Monographs on Endocrinology

Volume 14

J. H. Clark · E. J. Peck, Jr.

Female Sex Steroids

Receptors and Function

With 116 Figures

Springer-Verlag
Berlin Heidelberg New York 1979

James H. Clark, Ph. D.
Professor
Department of Cell Biology
Baylor College of Medicine
Texas Medical Center
Houston, Texas 77030 / USA

Ernest J. Peck, Jr., Ph. D.
Associate Professor
Department of Cell Biology
Baylor College of Medicine
Texas Medical Center
Houston, Texas 77030 / USA

ISBN-13: 978-3-642-81341-2 e-ISBN-13: 978-3-642-81339-9
DOI: 10.1007/978-3-642-81339-9

Library of Congress Cataloging in Publication Data
Clark, James H.
Female sex steroids. (Monographs on endocrinology; v. 14)
Bibliography: p. Includes index. 1. Steroid hormones--Receptors.
2. Estrogen--Receptors. I. Peck, Ernest J., joint author. II. Title.
QP572.S4C53 612.6'2 79-14030

Softcover reprint of the hardcover 1st edition 1979

Typesetting: R. Leipe, Hagenbach

2125/3020-543210

To Henry Koffler, who shaped our partnership.

And to our families, who have endured it.

Preface

It is the object of this series of monographs to present the experiments and interpretations of a given laboratory. We have attempted in this volume to present our view of receptor-steroid interactions and their relation to steroid-induced responses. This view is necessarily biased by our own experimental results. While we have tried to include the views and results of others, this volume is not meant to be comprehensive. Rather it is selective and examines those topics within the general field of steroid hormone action which we have addressed at the laboratory bench.

The work reviewed in this volume would not exist without the help and support of many associates. Our mentors included Jorge Awapara, Jack Gorski, Henry Koffler, Bill Ray, and Mike Zarrow. Associates instrumental in this work include George Barr, John Burgner, Sam Campbell, Constance Cardasis, Håkan Eriksson, Stan Glasser, Jim Hardin, Mohammed Kalimi, Bruce Lester, Barry Markavarich, Shirley McCormick, Tony Means, Dan Medina, Ann Miller, Bert O'Malley, Helen Padykula, Zigmund Paszko, Dale Snow, Susan Upchurch and Marian Walters. Students who have contributed to this effort include John Anderson, Joe DeLibero, Aaron Hsueh, Katrina Kelner, Debbie Metzger, Randy Richards, Jim Schaeffer, Mike Tytell, Lane von Brunt, and Cheryl Watson. Technicians without whose assistance this work could not exist include Toni Cetti, Wendy D'Attilio, Jeanie Haselby, Lily Hsueh, Helen Hyland, Jim Kovar, Kathy O'Connor, and Peggy Sansone. Most of the art work for this volume was done by Nancy Henn and the photography by Jim Kovar. This volume could not be without the tireless effort of Pat Williams, who has typed and retyped these words without complaint and Valerie Bolton who typed the references with her usual mastery.

Finally, we wish to thank the US Public Health Service, National Institutes of Health, American Cancer Society, Huntington's Chorea Foundation, and the Research Corporation for the financial support that made this work possible. Current grants from the NIH supporting this work include CA 20605, HD 00022, HD 08389, HD 08436, and NS 11753.

September 1979

J.H. Clark
E.J. Peck Jr.

Contents

Chapter I

Introduction

1. Historical Aspects of Steroid Hormone Action

Effects of drugs and hormones have long been thought to arise from binding interactions with receptors. From the simple observation that drugs and hormones act to modify the biochemistry or physiology of specific tissues and yet have no apparent effect on others, it was reasoned that "target" tissues, i.e., those that respond to a hormonal signal, must be able to detect or perceive the effector. This concept was first proposed by Langley (1905) and extended by Ehrlich (1913). Ehrlich's dictum, "drugs do not act unless they bind," constitutes the basic premise of the study of receptor-steroid interactions. These ideas were formalized by Clark (1937) and Gaddum (1937), who proposed that the intensity of a response is proportional to the number of receptor sites that are occupied. This relationship is given by the simple equation:

$$R + S \rightleftarrows RS \rightarrow \text{Effect}$$

where R is the receptor, S is a steroid hormone molecule, RS is the receptor-steroid complex, and the effect is the biochemical or physiological response to the steroid. This equation is analogous to the Michaelis-Menton formulation for enzyme kinetics and was used by Clark (1937) and Gaddum (1937) to analyze receptor-drug interactions. These authors proposed that the intensity of the effect was directly proportional to the number of receptors occupied by the drug. This occupancy theory of drug action is appealing; however, it does not explain the action of partial agonists or antagonists. Ariëns (1954) and Stephenson (1956) modified the occupancy theory by introducing a second step:

$$R + S \rightleftarrows RS \rightarrow [RS] \rightarrow \text{Effect.}$$

Step 2 represents a postulated transition state required for a given RS complex to produce an effect. In this model, not all compounds induce the transition and hence, though binding may take place, a response is not obligatory. The capacity of a given drug or hormone to promote this transition and effect a response has been called *intrinsic activity* by Ariëns (1954) and *efficacy* by Stephenson (1956). Thus, one can envision antagonists that have high affinity for the receptor, but no intrinsic activity, as well as partial agonists that have high affinity for the receptor, but reduced intrinsic activity.

Paton (1961) proposed that responses to drugs depend on the number of binding encounters between the receptor and the drug per unit of time. This rate theory does not differ formally from the occupancy theory of Paton and Rang (1966). These earlier concepts have been modified and extended by several investigators to explain cooperative effects and allosteric interactions (Koshland 1961; Belleau, 1964; Koshland et al., 1966; Koshland and Neet, 1968; Changeux et al., 1967; Wyman, 1967; Sherman, 1977).

Although the concepts presented above are useful and important, the relationship between receptor-steroid interactions and subsequent responses is not fully understood. This relationship is one of the basic concerns of this book.

2. General Consideration of the Steroid Receptor Concept

The preferential accumulation and retention of a steroid hormone by a target tissue was first described by Glasscock and Hoekstra (1959) and Jensen and Jacobson (1962). Jensen's laboratory succeeded in synthesizing ^{3}H-estradiol labeled to a high specific activity and showed that three tissues, the uterus, vagina, and pituitary, retained significant quantities of the hormone against a marked concentration gradient. Since nontarget tissues, such as diaphragm, muscle, and kidney, did not retain ^{3}H-estradiol, the assumption was made that specific molecules binding estradiol must be present in target tissues. The concept of steroid receptors was extended by King et al. (1965a), who proposed that ^{3}H-estradiol was localized in the nuclei of target cells and was in association with chromatin (King et al., 1965b, c, 1966; King and Gordon, 1966). Talwar et al. (1964) and Toft and Gorski (1966) verified the existence of specific steroid receptors when they showed that cytosol fractions of the rat uterus contained a macromolecule with the characteristics expected of an estradiol receptor. Toft and Gorski (1966) demonstrated that the putative receptor could be observed as a macromolecule with a sedimentation coefficient of about 8*S* on low ionic strength sucrose density gradients. Using sucrose density gradient analysis they were able to show: (a) that the 8*S* binding site could be saturated by adding excess ^{3}H-estradiol; (b) that it had a high affinity for estradiol ($K_d \sim 10^{-10}$ *M*); (c) that the binding of ^{3}H-estradiol was inhibited by the addition of the nonsteroidal estrogen, diethylstilbestrol (DES) but not progesterone or testosterone; and (d) that it was not found in nontarget tissues. These pioneer studies established the basic criteria for steroid receptor measurement and led to the subsequent demonstration of similar receptors for other steroid hormones.

The preferential accumulation of ^{3}H-progesterone by the reproductive tract of the rat was shown by Laumas and Farooq (1966), Falk and Bardin (1970) and Wiest and Rao (1971). Cytosol progesterone receptors have been extensively characterized (Sherman et al., 1970; O'Malley et al., 1971; Schrader and O'Malley, 1972); they possess characteristics similar to those of the estrogen receptor. These receptors appear as 6–8*S* macromolecules on low ionic strength sucrose density gradients and are proteins with molecular weights of ~ 200,000. This form may represent an aggregate of 4*S* subunits, which are observed when cytosol fractions are analyzed on sucrose gradients containing 0.3–0.4 *M* KCl (Korenman and Rao, 1968; Erdos, 1968). These points will be discussed in more detail below; for the present, suffice it to say that soluble macromolecules exist in the cytosol fraction of target tissue homogenates which manifest the binding characteristics of hormone receptors.

It has been proposed that, upon exposure to estradiol, the cytosol receptor forms a complex with estradiol and that this complex is translocated to the nucleus (Toft et al., 1967; Jensen et al., 1968; Shyamala and Gorski, 1967). This concept was based on the observation that very little receptor could be found in uterine nuclear fractions

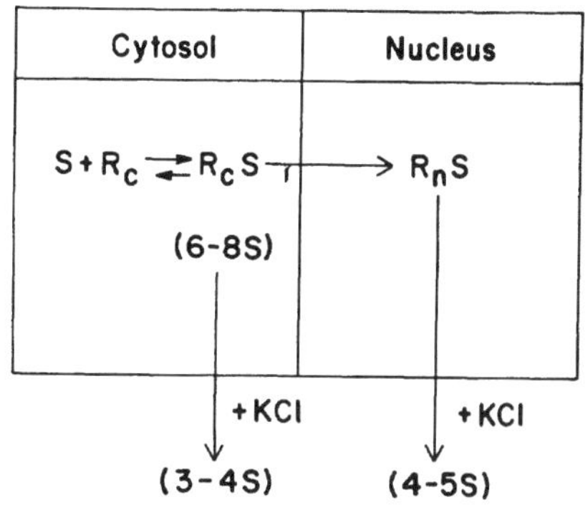

Fig. 1. Steroid hormone receptor interactions in the cytoplasm and nucleus of a target cell. The steroid hormone (*S*) binds to a soluble macromolecule (R_c) in the cytoplasm to form a receptor-steroid complex (R_cS). This complex undergoes translocation to the nucleus and is bound to nuclear sites (R_nS). Sedimentation coefficients are also shown in parantheses for low- and high-salt conditions

prior to exposure of the uterus to ^{3}H-estradiol, whereas after treatment with ^{3}H-estradiol the quantity of bound hormone increased in the nucleus, and the number of binding sites in the cytosol was depleted. The receptor-estradiol complex in the nucleus was partially extracted by 0.3–0.4 *M* salt and sedimented as a 4–5*S* moiety in sucrose density gradients. Studies by O'Malley et al. (1971) and Schrader and O'Malley (1972) have shown similar characteristics for the progesterone receptor.

From these studies a general picture of steroid binding and translocation of receptor-steroid complexes to the nucleus has evolved as shown in Fig. 1. Also shown are the relationships between low and high ionic strength forms of these receptor-steroid complexes. The binding of the R_nS complex to nuclear sites is assumed important to the mechanism by which this complex controls growth and metabolic activity of the tissue. This presumption is the basic premise of this book.

Chapter II

Steroid Receptor Characterization and Measurement

A complete understanding of the relationship between steroid receptor binding and the mechanism of hormone action depends on valid characterization and accurate measurement of steroid receptors. In this chapter the criteria and methods by which this can be accomplished are discussed. These provide the necessary background for the studies that will be presented in subsequent chapters and represent an expression of previous work published on this topic (Clark and Peck, 1977d, 1978a).

1. Receptor Criteria

The considerations discussed in Chap. I establish the basic criteria for steroid hormone receptors. These criteria are presented in this chapter in terms of simple systems that contain only one receptor site as well as of more complex systems made up of multiple receptor sites.

a) Finite Binding Capacity

The biologic response to an estrogen or progestin is a saturable phenomenon. If the formation of receptor-steroid complexes is obligatory for the production of a biologic response, then the quantity of steroid receptor should be limited to a finite number of binding sites. This criterion, limited capacity, should be met by demonstrating that

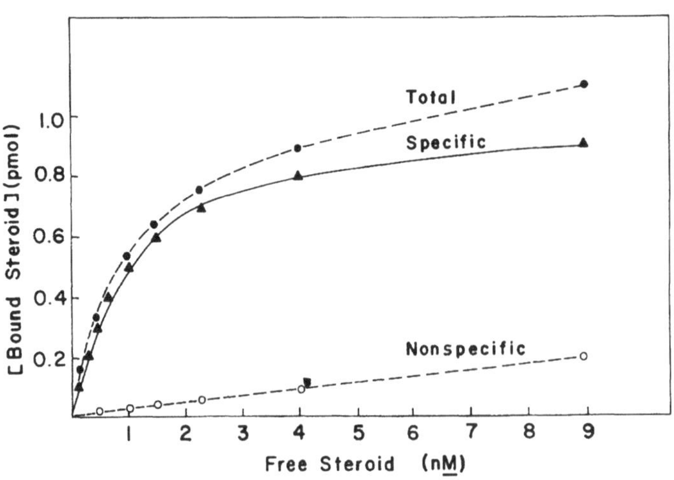

Fig. 2. Steroid binding to specific and nonspecific sites. The quantity of specifically bound steroid (▲) is determined by subtracting nonspecific (○) from total binding (●)

the binding activity of interest can be saturated by its specific ligand. This is usually accomplished by exposing a receptor source, for example, the cytosol or nuclear fraction of a target tissue, to various concentrations of ^{3}H-steroid until equilibrium is achieved. Subsequently the amount of bound ^{3}H-steroid is determined. Methods by which these data can be obtained are discussed later in this chapter; however, by way of example and for theoretical consideration, Fig. 2 is given as a typical binding curve for ^{3}H-estrogen to rat uterine cytosol.

The line designated *Total* in Fig. 2 represents ^{3}H-steroid bound to both specific receptor sites and nonspecific binding sites. Thus, this curve reflects both saturable and nonsaturable components. The saturable or receptor component is defined by the arithmetic difference between total and nonspecific binding. Nonspecific or nonsaturable binding is measured as the ^{3}H-steroid bound in the presence of excess unlabeled competitive ligand, for example, diethylstilbestrol (Fig. 2). Parameters required to define this molar excess will be discussed later in this chapter. Under these conditions the unlabeled steroid occupies essentially all high-affinity binding sites, but does not alter the binding of ^{3}H-steroid to low-affinity, nonspecific, or nonsaturable sites. This method assumes that nonspecific binding sites are of low affinity and high capacity relative to the receptor system. This has proven true for uterine, pituitary, and hypothalamic estrogen receptors. However, for careful quantitation this assumption should be validated by the demonstration of a *straight line* for nonspecific binding such as in Fig. 2 and/or by Scatchard analysis (Scatchard, 1949).

This analysis allows estimation of the number of receptor sites, i.e., *specific binding*, by subtracting *nonspecific* or nonsaturable binding from *total binding* as in Fig. 2. The calculated values for specific binding are used to estimate the total number of binding sites (n) and the dissociation constant (K_d) of the receptor-steroid complex (RS), a measure of the affinity of R for S. Two methods for such data analysis are shown in Fig. 3. In Fig. 3A, the method of Scatchard is employed, while in Fig. 3B a double reciprocal plot is shown. The number of high-affinity binding sites, n (B_{max} or R_t), is given by the X intercept in Fig. 3A and by the reciprocal of the Y intercept in Fig. 3B. Dissociation constants are obtained as described below.

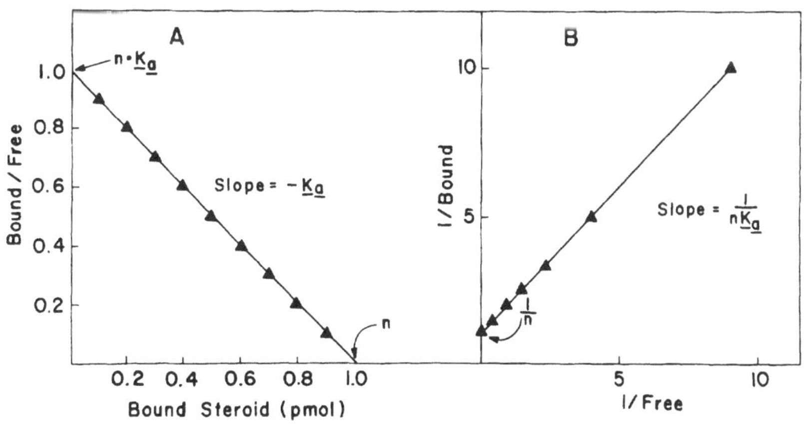

Fig. 3 A and B. Determination of receptor binding parameters. The data from Fig. 2 were used for construction of either a Scatchard plot (**A**) or a double-reciprocal plot (**B**)

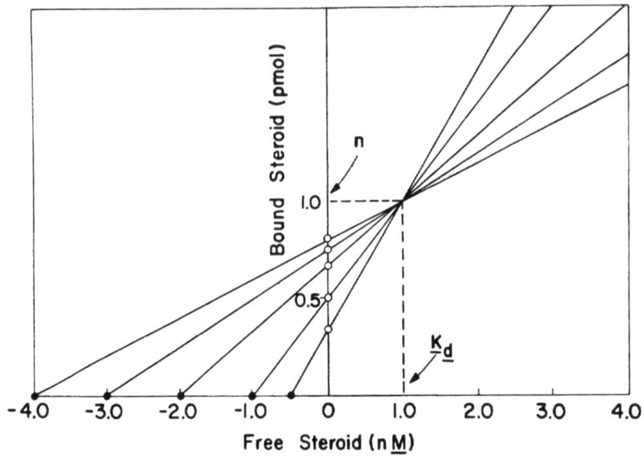

Fig. 4. Determination of receptor binding parameters by the direct linear plot. A *straight line* is drawn through the quantity of specifically bound steroid (○) and each of the negative values for free steroid (●). The *intersection* of these lines gives a direct estimate of the number of sites (n) and the dissociation constant (K_d)

An alternate method for determining the number of receptor sites is called the direct linear plot (Fig. 4). This method was originally developed for the analysis of enzyme kinetics by Eisenthal and Cornish-Bowden (1974) and has recently been used to estimate binding parameters for steroid receptors (Woosley and Muldoon, 1976; Walters and Clark, 1977b). This procedure is a simple graphic method, which is statistically sound and requires no calculation (Cornish-Bowden and Eisenthal, 1974). With this method n and K_d are given directly from the *intersection* of a series of lines defined by (−1) (free steroid concentration) and bound steroid at each of those concentrations (Fig. 4).

b) High Affinity

Steroid receptors should possess a high affinity for their respective hormones because the circulating levels of steroid are usually 10^{-10}–10^{-8} M. Thus, the existence of receptor-mediated responses of physiologic importance demands that the receptor have an affinity for the hormone that is in the range of the blood levels; otherwise the response would not occur. These considerations have proven true for a variety of target tissue receptors; however, they do not preclude receptor interactions of weaker affinity if blood or tissue levels of steroids or receptors are elevated (see Chap. X for discussion of this point).

Affinity or association constants (K_a), or their reciprocal, dissociation constants (K_d), are obtained from the analyses shown in Figs. 3 or 4. The *Y intercept* of the Scatchard Plot (Fig. 3A) equals nK_a or n/K_d, while the *X intercept* of the double reciprocal plot (Fig. 3B) equals $-K_a$ or $-1/K_d$. The K_d can be directly extrapolated from the intersection of the lines shown in Fig. 4. Although some investigators assume that the relative affinity of a receptor for a steroid measures the relative potency of the steroid for producing a response, this is not necessarily the case. A number of factors determine the ultimate expression of hormone action, and these considerations will be discussed throughout this volume.

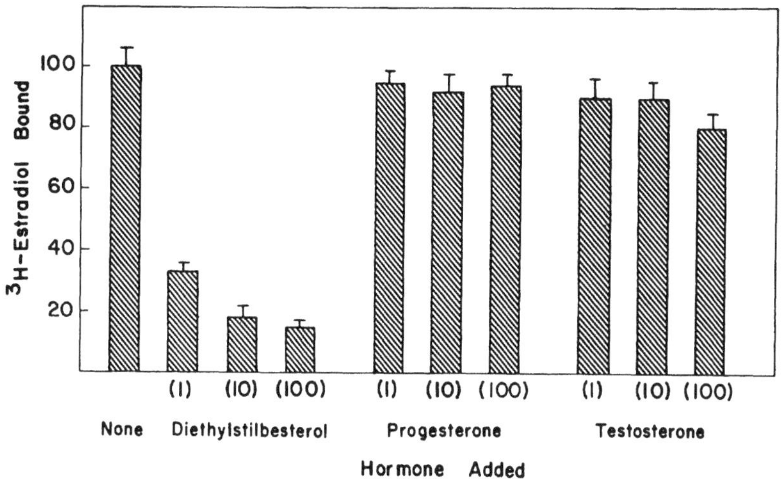

Fig. 5. Steroid specificity of the estrogen receptor. Effect of various hormones on ^{3}H-estradiol exchange in nuclear fractions of rat uterus. Rats were injected with 2.5 µg estradiol, and uterine nuclear fractions were prepared 1 h later. Nonlabeled hormone was mixed with $1.3 \times 10^{-8}\,M$ [^{3}H]-estradiol at the ratios in parentheses (nonlabeled hormone/[^{3}H]-estradiol). ^{3}H-estradiol bound in the absence of unlabeled ligand was taken as 100 percent or total binding

c) Steroid Specificity

Generally speaking, receptors are expected to display high affinities for a specific hormone or class of hormones. This "specificity" enables a given target cell to respond to a hormonal signal without interference from other signals. Thus, hormones of the same class as well as their agonists and antagonists should compete effectively for a given class of receptor while not affecting other receptor systems. This principle was used above to determine specific binding in Figs. 2 and 3. The inclusion of excess DES, which is an estrogen agonist bound by the receptor with an affinity similar to that of estradiol, prevents essentially all binding of ^{3}H-estradiol to the receptor. On the other hand, nonestrogenic steroids present at reasonable concentrations should not compete for the estrogen receptor, i.e., they should have little or no effect on the total binding observed in Fig. 2. Figure 5 is typical of the data obtained from an analysis of estrogen receptor specificity. As shown, DES very effectively reduces the amount of ^{3}H-estradiol bound to uterine nuclear fractions whereas progesterone and testosterone do not, even at relatively high concentrations.

It should be noted that steroid receptor sites do not display absolute stereo- and pharmacospecificity; i.e., the binding or recognition site on the receptor has a limited capacity for the recognition and differentiation of various steroid structures. Estrogen receptors have some affinity for all steroids, as has been shown recently for androgens (Ruh et al., 1975; Rochefort and Garcia, 1976; Garcia and Rochefort, 1977). These in vitro and in vivo studies have demonstrated that extremely high concentrations of androgen (3–10 mg in vivo; 1–10 µM in vitro) will promote the translocation of uterine estrogen receptor from cytoplasm to nucleus as well as stimulate both general protein synthesis and the synthesis of the so-called IP, induced protein. Under these conditions testosterone also binds to androgen receptors present in the uterus. These receptors are distinct from the estrogen receptor and do not appear to be involved in the observed estrogenic stimulation (Giannopoulos, 1971, 1973; Rochefort and Lignon, 1974). That androgens can interact at estrogen receptive sites was also demonstrated by Korach and Muldoon (1975). In an equilibrium and kinetic study of steroid binding, these authors demonstrated that large (1000-fold) excesses of unlabeled androgen competitively block the high-affinity estrogen receptor site and decrease the association rate of estrogen with this site. It seems doubtful that any of the effects discussed above are of physiologic importance since μM concentrations of *free* androgen would be required in vivo to produce these effects. Neither male nor female is likely to have these levels under normal circumstances.

The binding of progesterone to its receptor appears to be less specific than that of estrogen. Progesterone binds to both androgen (Bardin et al., 1975) and glucocorticoid receptors (Rousseau et al., 1972; Lippman et al., 1977a); in addition, glucocorticoids are known to bind to progesterone receptors (Walters and Clark, 1977b). Therefore, the progesterone receptor must be measured with particular attention to competitive binding analysis (see Sect. II. 2.e).

d) Tissue Specificity

Only certain tissues and organs appear to be stimulated by the sex steroids. Classically these have been termed target organs, e.g., uterus, vagina, mammary gland. It is generally accepted that if responses of these target organs result from receptor-steroid interactions, then the number of receptors in these tissues should be higher than that in nontarget tissues. Although there is no established minimum for the number of receptors required to define a target cell, tissue specificity should be part of any receptor validation scheme. Typical data are shown in Table 1. Note that the number of cytoplasmic estrogen receptors is very high in the uterus, vagina, and pituitary and low but not totally absent in other tissues. Some target tissues, such as the hypothalamus, possess low densities of receptor because of the heterogeneity of cell types within their anatomic boundary. Thus, not all cells within the bounds of the hypothalamus are estrogen targets, but those which are estrogen responsive probably possess receptors in numbers equivalent to other target cells. In those cases where receptors are fewer than 1 per cell, one must presume that the tissue is either not a target or that the target is more circumscribed than the tissue.

Table 1. Estrogen receptor content in target and nontarget tissues

Tissue	Receptor content (pmol/100 mg wet wt.)
Uterus	5.92
Vagina	2.15
Pituitary	1.43
Hypothalamus	0.5–0.10
Kidney	0.20
Diaphragm	0.06
Spleen	0.02

e) Correlation with Biologic Response

Implicit in all studies of macromolecules that bind steroids and meet the above criteria is the assumption that this binding results in a biologic response. Thus, binding of hormone to putative receptors must precede or accompany tissue responses, and the extent of response should relate to some function of receptor occupancy. This criterion, the demonstration of receptor-dependent hormonal response, is not often met and is the most difficult to establish. In some systems, e.g., the pseudohermaphroditic male mouse, the absence of the cytoplasmic receptor has been shown to preclude responses to the hormone. Thus, these animals do not respond to androgens, and this insensitivity correlates with an absence of androgen receptor (Bullock et al., 1971; Gehring et al., 1971; Milin and Roy, 1973). Sibley and Tomkins (1974) have shown that steroid resistance in lymphoma cells may result from deficiencies in a variety of receptor functions: (a) cytoplasmic receptor (R_c) may be reduced or altered such that the receptor steroid (RS) complex does not form; (b) the RS complex may form but fail to undergo translocation; and (c) the RS complex may form and undergo translocation but fail to stimulate nuclear responses. In addition, it is known that the uterus is refractory to estrogen stimulation at times when the quantity of R_c is low and returns to normal responsiveness as the quantity of R_c increases (Clark et al., 1973a; Anderson et al., 1973a). This criterion will be discussed more thoroughly in Chap. V.

2. Receptor Parameters: Theory and Practice

The preceding discussion of receptor affinity and capacity was based on data derived from a simple system that contained relatively high levels of one specific steroid receptor site and low levels of nonspecific steroid binding sites. Although this is

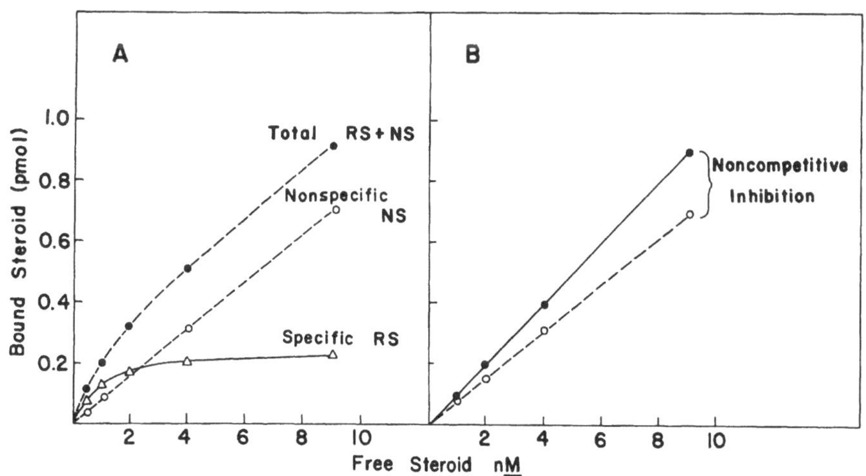

Fig. 6. A comparison of saturable and nonsaturable steroid binding under conditions of competitive (A) and noncompetitive (B) inhibition

sometimes the case, it most often is not. In this section we present some of the problems frequently encountered and provide methods by which they can usually be overcome.

a) Receptor Measurement when Nonspecific Binding is Large

Under ideal circumstances saturation analyses would be as shown in Fig. 2, and the binding capacity and affinity for such a receptor site could be determined with great confidence as in Fig. 3. In many hormone receptor systems, however, the quantity of nonspecific binding (NS) is great with respect to total binding (Fig. 6A), and thus the probability of accurately estimating specific binding (the difference between total bound steroid and NS) is very low. A careful analysis of saturation in the presence and absence of competing ligand is necessary to establish the existence of receptor sites under these conditions. Specific receptor sites, which are limited in number and possess a ligand binding site specific for a given hormone species, should be demonstrated by their saturable nature, not through the use of "one point" assays (see below).

Inhibition of ligand binding can occur by noncompetitive mechanisms to produce apparent "specific inhibition" of receptor-ligand interactions. If this occurs, the likely outcome is a *straight line* for both NS (nonspecific binding) and NS + RS (total bound S) as shown in Fig. 6B. Under such conditions "one point" assays (that is, assays using a single concentration of labeled ligand, such as 10 K_d plus and minus excess unlabeled ligand) can lead to erroneous conclusions. Depending on the tissue source of the receptor sites and the number of receptor sites per unit of tissue, the investigator may be forced to employ these "one point" assays for studies of receptor dynamics during altered physiologic states. However, the system must be characterized as to the saturable nature of the receptor site under study; in addition, any extremes in receptor number as measured with these assays (especially dramatic increases in receptor number) must be confirmed by Scatchard analysis.

b) Interpretation of Saturation Parameters by Scatchard Analysis

Other investigators have discussed many of the problems of Scatchard analyses (Rodbard, 1973; McGhee and von Hippel, 1974; Williams and Gorski, 1974; Chamness and McGuire, 1975; Buller et al., 1976). However, several points regarding their interpretation are sufficiently important that we wish to deal with them in this section.

1) Estimates of Receptor Number in the Presence of Nonspecific Binding

Typical data for a binding system that contains only one saturable receptor system (R) and one nonsaturable, nonspecific binding system (NS) are shown in Fig. 7. RS, NS and the sum of the two are analyzed by the method of Scatchard. In many cases, NS is not measured; that is, a plot of RS + NS is used to extrapolate a point on the *abscissa* for an estimate of *n*, the number of receptor sites in the system. Such analyses invariably overestimate the true value of *n*. An accurate estimate of *n* can be made either by accounting for nonspecific binding by competition studies, by parametric analysis using nonlinear curve fitting programs (Rodbard and Feldman, 1975), or by graphic methods of correlation as described by Rosenthal (1967) and Feldman (1972). These will be discussed in more detail later in this chapter.

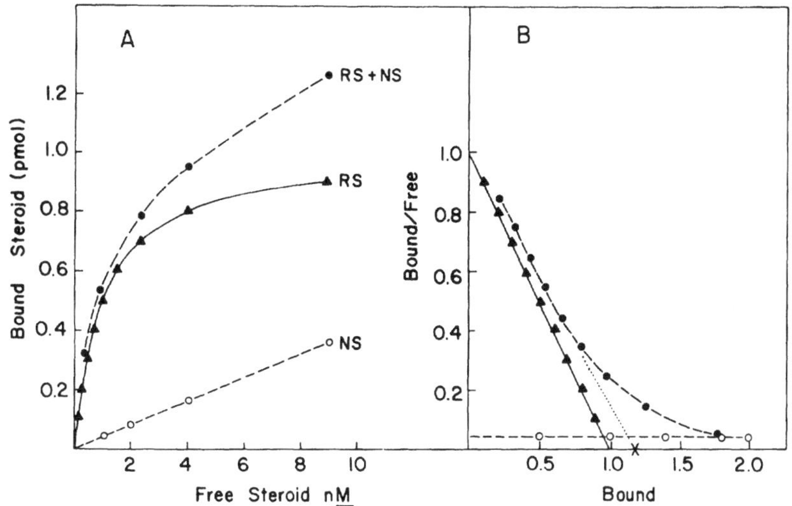

Fig. 7. Saturation (**A**) and Scatchard analysis (**B**) of one receptor component in the presence of nonspecific binding. Specific binding (▲) is determined by subtracting nonspecific binding (○) from total (●). The *dotted line* and *x* on the *abscissa* represent an incorrect extrapolation of the number of binding sites. See text for details

2) Two Specific Sites with Different Affinities

In many receptor systems two or more binding proteins exist which have similarly high affinities for the same steroid. This condition could produce the theoretical situation shown in Fig. 8. In this example the nonspecific binding component has been eliminated for convenience and will be discussed later. This mixture of binding sites produces a saturation curve (Fig. 8A) which does not appear to be composed of two binding components; however, the Scatchard analysis (Fig. 8B) clearly

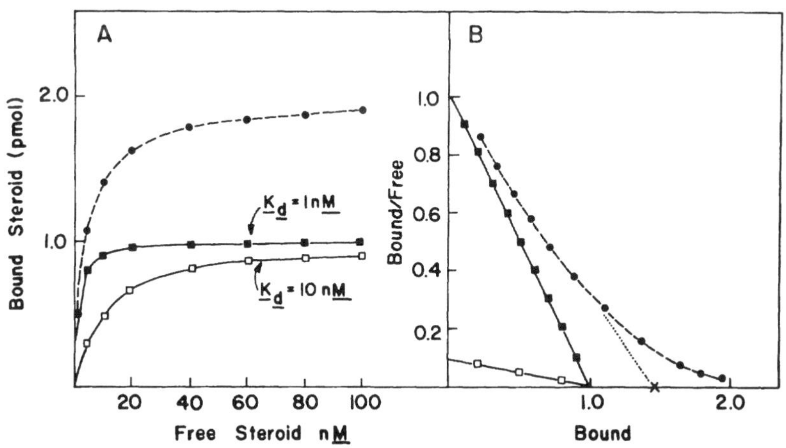

Fig. 8. Saturation (**A**) and Scatchard analysis (**B**) of two specific binding components of identical quantities but different affinities. Total specific binding (●); specific binding due to each component (■, □)

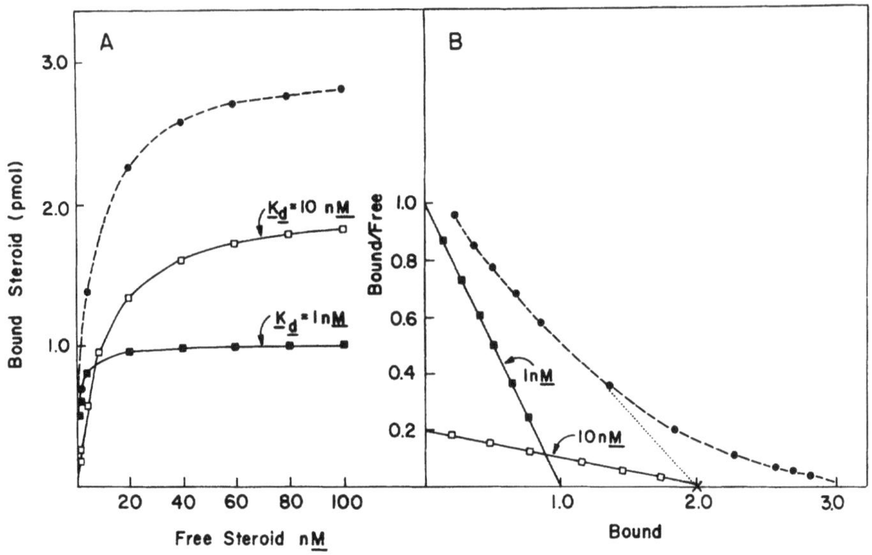

Fig. 9. Saturation (A) and Scatchard analysis (B) of two specific binding components of dissimilar quantities and affinities. Symbols are identical to those in Fig. 7

demonstrates their presence. Note that a usual saturation analysis might only include the lower range of steroid concentration; thus, extrapolation of an apparently straight line would yield an improper estimate of the number of binding sites (Fig. 8B). In addition, a false conclusion would be drawn — that only one specific binding component was present. Errors of this type are more exaggerated when the binding

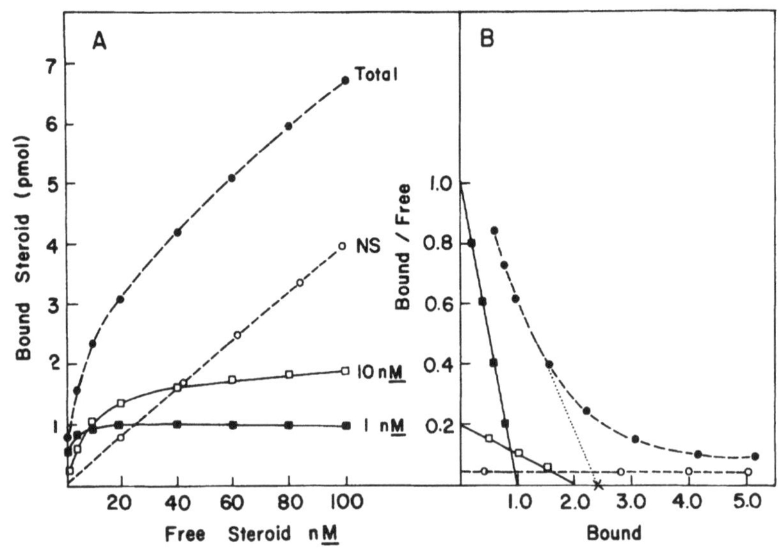

Fig. 10. Saturation (A) and Scatchard (B) analysis of two specific binding components of dissimilar affinities and quantities plus a non-specific binding component (O). Other symbols are identical to those in Fig. 8

component with the lower affinity (10^{-8} K_d) exceeds that with high affinity (Fig. 9). In such cases binding analyses at the low concentrations of steroid lead to gross overestimates of the number of a single class of sites and an underestimate of their affinity for steroids.

Inclusion of a nonspecific binding component which, of course, is usually present further complicates these analyses. Fig. 10 shows the contribution by nonspecific binding when present with two steroid-specific sites as discussed above. Again, the saturation curve (Fig. 10A) does not hint of the underlying complexity of the three-part system (Fig. 10B). A Scatchard analysis of data corrected for nonspecific binding reveals the presence of two or more binding components. In this case, ignorance of the presence of a complex system and extrapolation of what appears to be a straight line gives an estimate of a single binding class with a K_d of ~ 0.4 nM and a concentration of sites of 2.5 pmol/ml. Thus, careless analysis of complicated binding systems can lead to large errors in the estimation of receptor parameters. These examples are cited to demonstrate the necessity for pharmacologic (competitive ligand) differentiation or physical separation of different classes of binding sites to allow the precise measurement of all components.

c) Resolving Mixed Binding Systems

The obvious and ideal solution to the problem of mixed binding systems such as those above is the physical separation of the various components by purification procedures, thus allowing the study of each as an isolated system. However, this is usually not feasible due to limited quantities of tissue. Rosenthal (1967) and Feldman (1972) have proposed methods for resolving multiple binding components by the graphic analysis of curvilinear Scatchard plots. These have been reviewed recently by Munck (1976). Simply stated, curved Scatchard plots are resolved by finding straight lines that, when summed point by point in a vectorial manner, reproduce the

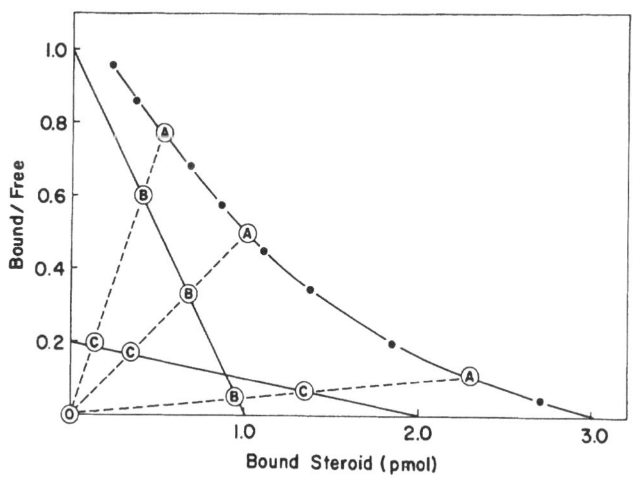

Fig. 11. Resolution of two binding components by vectorial analysis of a curved Scatchard plot. Each point A of the Scatchard plot is the vectorial sum of points B and C for each of the binding components. Thus, two linear components can be resolved by adjusting their slopes until $OC + OB = A$ for all lines drawn from the origin (O) to the Scatchard plot

original curve. This procedure is illustrated in Fig. 11. Note that vectors determined by two independent components must sum to the curve. A more complete description is available in the review of Munck (1976). Usually the data are limited and the precise curves so imprecisely determined that the resolution of more than two components is impossible.

Invariably these complex Scatchard plots require that the investigator use some iterative process to resolve the multiple components present in the system. This process may involve the use of an inexpensive ruler (e.g., for the vectorial analysis discussed above) or an extremely sophisticated computer. In the simplest case, the system is composed of one specific or saturable component and one nonsaturable component (as in Fig. 7). In most cases nonspecific binding is directly proportional to the free ligand concentration; therefore, the ratio of NS to S reaches a limit value on the Scatchard plot which remains unchanged once the high affinity receptor component is saturated (see Fig. 7; also see Davies and Ryan, 1973, and Chamness and McGuire, 1975). To resolve this simple system the investigator may use the graphic method of Rosenthal (1967), a modification of that procedure as discussed by Chamness and McGuire (1975), or the algebraic procedure derived by Hart (1965). Alternatively, competitive ligands as discussed for Fig. 2 may be employed to determine this component.

For complex systems involving two or more saturable components, the investigator either must rely on physical or pharmacologic techniques for distinguishing between sites or must use computer-assisted iterative processes, such as nonlinear curve-fitting programs (Rodbard and Feldman, 1975) or the statistical parameter-fitting procedures of Feldman (1972) or Priore and Rosenthal (1976).

Analytical methods employing geometric or parametric procedures, such as those discussed above, are useful if very complete and detailed ligand binding data can be obtained. Often this is not the case due to limitations in biologic material, and other methods must be found to cope with the problem. Differential inhibition of ligand binding has been employed effectively to measure a given component in a number of mixed systems. For example, in a three-component system, such as that in Fig. 10, let the binding site with a K_d of 10^{-8} M be a serum protein that binds estradiol and testosterone equally well. Such is the case for steroid-binding globulin, SBG, which is elevated during pregnancy in the human. The component with a K_d of 10^{-9} M might be the estrogen receptor (R), while nonspecific binding (NS) is attributable to numerous proteins and lipids. (In Chap. IV we will present data on such a mixed system which is found in the uterus.) The binding of ligand to receptor can be determined under these circumstances by taking advantage of the differential binding specificities of SGB and R. The relative binding of testosterone to SGB and R depends on the relative affinities of each for testosterone (see below for further discussion of this point). Such a system is outlined in Fig. 12 below. Using labeled estradiol in combination with unlabeled testosterone and/or DES, the investigator can determine total binding (R + SGB + NS), the binding of estradiol to specific and nonspecific sites (R + NS), and the binding of estradiol to nonspecific sites (NS) alone. *Part A* shows the binding of ^{3}H-estradiol to all three components, while *parts B* and *C* show the differential displacement of labeled estradiol from SGB and R by testosterone and DES, respectively. By subtraction it is possible to calculate the number of binding sites present for a given system. Note that one must be aware that the system contains three components and have considerable knowledge of the characteristics

MIXED BINDING SYSTEMS

A. $R + SBG + NS + E_2^{\bullet} \rightleftharpoons R\cdot E_2^{\bullet} + SBG\cdot E_2^{\bullet} + NS\cdot E_2^{\bullet}$

B. $R + SBG + NS + E_2^{\bullet} + T \rightleftharpoons R\cdot E_2^{\bullet} + SBG\cdot T + NS\cdot E_2^{\bullet} + NS\cdot T$

C. $R + SBG + NS + E_2^{\bullet} + T + DES \rightleftharpoons R\cdot DES + SBG\cdot T + NS\cdot E_2^{\bullet} + NS\cdot T + NS\cdot DES$

B - C = Receptor binding

Fig. 12. Determination of receptor binding in the presence of steroid-binding globulin: Receptor, R; steroid-binding globulin, *SBG;* nonspecific binding, *NS*; ^{3}H-estradiol, E_2^*; testosterone, *T*; diethylstilbestrol, *DES*

of each. Such an awareness implies that the investigator has carefully defined the pharmacology of the binding system, i.e., the degree of ligand binding specificity displayed by the various binding components at several concentrations of ligand. Armed with such information and/or knowledge of differential sensitivites of binding sites to extremes of pH, temperature, or physical stripping, it is possible to resolve complex mixtures of receptor sites.

A similar rationale can be used to assess the binding of estrogen to receptor in the presence of α-fetoprotein (α-FP). This protein is present in large quantities in the neonatal rat and has a high affinity for estradiol (K_d 10^{-9} – 10^{-10} *M*) which approximates that of the receptor. The receptor is measured by taking advantage of the fact that DES binds with low affinity to α-FP, but competes very effectively with estradiol for estrogen-receptor binding sites (Soloff et al., 1971; Plapinger et al., 1973; Kato et al., 1974; Attardi and Ruoslaht, 1976). This is illustrated in Fig. 13.

ESTROGEN RECEPTOR ASSAY IN PRESENCE OF α-FETOPROTEIN

A. $R + \alpha\text{-}FP + NS + E_2^{\bullet} \rightleftharpoons R\cdot E_2^{\bullet} + \alpha\text{-}FP\cdot E_2^{\bullet} + NS\cdot E_2^{\bullet}$

B. $R + \alpha\text{-}FP + NS + E_2^{\bullet} + DES \rightleftharpoons R\cdot DES + \alpha\text{-}FP\cdot E_2^{\bullet} + NS\cdot E_2^{\bullet} + NS\cdot DES$

A = Total $E_2^{\bullet}$ binding; B = $E_2^{\bullet}$ binding due to α-FP and NS sites

A - B = Specific binding of $E_2^{\bullet}$ due to R

Fig. 13. Determination of receptor binding in the presence of α-fetoprotein: Receptor, *R*; α-fetoprotein, *αFP*; nonspecific binding, *NS*; ^{3}H-estradiol, E_2^*; diethylstilbestrol, *DES*

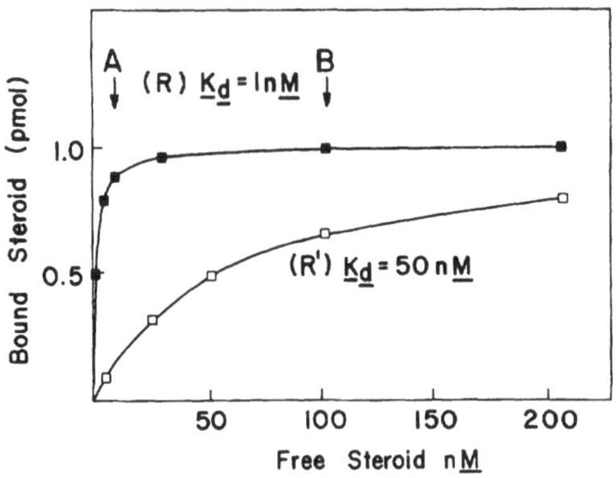

Fig. 14. Determination of a high-affinity specific-binding component in the presence of a lower-affinity specific-binding component. See text for details and abbreviations

Part A shows estradiol binding to R, α-FP, and NS, while *B* shows DES binding only to R and NS. Thus, the binding of labeled estradiol to R can be determined by subtracting *B* from *A*. Differential ligand specificity has been used by several investigators to resolve mixed binding systems (Rosner et al., 1969; Korsten and Persijn, 1972; Funder et al., 1973; Ratajczak and Hähnel, 1976; Philibert and Raynaud, 1977).

In the preceding examples mixed systems with binding components possessing different steroid specificities were discussed. It is possible to obtain reasonable estimates of a single-binding component in mixed systems with similar steroid specificities, providing their affinities for the steroid differ significantly. To illustrate, a system is shown in Fig. 14 in which two binding components, R and R′, have dissociation constants of 1×10^{-9} and 50×10^{-9} *M*, respectively. From the saturation curves of these two receptors, it can be seen that any attempt to measure R under conditions of saturation, i.e., 100 n*M* (Fig. 14, *point B*) would lead to the detection of significant quantities of R′. However, it is possible to perform assays at concentrations of steroid which are near saturation for R, between 5–10 n*M* and detect ~ 90 percent of the R sites while measuring only 10–20 percent of the R′ sites (Fig. 14, *point A*). A complete separation of two receptor sites by this method can be achieved when their K_d's differ by 100-fold. Note that such analyses presume a thorough investigation of the saturable nature and kinetic properties of the system under study. In addition, it should be obvious that such an estimate is invalid if R′ is in great excess of R.

d) Receptor Determination when the Concentration of Receptor Greatly Exceeds the Dissociation Constant of the Reaction

The analysis of binding systems as discussed above and utilized by many investigators applies only when the [R] is at or below the K_d of the reaction, $R + S \rightleftarrows RS$. When $[R] \gg K_d$, a condition exists which is not appreciated by many investigators. For the simplest system where the dissociation constant for the RS complex is defined by

$$K_d = \frac{[R] \cdot [S]}{[RS]} \tag{1}$$

and the total steroid (ligand) present in the system is defined by the conservation equation,

$$S_t = S + RS, \tag{2}$$

one can substitute the definition of RS from Eq. (1) and rearrange terms such that Eq. (2) becomes:

$$S = \frac{S_t}{1 + R/K_d} \tag{3}$$

If total R (R_t) is much greater than K_d (i.e., $R_t \gg K_d$), then free R $\gg K_d$ and therefore $R/K_d \gg 1$. Under these conditions Eq. (3) simplifies to:

$$[S] = \frac{[S_t] \cdot K_d}{[R]} \tag{4}$$

and solving for RS from Eqs. (1) and (4), RS $= S_t$. Thus, all of the steroid in the system will exist as an RS complex, at least to 90 percent saturation, and there will be little or no free ligand present.

Such a circumstance places severe limitations on competition studies for receptor sites and on the determination of thermodynamic constants. Under such conditions no determination of K_d or K_i is possible; however, if saturation is achieved, a value of R_t can be obtained. This analysis, which is applied to a single class of binding sites in this case, is further complicated if additional binding sites of reasonable affinity are present in the system. Such is often the case for steroid-receptor systems.

e) Steroid Binding Specificity

As discussed above, receptors are expected to display stereo-specific binding. This characteristic has been used extensively as a device for the detection of specific binding. Since the principles of competitive binding are of paramount importance for the valid measurement of hormone receptors, we feel that they deserve further consideration.

1) Competitive vs. Noncompetitive Interactions

Inhibition of steroid binding to receptor sites may occur by either competitive or noncompetitive means, i.e., by mechanisms involving mutually exclusive binding of ligands (competitive) or by mechanisms in which one ligand inactivates (either reversibly or irreversibly) the receptor (noncompetitive). This differentiation of inhibitory classes is important in the study of steroid receptors. Competitive inhibitors decrease ligand binding to receptor sites by combining with the receptor in such a manner that the ligand (hormone, substrate, "label") can no longer be bound, e.g., as when ligand and inhibitor compete for the same or adjacent and overlapping sites. The mutually exclusive nature of ligand and inhibitor binding in such systems results in the data shown in Fig. 15A which can be analyzed by double-reciprocal analysis as shown in Fig. 15B. Note that increasing concentrations of inhibitor alter the apparent K_d for the receptor-steroid complex but do not change the number of sites. Noncompetitive inhibition of ligand binding to receptors, as seen in Figs. 15C and D, may occur for several reasons. As examples, the inhibitor may precipitate or denature the receptor or its active site; alternatively the inhibitor might bind to a second

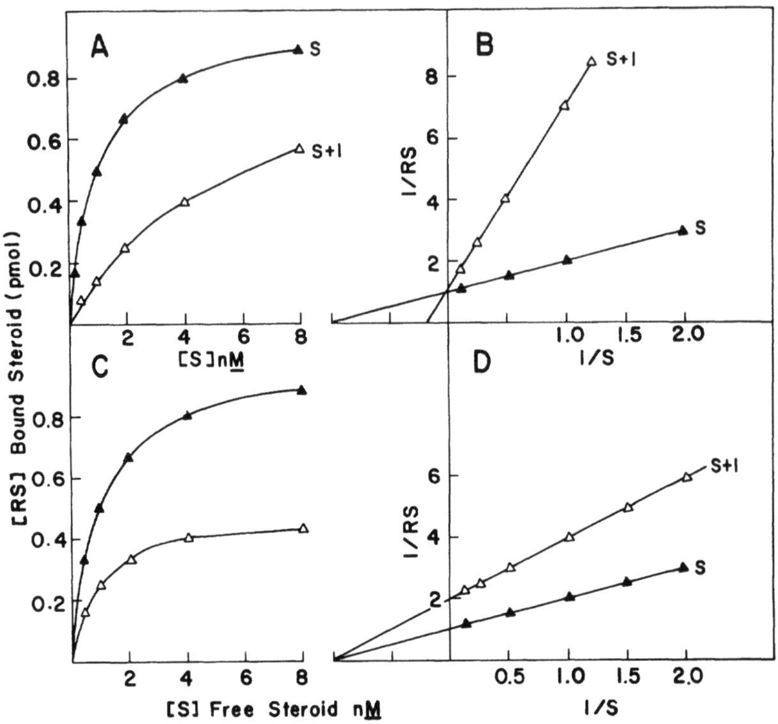

Fig. 15 A–D. A comparison of competitive and noncompetitive inhibition of hormone binding. Saturation (A) and double-reciprocal plots (B) of the competitive inhibition of hormone binding. Saturation (C) and double-reciprocal plots (D) of noncompetitive inhibition of hormone binding. Steroid, *S;* inhibitor, *I*

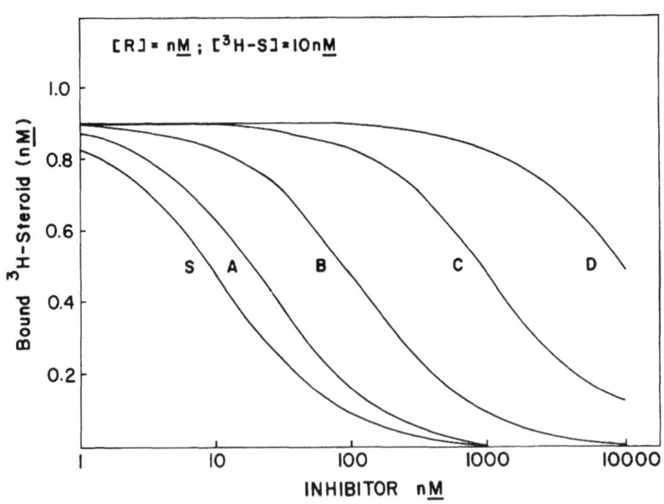

Fig. 16. Competitive inhibition of ^{3}H-steroid binding as a function of inhibitor concentration. Various competitive inhibitors are added to a reaction mixture containing 1.0 n*M* receptor and 10 n*M* labeled steroid

site on the receptor and in so doing alter the active site of the receptor. In this case (Fig. 15D) the number of "active" receptor species decreases with increasing concentrations of inhibitor. Thus, the simple demonstration of a "blockade" of RS formation by the addition of an inhibitor does not establish a competitive mechanism. Since a competitive inhibitor is often employed to define "specific binding", i.e., the binding of hormone to a given receptor site, it is important to perform the simple and classic analysis shown in Fig. 15 to validate the use of inhibitors.

2) Competitive Binding Analysis: Problems and Interpretations

Competitive inhibition is employed extensively to characterize the binding properties of hormone receptors. The classic use of such analyses has been to determine the relative binding affinities of various pharmacologic inhibitors. Thus, in Fig. 16, the affinity of inhibitors *A* and *B* for the receptor would be 1/2 and 1/10 that of S. The parallel nature of the displacement produced by the various inhibitors suggests that the inhibition of binding is competitive, although this should be examined by the double-reciprocal analysis discussed above.

Figure 16 also provides an explanation and clarification of problems introduced earlier concerning receptor binding analyses. Thus, when a nonlabeled steroid is added to an assay system at 100 × the concentration of labeled steroid, [S*], it will effectively compete for those receptor binding sites that would be occupied by S* in the absence of inhibitor if the two ligands have similar affinities for the receptor. This is the justification for the use of such additions for the determination of specific binding by difference (as discussed earlier in this chapter). It is insufficient simply to use 100 × [S*] for this correction. Rather one must consider the relative affinities of the receptor for the ligands. The concentration of unlabeled ligand necessary to displace greater than 99 percent of specific binding is defined by:

$$[\text{Unlabeled ligand}] = K_i/K_d \cdot [S^*] \cdot 100.$$

Estrogen antagonists for which R has low affinities (high K_d's) have been employed for the determination of estrogen receptors (Jensen and De Sombre, 1972). Although useful for qualitative assessments, quantitative measurements with such low-affinity inhibitors are inferior to measurements that utilize inhibitors with K_d's similar to or greater than that of the labeled steroid. For example, the use of Nafoxidine as a competitive inhibitor of estrogen receptor binding in quantitative assays is inappropriate. The receptor-Nafoxidine complex has a K_d of 10^{-7} *M*; therefore, 10,000 n*M* Nafoxidine is required to compete with estradiol for the majority of R sites. One should realize that the addition of large excesses of inhibitor may alter nonspecific binding, decrease receptor binding by noncompetitive inhibition, and/or increase apparent receptor binding by competition with a second saturable (or nonsaturable) component. In addition, equilibrium conditions may be quite different in the presence of large excesses of steroid.

Any compound thought to be a competitive inhibitor of receptor-ligand interactions should be examined over a wide concentration range. Otherwise its inhibitory capacity may go undetected. The use of inhibitors at concentrations near the K_d of the RS* complex may yield erroneous results. If compound *B* (Fig. 16) was examined under such circumstances, very little competition would be observed, and the wrong conclusion would be drawn, i.e., that compound *B* does not interact effectively with the receptor. In fact, under physiologic circumstances compound *B* may be present in the blood and extracellular spaces at very high concentrations when compared to S and thus be a physiologically active ligand.

One common fault with analyses such as those in Fig. 16 is the use of IC_{50} values (the concentration of a given unlabeled ligand required to displace or inhibit labeled ligand binding by 50%) as estimates of K_i. In fact, only under special circumstances can the IC_{50} value be employed as an estimate of K_i, i.e., when the concentration of labeled ligand, [S*], is much less than its K_d ($[S^*] \ll K_d$). In addition, as discussed previously, the concentration of receptor sites, $[R_t]$, must also be less than the dissociation constant of the RS complexes (i.e., $[R] \ll K_d$). The necessity of these conditions should be apparent from the following derivation.

In the absence of an inhibitor, the amount of bound steroid, RS_a^*, is defined by

$$[RS_a^*] = \frac{[R_t] \cdot [S_{Free}^*]}{[S_{Free}^*] + K_d} \tag{5}$$

whereas in the presence of a competitive inhibitor, I, the amount of bound steroid, RS_b^*, is defined by

$$[RS_b^*] = \frac{[R_t] \cdot [S_{Free}^*]}{[S_{Free}^*] + K_d + K_d[I_o]/K_i} \tag{6}$$

and the ratio, RS_b^*/RS_a^* is

$$\frac{[RS_b^*]}{[RS_a^*]} = \frac{[S_{Free}^*] + K_d}{[S_{Free}^*] + K_d + K_d[I_o]/K_i} \tag{7}$$

For the IC_{50} of I to equal K_i, it is necessary that $[I_o]$ equals K_i ($I_o/K_i = 1$) and that RS_b/RS_a equal 0,5 in Eq. (7). For ($S_{Free}^* + K_d$) divided by ($S_{Free}^* + 2\ K_d$) to equal 0.5, $[S_{Free}^*]$ must be much less than K_d ($[S_{Free}^*] \ll K_d$). Obviously no determination of the number of binding sites is possible under conditions of subsaturating labeled ligand; rather saturation analyses as discussed above are required to estimate R.

f) Hooks and Curves in Scatchard Plots

When using the Scatchard plot to analyze ligand-binding data, it is important to remember the basis of this analysis and the conditions under which it is valid. Values of bound and free ligand must be determined at equilibrium, and the plot must either be linear or readily resolved into linear components as described above. The presence of hooks, convexities, or concavities in Scatchard plots should not be taken as a priori evidence of cooperativity. Such nonlinearity indicates that the data do not fit a simple two-state model in which ligand and receptor exist at equilibrium as bound (occupied) or free. In addition to multistate systems, a number of methodologic artifacts can lead to nonlinear Scatchards. A few of these are discussed below.

1) High Concentration of Ligand or Receptor Sites

Although estimates of affinity by Scatchard analysis are valid even when ratios of ligand or receptor site to K_d are high, several artifacts hamper estimates at extremes of either component. Thus, when labeled ligand concentrations are very high (i.e., only a small fraction, 0.5%–5%, of the total ligand is bound), errors in the estimate of bound ligand can be large. Similarly when receptor site concentration is high, the situation discussed in Sect. II.2.d applies, and an accurate estimate of free ligand is not possible. Over- or underestimates of either parameter, bound or free, are magnified on Scatchard plots to produce bizarre and deceiving nonlinearities.

2) Nonequilibrium Conditions

Although this should be obvious to all, many studies are performed under conditions that are sufficient for equilibrium at saturating concentrations of ligand but which are not sufficient for equilibrium *at subsaturating concentrations*. Such a circumstance invariably leads to an underestimate of "bound at equilibrium" at low ligand concentrations, an overestimate of free, and a "hook" in the Scatchard suggestive of "positive cooperativity." An increase in equilibration time or temperature often straightens these hooks.

3) Instability of Ligand or Receptor Sites

If labeled ligand is sufficiently different from unlabeled ligand with respect to affinity for the receptor or with respect to chemical stability, there is no straightforward way to estimate the concentration of free ligand at equilibrium. Separation of total free ligand from bound ligand will not suffice since total free will be composed of subsets of ligand species. Heterogeneity in ligand species due to chemical instability can produce nonlinear Scatchard plots.

Similarly, receptor sites may show heterogeneity with respect to stability, a situation resulting in a nonlinear Scatchard. Instability of steroid receptors in the absence of ligand is well known, especially at elevated temperatures. This is true for the uterine estrogen receptor even though the uterus is kept intact (Peck et al., 1973b). Therefore, stability can be a significant problem during the assay of hormone receptors and further complicates their measurement. One example of a ligand-dependent trend in binding data which invariably results in Scatchards suggestive of positive cooperativity arises if receptor sites are stable when occupied but prone to inactivation when unoccupied. This is often the case at low concentrations of protein.

Thus, during equilibration with ligand, receptor sites at saturation are stable while those at subsaturating conditions are constantly changing in number throughout the incubation. This situation has probably produced many of the reports of cooperativity in receptor-steroid interactions. Procedures for avoiding or detecting this artifact have been discussed by Chamness and McGuire (1975).

4) Dilution of Radioactivity

Many cytosol and nuclear preparations contain significant levels of endogenous steroid, usually the result of some physiologic or pharmacologic manipulation. These levels may vary from treatment group to treatment group and between various subcellular fractions. If these fractions are employed for Scatchard analyses, the endogenous steroid will dilute the specific activity of the ligand. If endogenous levels are similar to those employed in the ligand binding analysis, the final specific activity at low concentrations of labeled ligand will be markedly reduced relative to that at higher concentrations of labeled ligand. These effects can alter the shape of Scatchard plots. One can either pretreat such fractions with charcoal to strip them of endogenous steroid or correct all Scatchard plots for this dilution by endogenous ligand (Katzenellenbogen et al., 1973).

5) Interacting Species

A number of other conditions exist under which Scatchard plots will not be linear. These include multiple-binding sites, as discussed above, ligand-ligand interactions, interactions between receptor sites, and cooperative states in which the interaction of ligand with receptor sites is not simply R + S $\rightleftarrows$ RS. Some of these are discussed in detail by Hollenberg and Cuatrecasas (1976) and by Rodbard and Feldman (1975). Anyone faced with nonlinear Scatchard plots after considering the points discussed above (Sects. f. 1–5) should use these sources for a thorough analysis of interacting systems.

3. Receptor States and ^{3}H-Steroid Exchange

Most biologic systems contain both occupied and unoccupied receptors. The assessment of both forms is important if an accurate picture of the relationship between steroid binding and physiology is to emerge. We have developed a method for the quantitation of both forms of the estrogen receptor which is based on the exchange of ^{3}H-estradiol for nonlabeled estrogen (Anderson et al., 1972a). This assay was first developed for measurement of estrogen receptors in uterine nuclei (Anderson et al., 1972a) and the cytosol progesterone receptor in the guinea pig uterus (Milgrom et al., 1973a). Subsequently the ^{3}H-steroid exchange technique has been widely employed (for references, see Sects. a and b below). Although the process cannot be applied to all steroid receptors, this method has allowed estimation of receptor-steroid complexes in a number of systems under a variety of experimental conditions. In the following sections we will discuss theoretical as well as practical aspects of the estimate of occupied and unoccupied nuclear and cytoplasmic receptor species.

a) Exchange Techniques in Theory and Practice

The theoretical considerations involved in the measurement of estrogen receptors differ from those for the progesterone receptor. The general conditions for the measurement of estrogen receptors is given in Fig. 17; those for the progesterone receptor

RECEPTOR MEASUREMENT

$$R_c + R_cS + S^{\star} \xrightarrow[4^\circ C]{\text{Assay}} R_cS^{\star} + R_cS \qquad \text{Parameter Unoccupied Sites}$$

$$R_c + R_cS + S^{\star} \xrightarrow[30^\circ\text{-}37^\circ C]{\text{Assay}} R_cS^{\star} + R_cS^{\star} \qquad \text{Total Sites}$$

$$R_nS + S^{\star} \xrightarrow[30^\circ\text{-}37^\circ C]{\text{Assay}} R_nS^{\star} \qquad \text{Total Nuclear Sites}$$

Fig. 17. Measurement of occupied and unoccupied receptor sites in cytosol and nuclear fractions by ^{3}H-steroid exchange. Cytoplasmic receptor, R_c; cytoplasmic receptor-steroid complex, R_cS; labeled steroid, S^*; cytoplasmic receptor-labeled steroid complex, R_cS^*; nuclear receptor-steroid complex, R_nS; nuclear receptor-labeled steroid complex, R_nS^*

will be discussed later in this chapter. The basic rules for exchange studies with the estrogen receptor are simple. Receptor-estrogen complexes dissociate slowly at low temperature but rapidly at elevated temperature. The estrogen receptor is also inactivated at elevated temperature. To effect an exchange between unlabeled ligand bound to receptor and free labeled ligand, as in the estimation of R_nS or R_cS, one must incubate the appropriate subcellular fraction at an elevated temperature in the presence of labeled ligand. Because these conditions of elevated temperature may compromise the estimate of the occupied receptor via its inactivation, a number of control studies are necessary to determine maximal conditions for exchange (see below).

Figure 17 shows that unoccupied sites can be measured in the presence of previously occupied sites by exposure of the appropriate subcellular fraction to labeled estradiol at low temperature. Under these conditions the labeled ligand rapidly associates with the receptor; incubation for 30 min – 1 h is usually sufficient for reliable estimates. Following this period, unbound steroid must be separated from bound for the estimate of receptor-estrogen complexes. A variety of techniques may be used for this purpose. Currently we use the rapid adsorption of R_cS^* complexes to hydroxylapatite (HAP), their precipitation by protamine sulfate, or the removal of unbound ligand by charcoal. The charcoal procedure relies on the adsorption of free steroid by charcoal followed by separation of charcoal from a supernatant containing the R_cS^* complex. The HAP procedure is quite the opposite; small molecular weight ligands do not associate readily with HAP, whereas proteins bind rather tightly to this matrix. Thus, HAP is used to adsorb R_cS^* complex at low temperature. The HAP is then washed and extracted with absolute ethanol to determine bound S^*. Alternatively, protamine sulfate may be employed to precipitate R_cS^* or R_nS^* complexes to effect their separation from free S^*. Each of these procedures should be carried out under cold conditions to avoid exchange and loss of label during manipulation. The protocols for these assays are given later in this chapter.

Also shown in Fig. 17 is the assay of total cytoplasmic sites ($R_cS + R_c$) and nuclear-receptor estrogen complexes (R_nS). These are measured by exposing the appropriate subcellular fraction to labeled steroid at elevated temperature for a short period. Exchange occurs during this period, and all receptor sites are occupied by labeled hormone. The original protocol for nuclear sites employed 37°C for 30 min for exchange (Anderson et al., 1972a). However, depending on the system under consideration, an appreciable number of receptor sites may be inactivated under these conditions. Therefore, lower temperatures and longer periods of incubation can be used to effect exchange without significant degradation. A more complete discussion of this problem follows.

b) Time and Temperature of Steroid Exchange

Exchange is a method for the determination of previously filled high-affinity receptor sites. As stated above, the procedure depends on the temperature dependence of the dissociation of receptor-estrogen complexes. Thus, low temperatures are employed to "trap" receptor-steroid complexes; the extent of exchange can be manipulated by the time and temperature at which the system is exposed to excess labeled steroid. Figure 18 shows the extent of exchange (bound ³H-steroid as a function of incubation time) for a hypothetical receptor-ligand complex assuming that the half-life of

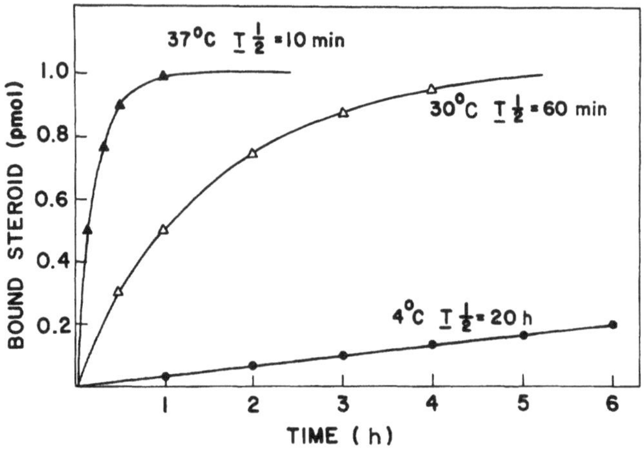

Fig. 18. Steroid exchange as a function of temperature

this complex ($T_{½}$) varies with temperature. The $T_{½}$ values refer to the half-time for dissociation of the complex in the presence of excess steroid at the given temperature. Note that ~ 1 h is adequate for complete exchange at 37°C, but days would be required at 4°C.

Unfortunately the system is seldom as simple as that shown in Fig. 18, but is usually like that shown in Fig. 19. In this case two processes occur during the exchange reaction. One is the temperature-dependent exchange shown in Fig. 18; the other is a temperature-dependent loss of receptor sites. The sum of these processes is depicted by the *dashed lines* in Fig. 19. This combination of reactions, exchange tending to increase measurable RS* while inactivation decreases measurable RS*, was first noted when comparing 37°C and 30°C as conditions for exchange of R_nE complexes in uterine tissue. However, it was most dramatically demonstrated in our

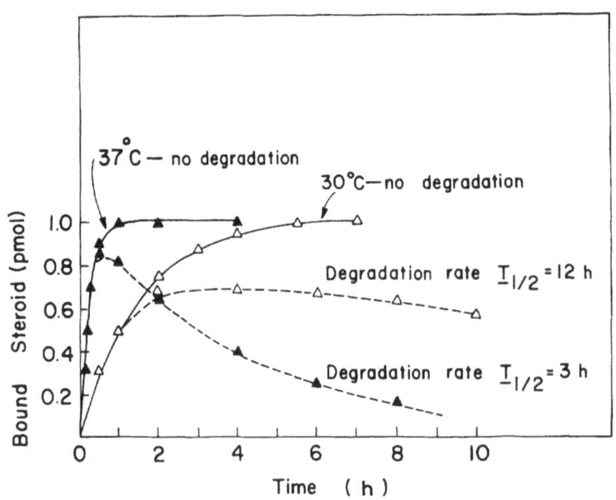

Fig. 19. Effect of receptor loss due to degradation on the quantity of receptor and apparent rate of steroid exchange

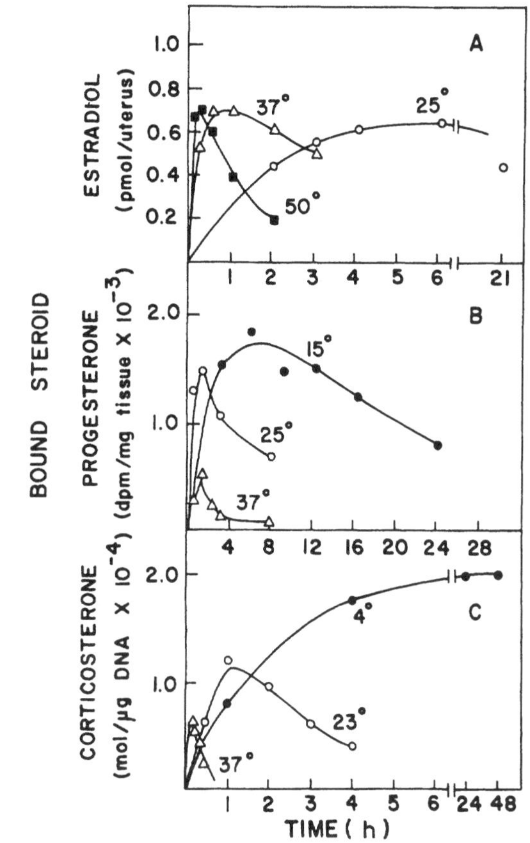

Fig. 20 A–C. Effect of time and temperature on the quantity of receptor-steroid complex measured with the [^{3}H]-steroid exchange assay. **A** R_nE complexes in uterine nuclei measured by exchange at the times and temperatures indicated. **B** R_nP complexes in uterine nuclei measured by exchange. **C** Assay of nuclear receptor corticosterone complexes in liver nuclei by exchange

laboratory in attempts to measure progesterone and corticosterone receptors by exchange (Hsueh et al., 1974; Walters and Clark, 1977a). Data for three experimental systems are shown in Fig. 20. Exchange of R_nE occurs without substantial inactivation except at very high temperatures, whereas complexes of progesterone and corticosterone with their receptors exhibit complex kinetics for exchange. These studies emphasize the need for a series of time and temperature studies before choosing a procedure for an exchange assay. Otherwise, an artificially low number of nuclear-receptor steroid complexes is apt to be measured.

Both cytoplasmic and nuclear progesterone receptors of the rat uterus can be measured by exchange at 4°C (Walters and Clark, 1977a, b; 1978). Exchange is complete by 10 h, and no degradation of receptor is detectable by 50 h (Fig. 21, see next page). Therefore, this assay is not complicated by a degradation factor; however, it is not possible to distinguish between occupied and unoccupied binding sites.

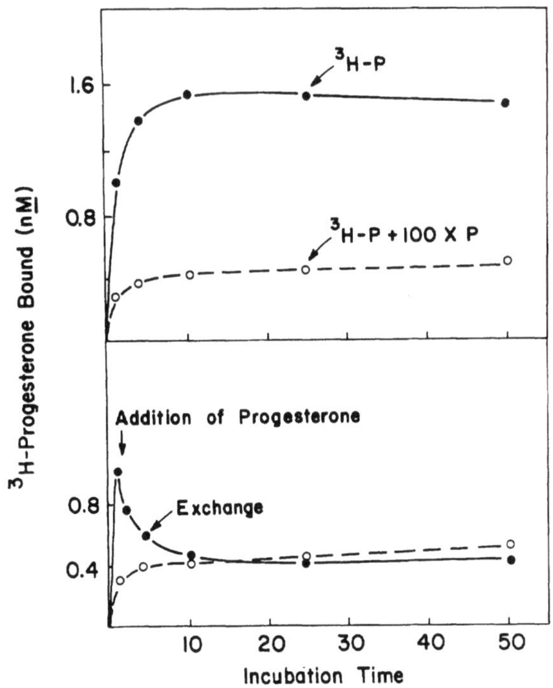

Fig. 21. Binding and exchange of progesterone. Uterine cytosol was incubated with 20 n*M* ^{3}H-progesterone (●) or ^{3}H-progesterone plus 100 × excess of progesterone (○) at 4°C as function of minutes (*top panel*). In the *lower panel* 2 μ*M* progesterone was added to uterine cytosol after 60 min exposure to 20 n*M* ^{3}H-progesterone

c) Determination of the Binding State of Receptor Sites

The actual binding state in situ and relationship between occupied and unoccupied estrogen receptor sites was first demonstrated by Williams and Gorski (1971) in their study of the cytoplasmic estrogen receptor. This investigation, since confirmed by our own laboratories and others, demonstrated that a constant and very low percentage of total occupied estrogen receptors remain in the cytoplasmic compartment of uterine cells despite the level of estrogen present. Furthermore, the number of filled cytoplasmic sites could be artifactually high unless precautions were taken to prevent the association of estrogen (which was free in interstitial spaces and loosely associated with nonspecific sites) with unfilled cytoplasmic estrogen receptors following tissue disruption. The "true" and "artifactual" states are represented in Fig. 22. Note that a target tissue has several pools of estrogen. Among these are free estrogen in both inter-and intracellular spaces as well as estrogen tightly associated with high-affinity receptor sites and that loosely associated with nonspecific sites (e.g. lipid, hydrophobic proteins, other estrogen specific sites). Occupied receptor sites are present in both the cytoplasmic and nuclear compartments, whereas unoccupied sites are present primarily in the cytoplasmic pool (see below for exception). If tissues are homogenized, estrogen, previously in the intercellular space and unavailable for interactions with the cytoplasmic receptor, is free to occupy previously unoccupied sites. In addition, estrogen previously bound to nonspecific, low-affinity sites is likely to dissociate and be available for interaction with unoccupied receptor sites. Thus, the act of homogenization with its attendant disruption of compartments and dilution of pools results in artifactually high estimates of occupied

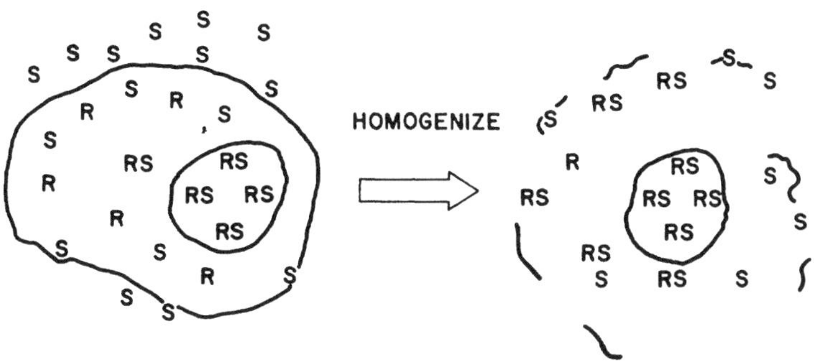

Fig. 22. Artifactual production of receptor-steroid complexes as a result of cell breakage by homogenization

cytoplasmic sites. The correction introduced by Williams and Gorski (1971) to avoid this artifact consisted of homogenization in the presence of excess unlabeled ligand (estrogen). This manipulation allowed the estimation of receptor sites occupied with labeled estrogen under various in vitro incubation conditions.

Our laboratory has modified this technique to allow the estimation of occupied and unoccupied estrogen receptors that exist in tissues of normally cycling animals. The procedure, outlined in Fig. 23, represents the integration of the Williams and Gorski homogenization procedure with the steroid exchange assay (Anderson et al., 1972a). Briefly, tissues previously exposed to endogenous or exogenous steroids in vivo are homogenized in the presence of excess labeled steroid (S*) at 4°C (*A*). This treatment results in the occupation of previously unoccupied sites by S* and allows

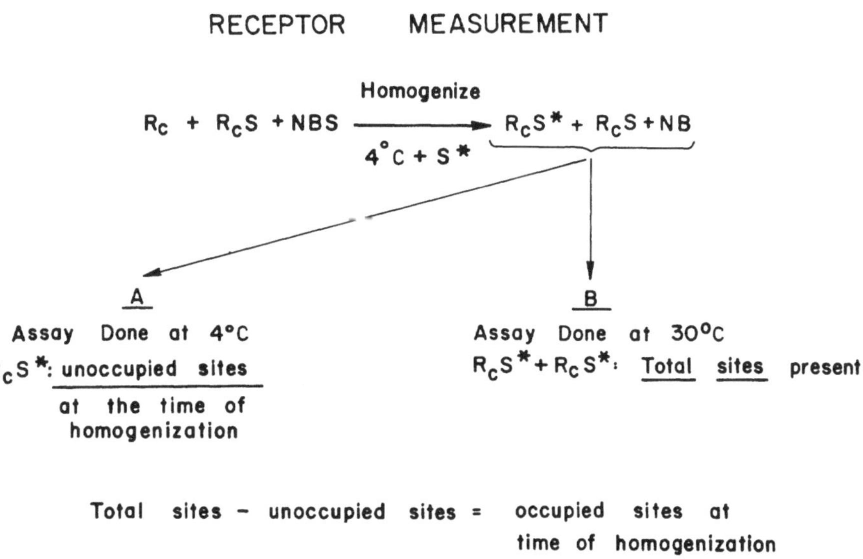

Fig. 23. Determination of receptor state at the time of homogenization and measurement of unoccupied and occupied sites. Unoccupied cytosol receptor, R_c; occupied cytosol receptor, R_cS; occupied nonspecific binding sites, *NBS*; ³H-steroid, *S**; ³H-steroid bound to receptor, R_cS^*; nonlabeled steroid bound to receptor, R_cS; unoccupied nonspecific binding sites, *NB*

the direct estimate of their number (R_c) by charcoal or hydroxylapatite adsorption techniques. Occupied sites (R_cS) do not dissociate during the assay period due to the slow rate of dissociation of estrogen at low temperature (see above). To assess those sites occupied by physiologic or pharmacologic estrogen prior to homogenization, a duplicate sample of cytosol is warmed at 30°C for 1 h or room temperature overnight (*B*). This elevated temperature allows the exchange of labeled steroid for the unlabeled ligand previously bound to receptor sites. The estimate of bound steroid in this case (i.e., by charcoal or hydroxylapatite adsorption) is the total of receptor sites present in the cytosol ($R_c + R_cS$). Occupied sites present at the time of homogenization (R_cS) can then be determined by difference using the value obtained above for unoccupied sites. Note that in this case, as in all studies of receptor-estrogen interactions, controls must be maintained for nonspecific binding. Thus, all incubation and homogenization manipulation should include samples with and without excess unlabeled estrogen (i.e., DES).

As mentioned earlier, it is not possible to differentiate between occupied and unoccupied progesterone receptors by exchange since this reaction takes place at 4°C. Therefore the method outlined here for the determination of the actual binding state of a receptor is not applicable to the progesterone receptor of the rat uterus.

In summary, the exchange procedure can be used to measure occupied and unoccupied estrogen receptor sites in various subcellular compartments. Two conditions must be established to maximize exchange and minimize receptor loss — the time and the temperature of incubation in the presence of labeled steroid. Additionally, one must establish that receptor-steroid complexes do not dissociate to an appreciable extent at low temperatures during subsequent manipulation. Given that these conditions are met, one should be able to use this procedure in relating receptor occupancy to hormone-induced responses.

4. Methods of Receptor Assay

a) Cytoplasmic Receptors

The methods to be discussed in this section represent some of the assays currently used for interactions of steroid hormones with receptors. We will consider only rapid procedures based on adsorption of free steroid or receptor-steroid complexes for the separation of bound and free hormone. Other procedures are used routinely by clinicians and researchers alike (Fishman and Fishman, 1974; Ritzen et al., 1974; Castaneda and Liao, 1975; Floridi et al., 1975; Cochet et al., 1976; Sherman et al., 1976; Wagner and Jungblut, 1976). That they are not included does not imply that they are undesirable, but rather that we have had little or no experience with them. Among the techniques not mentioned are equilibrium dialysis, discussed at length by Baulieu and Raynaud (1970), and gel filtration (Giannopoulos and Gorski, 1971a). We consider these methods lengthy and difficult to quantitate. Thus, a major advantage of adsorption procedures over equilibrium dialysis or gel filtration is their rapidity and ease of performance. All depend on the rapid separation of free S from bound S at 0–4°C, a temperature at which dissociation of steroid from the receptor is slow. The first approach, involving the adsorption of free S with charcoal and the measurement of R_cS complexes, was first described by Korenman et al. (1969) and modified by Mester et al., (1970) and Sanborn et al., (1971). Alternative procedures depend on the adsorption of RS complexes to charged surfaces as first described by Clark and Gorski (1969) for glass pellets and modified for hydroxylapatite and DEAE filters by Erdos et al., (1970) and Santi et al., (1973) or the precipitation of RS complexes with protamine sulfate (Steggles and King, 1970; Chamness et al., 1975a). Although these procedures are rapid and generally effective for the separation of free from bound S, they are not without fault. One should employ them cautiously, always bearing in mind the dangers inherent in determining equilibrium values under non-equilibrium conditions.

As discussed earlier, both occupied and unoccupied sites can be measured by ^{3}H-estradiol exchange. The ^{3}H-estradiol exchange assay for nuclear receptor (Anderson et al., 1972a) was modified for the measurement of cytosol-estrogen receptors by Katzenellenbogen et al. (1973). This methodology has subsequently been used to measure (1) cytosol estrogen receptors in the rat ovary (Saiduddin and Zassenhaus, 1977a), mammary gland, normal and cancerous (Watson et al., 1977), pituitary and hypothalamus (Anderson et al., 1973b; Peck et al., 1978); (2) cytosol progesterone receptors in the rat uterus (Walters and Clark, 1977a, b), mouse uterus (Philibert and Raynaud, 1977) and mammary tumors (Daehnfeldt and Briand, 1977); (3) cytosol androgen receptors in the rat prostate (Bonne and Raynaud, 1976); (4) nuclear androgen receptors in the rat testes (Sanborn and Steinberger, 1975) and ventral prostate (Davies et al., 1976); (5) nuclear estrogen receptors in the ovary (Richards, 1974), avian Müllerian duct (Teng and Teng, 1976), chick oviduct (Kalimi et al., 1976); (6) nuclear progesterone receptors in the uterus (Hsueh et al., 1976; Walters and Clark, 1977a, b); and (7) cytosol and nuclear estrogen receptors in mammary tumors (Nicholson et al., 1976; Watson et al., 1977).

1) Charcoal Adsorption Procedure

This method has been adopted by many investigators because of its inexpensive nature and ease of application. In addition, charcoal is a sufficiently strong adsorbent to remove or strip hormone from nonreceptor proteins; thus, the "nonspecific" term in experiments is often low when this method is used. However, the method should be confirmed by an alternate procedure since charcoal will adsorb receptor at low protein concentrations and will strip ^{3}H-steroid from high-affinity sites such as 4*S* uterine estrogen receptors (Peck and Clark, 1977).

The desired protocol varies greatly with the receptor system under study. In general, the more rapid the rate of interaction of hormone with receptor, the more labile the complex is to charcoal exposure. To counteract this tendency various labs have shortened the exposure time to charcoal. In this way the technique has been used to measure chick progesterone receptors (Schrader and O'Malley, 1972; Hansen et al., 1976), mammalian progesterone receptor (Feil et al., 1972; Walters and Clark, 1977a, b, 1978), and glucocorticoid receptors (Santi et al., 1973).

The assay methods for estrogen and progesterone receptors differ in some respects but are essentially identical in others. Therefore, only when differences exist will they be pointed out; otherwise the methods are identical for both receptors and are presented in the following steps:

1. For the assay of estrogen receptors, tissue should be removed, weighed and placed in ice-cold TESH buffer (10 m*M* Tris, 1.5 m*M* EDTA·4Na, 1 m*M* dithiothreitol, pH 7.4) as quickly as possible. Tissue/volume = 25–30 mg/ml for

uterine tissue (1 uterus/ml or 4–5 mg protein/ml; 2–3 hypothalami/ml; 2–4 pituitaries/ml).

For the measurement of progesterone receptors tissue should be placed in ice-cold TG buffer (10 m*M* Tris, 30% glycerol v/v, pH 7.4). Tissue to volume ratios are similar to those above.

2. Homogenize tissue in ground-glass homogenizer with motor-driven pestle. Use short periods of homogenization with alternate cooling in an ice bath.

3. Centrifuge the homogenate at 800–900 *g* for 20 min.

4. Separate cytosol from pellet.

5. Centrifuge cytosol at 48,000 *g* for 20 min. If a nuclear pellet is not desired, steps 3 and 4 may be omitted, and the initial homogenate centrifuged at 48,000 *g* for 20 min.

6. The following is for "one point" assays only.

Estrogen receptors

Set up assay tubes (12 × 75 mm disposable culture tubes) as follows:

Series A: 20 μl $^3H\text{-}E_2$ (1.5 ng/20 μl) and 20 μl TESH buffer

Series B: 20 μl $^3H\text{-}E_2$ (1.5 ng/20 μl) and 20 μl DES 150 ng/20 μl)

These will allow final concentrations of estrogen (after Step 8 below) of 10 n*M* plus or minus 1 μ*M* DES.

Progesterone receptors

^{3}H-Progesterone is used as the labeled ligand (20 n*M*) and unlabeled R5020 as the competitor (2 μ*M*). R5020 is a synthetic progestin that binds to the progesterone receptor with high affinity and does not bind to corticoid binding globulin (CBG) (Philibert and Raynaud, 1974). Since progesterone does bind to CBG under the conditions used in this assay the use of nonlabeled progesterone as a competitor could lead to errors.

7. The following applies when saturation analysis is desired:

Tubes should be arranged as above with the concentration of ^{3}H-steroid varying from 0.1–10.0 n*M*. DES and R5020 concentrations should be 100 × that of each labeled ligand. (Note: The use of competitors in large excess may result in invalid estimates of specific binding, as discussed in Sect. 2.e of this chapter.) Assays in this concentration range measure primarily the classic estrogen receptor; however, it is possible that secondary sites may interfere, and thus, the data should always be analyzed by the methods discussed earlier for resolving mixed binding systems (Sect. 2.e, also see Chap. IV for discussion of secondary estradiol binding sites).

8. Place 500 μl cytosol in each assay tube and mix.

9. Incubation of cytosol with ^{3}H-estradiol can be performed in several ways (two of which follow) depending on the desired end point:

Measurement of unoccupied receptor sites: Incubate cytosol + hormone for 15 min to several hours at 4°C.

Measurement of total sites (occupied and unoccupied): Incubate cytosol + hormone for 18–24 h at room temperature or for 60 min at 30°C.

For the estrogen receptor assay, tubes should be placed in an ice water bath for 5–10 min. Add 500 μl 0.2 percent charcoal and 0.05 percent dextran suspension to each tube and leave at 4°C for 15 min with vortexing every 5 min. Centrifuge assay tubes at 5000 *g* for 10 min.

Incubation of cytosol for progesterone receptor assay should be terminated by a brief exposure (~ 30 s) to a charcoal-dextran solution (1.0 g charcoal, 0.05 g dextran/100 ml of TG buffer). Add 250 µl charcoal-dextran suspension to tubes with shaking and centrifuge immediately at 2500 *g* for 5–6 min. This entire procedure should be done as rapidly as possible.

10. Add a 500 µl aliquot of the supernatant to 4.0 ml aqueous counting scintillant (Caution: avoid getting charcoal in the 500 µl aliquot).

2) Hydroxylapatite Procedure

This procedure, first described as a column method by Erdos et al. (1970), has recently been modified to a batch procedure (Pavlik and Coulson, 1976a; Williams and Gorski, 1974; Clark et al., 1977b; Peck and Clark, 1977; Walters and Clark, 1977b). The batch process employed in our own laboratory will be described.

This method is excellent for the determination of multiple forms of the estrogen receptor in the presence or absence of elevated KCl. All known forms of steroid receptor adsorb to hydroxylapatite (HAP) and can be assessed with this procedure. An additional advantage of this technique is that receptors can be eluted from HAP if desired by using 0.5 *M* phosphate. Thus, the method allows quantitative analysis via scintillation spectrometry and qualitative assessment via elution and either sucrose gradient centrifugation or gel filtration. The method is gentle enough to use with highly purified receptors (Schrader et al., 1975).

The procedure outlined below is for the assay of estrogen receptors. For the progesterone receptor assay, TG buffer should be substituted for TE. Otherwise, the procedures are identical (Walters and Clark, 1977b):

1. Prepare cytosol and assay tubes as described in the charcoal assay (steps 1–7).
2. Place 250 µl cytosol and 250 µl TESH buffer in each assay tube and mix.
3. Incubation is carried out under any of the conditions described in step 9 of the charcoal procedure.
4. Place the tubes in an ice water bath and leave them for 5–10 min.
5. Add 500 µl of HAP suspension (see below) to each tube and leave on ice for 15 min with vortexing every 5 min.
6. Centrifuge the tubes at 5000 *g* for 5 min to pack the HAP.
7. Wash the HAP pellets three times by removing the supernatant, resuspending with 1.5 ml aliquots of Tris-EDTA buffer, and centrifuging at 5000 *g* for 5 min.
8. Add 1 ml 100% EtOH to each tube and vortex.
9. Place tubes in a 30°C water bath for 10 min or at room temperature for 30 min.
10. Vortex the tubes and then centrifuge them at 5000 *g* for 10 min.
11. Add the EtOH extract (supernatant) to 4 ml aqueous counting scintillant.

Preparation of HAP suspension: A HAP suspension is made by simply washing Biogel HTP (Biorad) three times with TE buffer and resuspending to a final concentration of 60 percent HAP.

3) Protamine Sulfate Precipitation Method

Protamine sulfate, used to measure unoccupied estrogen receptor sites by Steggles and King (1970), has been modified by Chamness et al., (1975), to measure occupied and unoccupied sites by exchange. We have recently employed the procedure to measure cytoplasmic and nuclear receptors in uteri, pituitaries, and hypothalami (Kelner and Peck, unpublished observations).

1. Prepare cytosol and assay tubes as described in steps 1–7 of the charcoal assay procedure.
2. Add 250 µl cytosol to each assay tube and vortex.
3. Incubate under any of the conditions described in step 9 of the charcoal assay procedure. Total or unoccupied receptor may be measured.
4. If the incubation was at room temperature or 30°C place the tubes in an ice water bath for 5–10 min to terminate the incubation.
5. Add 250 µl protamine sulfate (1 mg protamine sulfate/ml TESH buffer), vortex, and let stand on ice (0°–4°C) for 10–15 min.
6. Add 1 ml cold TESH buffer to each tube and centrifuge at 5000 *g* for 15 min.
7. Pour off supernatant and wash the pellet twice with 1 ml ice-cold TESH, centrifuging at 5000 *g* for 10 min after each wash.
8. Pour off supernatant and add 1 ml 100% EtOH to each tube; vortex.
9. Let stand for at least 1 h at room temperature or at 30°C for 15 min.
10. Vortex tubes and pour directly into scintillation vials together with 4 ml scintillation fluid.

4) DEAE Filter Method

This method, developed by Santi et al. (1973) for the study of the glucocorticoid receptor, involves the adsorption of R_cS to DEAE cellulose filters with removal of unbound ^{3}H-steroid by washing with low ionic strength buffer. It is rapid and reproducible. One of its major advantages is the means to measure ^{3}H-steroid bound to dilute receptor species in large volumes. Major drawbacks include the adsorption of free ^{3}H-estradiol and ^{3}H-testosterone to the ion exchange filter at low ionic strength and the inability of receptors to bind to DEAE in high ionic strength media. These limitations compromise its use in the assay of estrogen or androgen receptors or of receptor samples in high ionic strength media.

In this procedure cytosol concentrations are usually 1–2 uteri/ml, and incubation volumes are small (50–100 µl). After equilibration with ^{3}H-steroid (0.5–13 n*M*) with or without excess unlabeled steroid (competitor), 50–100 µl aliquots of labeled cytosol are applied to DEAE filters that have been presoaked in Tris-EDTA buffer. Cytosol is filtered after at least 1 min of exposure to the filter. Filters are washed with five 1.0 ml portions of ice-cold Tris-EDTA buffer with gentle vacuum and excessive moisture removed with increased suction. Damp filters are placed into a toluene-based scintillation fluid, mixed by vortexing, and analyzed via liquid scintillation spectrometry.

b) Nuclear Receptors

The ^{3}H-estradiol exchange assay has proven useful for measuring nuclear receptor-estrogen complexes in the uterus (Russell and Thomas, 1974; Anderson et al., 1975; Lan and Katzenellenbogen, 1976), pituitary and hypothalamus (Anderson et al., 1973b), lactating mammary gland (Hsueh et al., 1973), corpus luteum (Richards, 1974), chick oviduct (Kalimi et al., 1976), Leydig cell tumors (Samuels et al., personal communication), and testes (de Boer et al., 1977). Recently the method has been modified in our laboratory for the measurement of progesterone and glucocorticoid receptors in the nuclear and cytoplasmic compartments of the rat uterus (Hsueh et al., 1974; Walters and Clark, 1977a, b) and in liver nuclei from both rat and chicken (Eriksson, Hardin, and Clark, unpublished observations). Other laboratories have used exchange to measure androgen receptors in nuclei from testes (Sanborn and Steinberger, 1975). We have also used this method to quantify the number of receptor-estrogen complexes that are bound to chromatin (Hardin et al., in press). Thus, it appears generally applicable. In this section we present the method currently in use for estrogen receptors and then indicate modifications that apply to other steroids.

1) ^{3}H-Estradiol Exchange of Nuclear Receptor-Estrogen Complexes

Tissues are removed, weighed, and placed in ice-cold TESH buffer (0.01 *M* Tris, 0.0015 *M* Na_2 EDTA, 0.001 *M* dithiothreitol, pH 7.4) as quickly as possible. Unless stated otherwise all procedures should be performed at 2–4°C. Tissue/volume = 10–50 mg/ml for uterine tissue (1–2 immature uteri/ml or 4–8 mg protein/ml) and 100–200 mg/ml for tissues containing lower quantities of receptor (e.g., hypothalamus or mammary gland).

Tissue is homogenized in a ground-glass homogenizer with a motor-driven pestle. Short periods (10–30 s) of homogenization are used with alternate cooling in an ice bath. For easily homogenized tissues such as hypothalamus or pituitary, a glass homogenizer with a Teflon pestle should be used.

The homogenate is centrifuged for 20 min at 800 *g* to obtain the crude nuclear pellet which is then washed three times. Washing is accomplished by adding ice-cold TESH buffer to the pellet (1 ml per 50 mg of uterus), vortexing thoroughly, and centrifuging at 800 *g* for 10 min. The assay may also be performed on purified nuclei or chromatin (see below). However, we usually use crude nuclear preparations to avoid losses in the preparation of these components.

The washed nuclear pellet is resuspended by gentle homogenization in TESH buffer (same ratios as above) and 250 μl aliquots of the nuclear suspension are dispensed into 12 × 75 mm disposable culture tubes, which contain either ^{3}H-estradiol or ^{3}H-estradiol plus a 100-fold excess of diethylstilbestrol in 250 μl of TESH buffer (final vol = 500 μl). 2,4,6,7-^{3}H-Estradiol (90–115 Ci/mmol) should be used in most cases; however, 6,7-^{3}H-estradiol (40–50 Ci/mmol) may be used in assays that require only higher ^{3}H-estradiol concentration ranges.

Assays to *estimate* the total number of receptors may be done using 10 n*M* ^{3}H-estradiol with or without 1 μ*M* DES. If a saturation analysis is desired, the ^{3}H-estradiol concentration (final) should range from 0.1 n*M* to 10.0 n*M*, and the DES concentration should be 100 × that of ^{3}H-estradiol. Most assays, especially at the beginning of a study, should utilize saturation analysis. This will ensure accurate determinations of appropriate binding parameters. *Note*: The use of DES concentrations higher than 100 × that of ^{3}H-estradiol may result in invalid estimates of specific binding (see Sect. 2. e 2 of this chapter for further discussion).

Assay tubes are incubated at 37°C for 30 min or 30°C for 60 min with occasional vortexing. This step should be varied to optimize assay conditions for other receptors, i.e., temperature can be lowered and/or time of incubation can be extended. Thus, for the glucocorticoid and progesterone receptors in the rat, an overnight incubation period at 4°C is appropriate (Fig. 18). Exchange is terminated by the addition of 1 ml ice-cold TESH buffer to each tube followed by centrifugation at 800 *g* for 10 min.

The nuclear pellet is washed three times by resuspending the pellet in 1.5 ml ice-cold TESH buffer, vortexing, and centrifuging as described above.

1.0 ml 100% ethanol is added to the washed nuclear pellets and assays tubes are placed in a 30°C water bath for 30 min. Tubes are vortexed and centrifuged at 800 *g* for 10 min. The total ethanolic extract is added to 5 ml scintillation cocktail (21 g Permablend [Packard] per gallon of toluene) and samples are counted.

The exchange method can be used for the determination of unoccupied receptor sites (R_n) as well as occupied complexes (R_nE). Incubation of nuclear preparations at 4°C with ^{3}H-estradiol for 15 min – 2 h will result in the binding of labeled steroid to R_n sites but will leave R_nE sites unaffected. Thus, a comparison of assays run at 4°C with those run under exchange conditions, i.e., elevated temperatures, provides a method for the determination of both states of the receptor. This method is not suitable for the differentiation of occupied and unoccupied progesterone receptor sites since exchange takes place at 4°C.

The quantities of receptor-steroid complex (RS, Fig. 6) are calculated by subtracting nonspecific binding (Fig. 6B, NS) from the total steroid bound (Fig. 6A, RS + NS). These values may then be plotted according to the method of Scatchard (1949) for the evaluation of the total number and dissociation constant of RS complexes.

Although it is not necessary to plot total and nonspecific binding to make the above determinations, these data give a complete picture of the influence of these parameters on the accuracy of estimates.

One of the common problems incurred when using the ^{3}H-steroid exchange assay is the presence of high levels of nonspecific binding. This situation decreases the probability that the receptor can be measured accurately. An example is shown in Fig. 5A; as discussed earlier, a careful analysis of saturation is necessary to establish the existence of competitive and saturable binding under these conditions. To overcome this problem, Zava et al. (1976) have used protamine sulfate to precipitate receptor that has been solubilized by 0.6 *M* KCl from nuclear pellets. ^{3}H-Estradiol exchange is then performed on the precipitate. This method reduces possible losses due to degeneration of the receptor and decreases the level of nonspecific binding. However, this method fails to account for the residual receptor that is not extracted by the salt solution. As discussed later in this chapter, this salt-insoluble fraction may be an important contributor to the overall mechanism by which R_nE complexes stimulate uterotropic responses. Other methods to decrease nonspecific binding include washing nuclear pellets with detergent and/or dilute ethanol solutions.

2) Modifications of ^{3}H-Steroid Exchange Assay

The measurement of the progesterone receptor by steroid exchange differs slightly from the procedure employed for estrogen receptors as discussed above (Hsueh et al., 1974; Walters and Clark, 1977a, b). The major difference stems from the tendency of mammalian progesterone receptors to undergo degeneration or rapid dissociation. Milgrom et al. (1972b) developed this procedure for the measurement of progesterone receptors in the cytosol of the guinea pig uterus. Measurement of mammalian progesterone receptor by exchange requires lower temperatures and therefore longer periods for complete exchange. The modifications described below permit exchange to occur with minimal degradation of the receptor.

Tissues are weighed and washed in cold Tris-glycerol (TG) buffer (10 m*M* Tris-HCl, 30% glycerol (v/v), pH 7.4). All subsequent steps are performed at 4°C unless otherwise indicated. The crude nuclear fraction is obtained and washed as before.

The washed pellet is suspended in TG buffer to a concentration equivalent to 50 mg tissue/ml. Portions (0.5 ml) of this suspension are dispensed into two series of tubes, A and B, containing 40 μl buffer. Series A contains 20 nM ^{3}H-progesterone and is used to determine the total amount of ^{3}H-progesterone exchanged. Series B contains ^{3}H-progesterone as in series A plus a 100-fold excess of unlabeled R5020. Saturation analysis is performed as described above for the ^{3}H-estradiol exchange assay. The nuclear fractions are incubated with shaking at 4°C for 18–24 h. Following incubation, 1.5 ml TG buffer is added, and the samples are centrifuged at 800 *g* for 10 min. Pellets are washed three times with TG buffer and extracted with 1.0 ml ethanol. The radioactivity in the ethanol extract is determined as explained for the ^{3}H-estradiol exchange procedure.

3) Nuclear Binding and Retention of Receptor-Hormone Complexes

The interactions between R_nS complexes and nuclear binding sites are poorly understood. However, we have demonstrated that one parameter of importance to physiologic response is the retention of R_nS complexes by nuclei. Therefore the measurement of R_nS complexes as a function of time during or after hormone exposure is of utmost importance. In our laboratory this has been done with the ^{3}H-steroid exchange assay under a variety of conditions as described below.

Nuclear and chromatin preparation. The ^{3}H-steroid exchange procedure can be used to measure hormone-receptor complexes bound to purified nuclei or to chromatin. Several different methods exist for the preparation of purified nuclei from a variety of tissues. Many are laborious and give relatively poor recoveries. However, a new method has been proposed by Wray and Stubblefield (1970) and modified by Conn and O'Malley (1975) and Hardin et al. (1976) which utilizes buffered hexylene glycol. This method has proven to be simple and reliable and has been used successfully in nuclear isolations from a variety of cells and tissues (Conn and O'Malley, 1975; Hardin et al., 1976).

For studies on receptor steroid binding to chromatin, this nuclear component can be prepared according to the method described by Spelsberg et al. (1971a, b). However, since the majority of receptor complexes are extracted with high ionic strength solutions, the final salt wash of the isolated chromatin must be omitted from this procedure.

Nuclear retention and acceptor sites. The long-term retention of 1000–3000 receptor estrogen complexes/cell is a requirement for uterine growth (Anderson et al., 1975; Clark and Peck, 1976a; see Chap. V for further discussion). Since uterine cells contain 15–20,000 receptor sites in the cytoplasm, only a fraction of these are required for maximal growth responses. We have suggested that the long-term retention of receptor-estrogen complexes in the nucleus results from the binding of these complexes to a limited number of nuclear-acceptor sites and that retention at these acceptor sites for greater than 4–6 h is required for the production of true uterine growth (Clark et al., 1973a; see Chaps. V and X for further discussion of these points).

To test this hypothesis we employed salt extraction of uterine nuclei to examine the differential extractability of receptor-estrogen complexes. The rationale for this technique was based on the observation that extraction of nuclei with 0.3–0.4 M KCl does not remove all of the nuclear-bound estrogen, suggesting that some receptor-hormone complexes are bound more tightly than others. With this technique it can be shown that R_nE exists in two forms: (a) KCl-extractable, defined as any R_nE complex that can be extracted by 0.4 *M* KCl or less and (b) KCl-resistant, defined as any R_nE that resists extraction by 0.4 *M* KCl. The stoichiometric and temporal relationship between these forms suggests that R_nE complexes that resist KCl extraction may be bound to nuclear acceptor sites (Clark and Peck, 1976a).

KCl extraction of nuclear R_nE is a simple procedure consisting of exposing nuclei to various concentrations of KCl. Nuclear fractions are washed three times with cold TE buffer (0.01 *M* Tris, EDTA 0.0015 *M*, pH 7.2, 4°C). Salt extraction is performed by adding various concentrations of TK buffer (0.01 *M* Tris and 0.1–0.6 *M* KCl, pH 8.0, 4°C) to the nuclear pellet with the mass to volume ratio maintained at 30–50 mg tissue/ml (300–400 μg DNA/ml). The nuclear pellets are resuspended in the TK buffer and left on ice for 15 min with occasional mixing. This is followed by centrifugation at 800 *g* for 10 min, and the KCl extract is carefully decanted. TE buffer is added to the nuclear pellet, and this pellet is resuspended. If necessary, resuspension is accomplished by rehomogenization of the pellet with a glass-Teflon homogenizer. After the pellet is resuspended it is washed three times with TE buffer. The salt-extracted pellet is used in a ^3H-estradiol exchange procedure for the determination of KCl-resistant complexes. The quantities of receptor steroid complex in the KCl extract are determined by the hydroxylapatite exchange assay. This assay technique is an integration of the ^3H-estradiol exchange procedure and the hydroxylapatite assay, both of which are described above. In this procedure salt extracts are centrifuged at 18,000 *g* for 20 min, exposed to labeled steroid for exchange, and subsequently adsorbed to HAP. HAP pellets are centrifuged, washed in cold buffer, and bound label extracted as before. Numbers of complexes measured are reported as a function of the salt concentration required for their extraction. It is very important to determine the actual salt concentration in the extract since the volume of the nuclear pellet will dilute the original salt solution. This may be done by measuring the conductivity of the extract.

In our laboratory all measurements of KCl-extractable vs. nonextractable sites have employed 0.4 *M* KCl extraction followed by application of exchange techniques to both soluble and insoluble fractions. The direct measurement of extracted radioactivity without use of exchange procedures is very likely to give different results, especially considering the variety of loosely bound pools of estrogen present in these subcellular compartments (see Chap. X for discussion).

Chapter III

Cellular Compartments and Translocation of Receptor-Steroid Complexes

1. Cytoplasmic or Nuclear Localization of Receptors

As discussed in the preceeding chapters, steroid-hormone receptors are generally considered to exist as soluble macromolecules in the cytoplasm of target cells (Fig. 1). When these receptors (R_c) bind to steroid hormones (S) they form receptor-steroid complexes (R_cS), which undergo translocation to the nucleus (R_nS). These

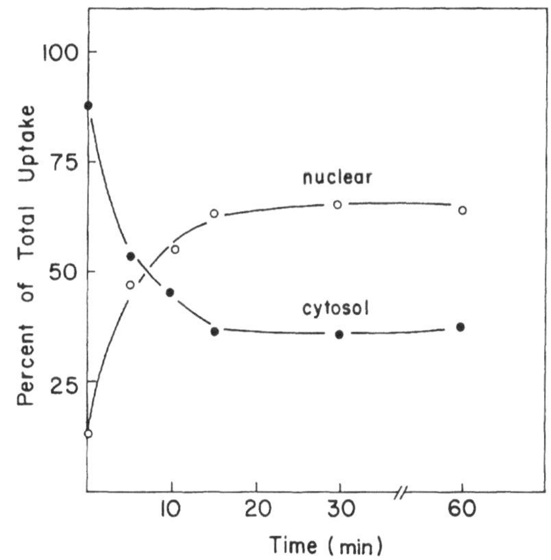

Fig. 24. The cellular distribution of ³H-estradiol in the uterus during in vitro incubation. Uteri were exposed to ³H-estradiol for 1 min, placed in fresh medium, and incubated at 37°C for various time periods (from Gorski et al., 1968)

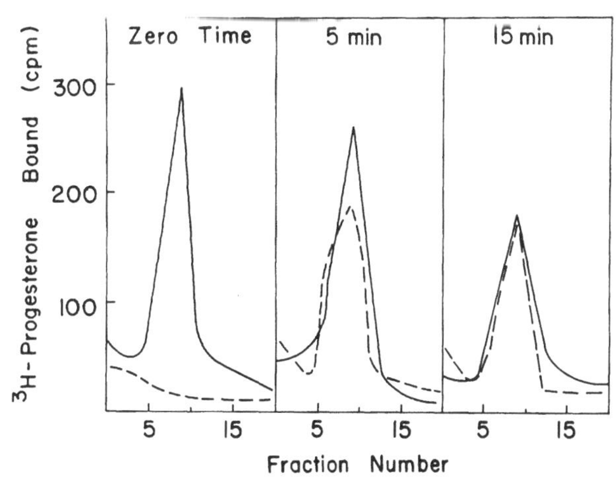

Fig. 25 Distribution of bound ³H-progesterone in the nuclear and cytosol fractions of the chick oviduct. Chick oviducts were incubated with ³H-progesterone, and bound ³H-progesterone was measured in the cytosol (*solid line*) and nuclear fractions (*broken line*) by sucrose density gradient analysis (O'Malley et al., 1970)

conclusions were reached from experiments by Jensen et al. (1968) and Gorski et al. (1968). Figure 24, taken from an experiment by Gorski et al. (1968), demonstrates the rapid shift of ^{3}H-estradiol from the cytoplasmic to the nuclear compartment of the rat uterus in vitro. Cytoplasmic receptor declines after either in vivo or in vitro exposure to estradiol and, as if in a concerted or ordered process, there occurs a concomitant increase in the quantity of a salt-extractable 5*S* complex in the nucleus (Jensen et al., 1968; Gorski et al., 1968). Similar results have been obtained by O'Malley et al. (1970) for the chick oviduct and are shown in Fig. 25.

This translocation of receptor from cytoplasm to nucleus is a temperature-dependent process. Thus, in vitro nuclear accumulation of receptor-steroid complexes is not observed at 4°C, whereas elevating the temperature of incubation causes rapid translocation (Gorski et al., 1968; Jensen et al., 1968; Rochefort and Baulieu, 1969). Metabolic inhibitors have no effect on receptor translocation in the uterus (Shyamala and Gorski, 1967), which suggests that energy utilization is not involved. Translocation does not appear to depend on an intact microtubule or micro-

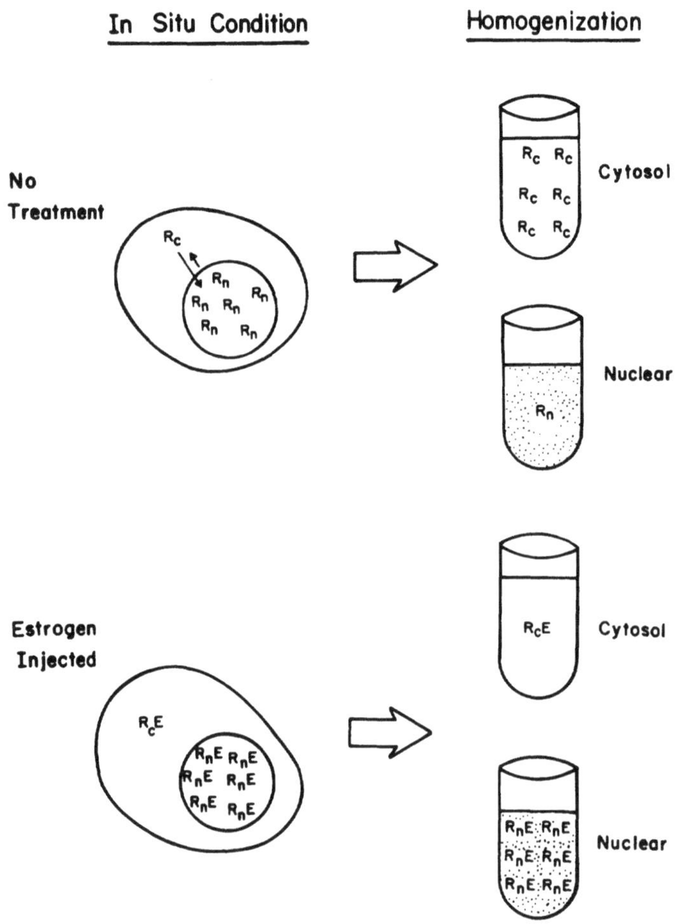

Fig. 26. Hypothetical model of steroid receptor distribution in situ and after homogenization

filament network since cytochalasin or vinblastine do not block the process (Gorski and Raker, 1974). Likewise, inhibitors of RNA and protein synthesis fail to interfere with the translocation process (Shyamala and Gorski, 1967).

The experiments cited above suggest that R_cS complexes undergo translocation to the nucleus of target cells; however, they do not prove that such translocation exists. Indeed, it has been suggested that steroid receptors reside predominantly in the nucleus in vivo and that R_c may represent an artifact found only after cellular disruption (Williams and Gorski, 1971; Sheridan, 1975). According to this hypothesis, which is illustrated in Fig. 26, R_c is observed in cytoplasmic fractions after homogenization because it is readily solubilized from the nucleus where it resides as unoccupied nuclear receptor in vivo. Steroid treatment would result in apparent translocation because R_nS complexes are tightly bound or stabilized in the nucleus and are not readily dislodged upon homogenization. In this manner exposure to steroid would appear to decrease the quantity of R_c while increasing the quantity of R_nS.

Other investigators have shown, in fact, that nuclear localization of unoccupied receptor is the predominant condition in some tissues. This localization varies with the source of the target tissue and the conditions of homogenization. As discussed in Chap. V, the chicken liver appears to contain no R_c for estrogen, while nuclear estrogen receptors are readily observed. In addition, Zava et al. (1977) and Panko and MacLeod (1978) have observed unoccupied nuclear receptors in certain mammary tumors. Thus, the precise subcellular distribution of unoccupied receptors re-

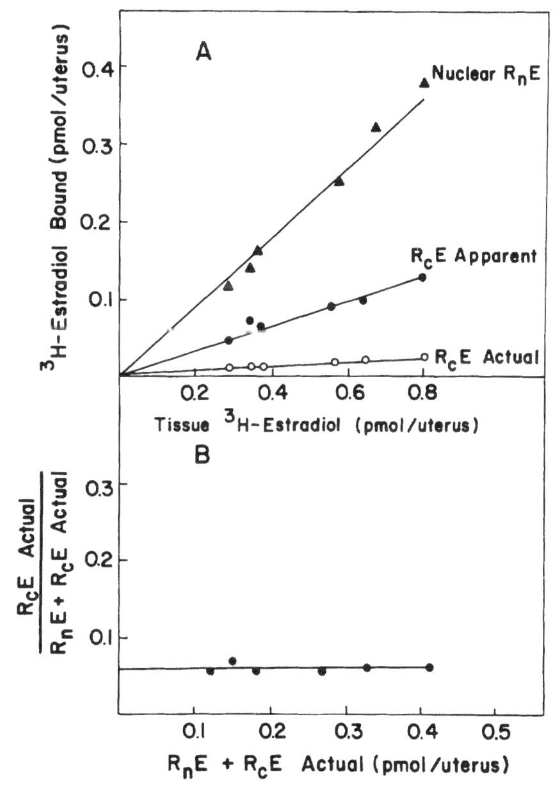

Fig. 27 A and B. The actual and apparent distribution of receptor estradiol complexes in the rat uterus. Uteri were incubated in varying concentrations of ^{3}H-estradiol and homogenized in buffer with or without excess DES. **A** The quantity of nuclear R_nE complexes was not different when homogenization was in the presence or absence of excess DES (▲); however, apparently occupied cytosol receptor (●) was reduced by homogenization in DES (○). **B** The quantity of cytoplasmic receptor sites actually occupied in situ remains a constant percentage of the total number of occupied sites (●)

mains a mystery. Certainly the dogma relating to localization of occupied and unoccupied receptors must be tempered by the observations discussed above.

Only a small percentage of the total bound hormone present in the cell at any time is in the R_cS form. Williams and Gorski (1971) showed this by incubating uteri in vitro with labeled steroid and subsequently homogenizing uteri in buffer containing excess unlabeled estradiol. This procedure, discussed below and in detail in Chap. II, permits the determination of the quantity of receptor-steroid complex in both cytoplasmic and nuclear compartments at the time of homogenization (see Figs. 22 and 23). Any R_cE complexes that are formed as a result of homogenization are readily detected by this procedure; in addition, the quantity of R_cE^* which existed at the time of homogenization can be determined. With this technique only a small quantity of R_cE (actually a constant percentage of total occupied receptor) exists either in vivo or in vitro, whereas 95 percent of the bound estrogen is in the R_nE form (Fig. 27).

The artifactual formation of R_cE^* as a result of homogenization is caused by the release of E^* from various extra- and intracellular pools. Among these pools are the free estrogen in both inter- and intracellular spaces as well as estrogen associated with various "nonspecific" or "nonreceptor" sites (such as lipid, hydrophobic proteins, steroid-binding globulins, etc.). Upon homogenization, estrogens previously localized in these pools (and not available for binding to receptor sites) are free to occupy previously unoccupied cytoplasmic receptors. In summary, homogenization of target tissues with its attendant disruption and dilution of compartments or pools can result in artifactually high estimates of R_cE.

2. Evidence for Nuclear Preference

As discussed above, one should not be dogmatic about the compartmentalization of unoccupied steroid receptors. However, it is possible to show that receptor-steroid complexes show preferential nuclear binding. The results shown in Fig. 28 demonstrate that greater quantities of receptor-estrogen complex accumulate in the nuclear fraction than in the mitochondrial-microsomal fraction. This observation was first made by Noteboom and Gorski (1965); other investigators have shown similar results for other steroid receptors (see Chap. IV and V for references). However, it is necessary to show that this apparent nuclear preference does not result from the large size of the nuclear fraction relative to that of the mitrochondrial-microsomal fraction as well as from the "sticky" nature of receptor-estrogen complexes (Clark and Gorski, 1969). Under certain conditions receptor-estrogen complexes will adhere to many surfaces including glass and hence might bind equally well to many cell organelles (see Chap. IV for further discussion of this point). Thus, when the uterus is homogenized, more R_cE complexes might bind (stick) to the large mass of the nuclear fraction than to the smaller mitochondrial-microsomal pellet and yield the results shown in Fig. 28A. When these data are expressed as a function of the dry weight or protein content of these fractions, however, nuclear preference is clearly demonstrated (Fig. 28B). Preferential nuclear accumulation probably accounts for the high nuclear/cytoplasmic ratio of bound ^{3}H-steroid observed by autoradiography (King et al., 1965a; Jensen et al., 1968; Stumpf, 1968, 1969). Although it has been suggested that R_nE may associate with the nuclear membrane (Jackson and Chalk-

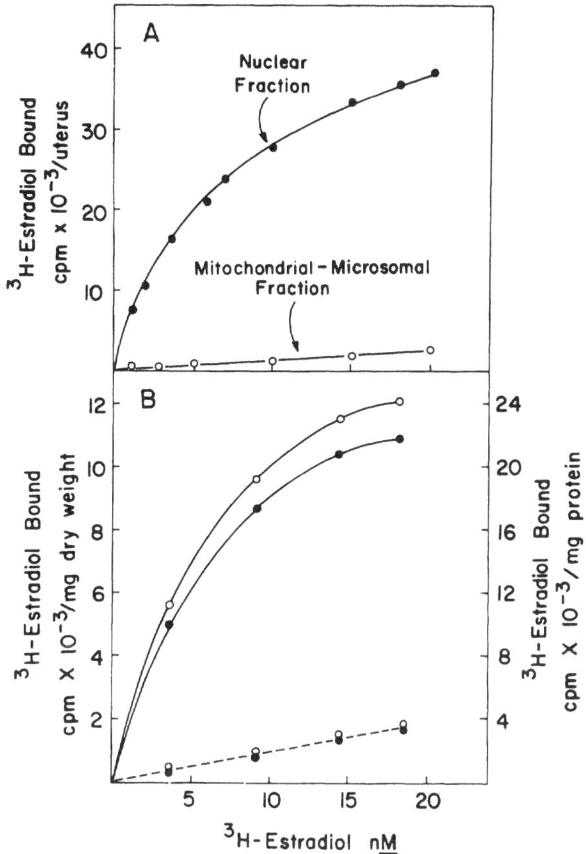

Fig. 28 A and B. Binding of ^{3}H-estradiol to various cellular compartments. Immature rat uteri were incubated with various concentrations of ^{3}H-estradiol at 37°C for 60 min. **A** The nuclear (●) and mitochondrial-microsomal (○) fractions were prepared by differential centrifugation, and specific binding of hormone was determined. **B** These data are expressed in terms of the dry weight (●) and protein content (○) for the nuclear fraction (*solid line*) and mitochondrial-microsomal fraction (*dotted line*)

ley, 1974), the majority of nuclear hormone-receptor complexes are bound primarily to chromatin (King et al., 1965b; 1969; Maurer and Chalkley, 1967; Teng and Hamilton, 1969). This is discussed more fully in Chaps. II and IV.

3. Are Cytoplasmic Receptors Necessary?

Since Terenius (1968) studied the uptake and retention of ^{3}H-estradiol and ^{3}H-hexestrol into tissue slices from mouse uterus and vagina as well as diaphragm, heart, and kidney, it has been widely assumed that steroids enter target and non-target cells by passive diffusion. Terenius concluded that high estrogen concentrations are maintained in target tissues by noncovalent association with binding sites (presumably receptors), not via a pump mechanism. These in vitro studies amplified the observations by Jensen and Jacobsen (1962) some years before that labeled estrogens are rapidly accumulated by target and nontarget tissues alike. In addition, where differences in retention of steroid were observed, these tissues invariably possessed receptors or high-affinity binding sites for steroids.

Despite these elegant early demonstrations of nonspecificity with respect to steroid uptake, from time to time it has been suggested that a membrane-bound carrier mediates the cellular uptake of steroid. Most of these studies have utilized in vivo or in vitro exposure of target and nontarget tissues to labeled steroids for extended periods of time (Jensen and Jacobsen, 1962; Gorski et al., 1968; Milgrom et al., 1972a, 1973b). Such studies invariably conclude that a saturable component is involved in steroid "uptake"; however, these more accurately reflect the retention of steroid via its interaction with receptors after entry, not the rate of entry of steroid itself.

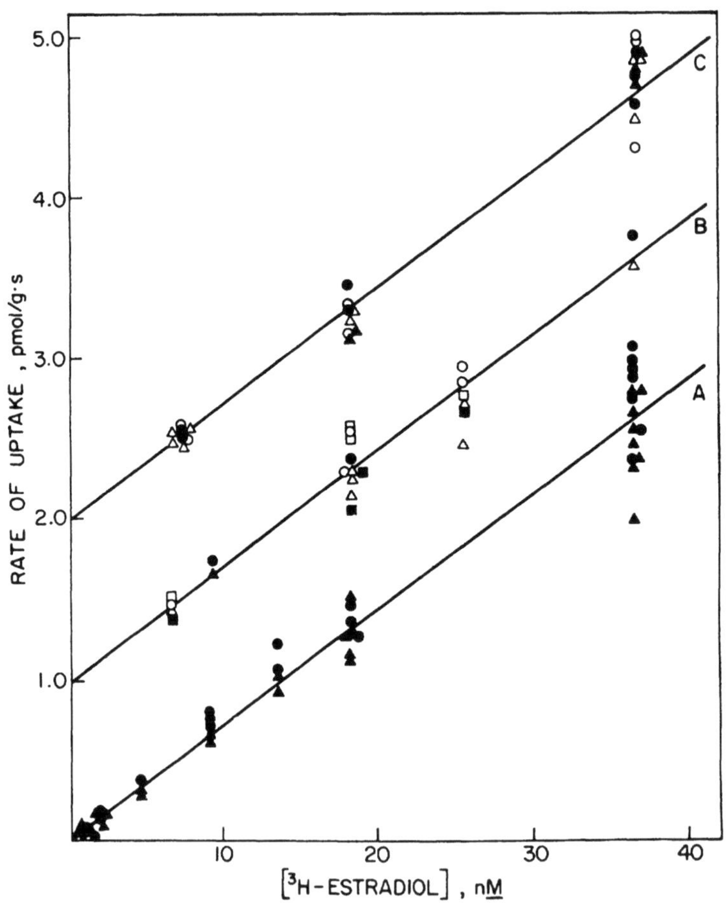

Fig. 29. Initial rate of uptake of ^{3}H-estradiol by the uterus and diaphragm. Tissues were incubated for 15 and 30 s with ^{3}H-estradiol, washed, and content of ^{3}H-estradiol determined. *Note:* The data for *B* and *C* have been displaced along the vertical axis by 1 and 2 pmol/g · s, respectively: *A*: uteri (●), diaphragm (▲); *B*: control uteri (○), uteri plus *N*-ethylmaleimide (●); uteri plus 2,4-dinitrophenol (□), control diaphragm strips (△), diaphragm strips plus *N*-ethylmaleimide (▲), and diaphragm strips plus 2,4-dinitrophenol (■); *C*: initial rates were determined for uteri (*circles*) and diaphragm strips (*triangles*) in the presence (*open symbols*) and absence (*closed symbols*) of DES (Peck et al., 1973a)

Two experimental approaches have been employed to counter the difficulty of entry versus retention. In an elegant series of experiments employing steady-state perfusion of two isotopes, Gurpide and co-workers (Gurpide and Welch, 1969; Gurpide et al., 1971) showed that various steroids can enter target and nontarget cells with equal facility at rates linearly related to their steady-state concentration over a rather wide range of concentrations (0.2–5000 ng/ml). Although it is not possible to completely rule out a carrier-mediated process, these results are most easily interpreted in terms of the diffusion of steroid across cell membranes.

In an attempt to differentiate between entry and retention, we examined the uptake of ^{3}H-estradiol by uteri in vitro under initial velocity and equilibrium conditions, subdividing the uteri into cytoplasmic and crude nuclear compartments after the termination of incubation (Peck et al., 1973a). Such an investigation of uptake either must use steady-state kinetics as described above (via constant perfusion) or must employ very short-time intervals to determine the initial velocity of uptake. Only these methods can adequately study catalyzed processes such as active transport or facilitated diffusion; any deviation from short-time periods or rapid kinetic measurements will result in confusion due to translocation of R_cE to the nuclear compartment. In these studies it was demonstrated that R_c is not involved with the entry of estradiol into uterine tissue. As shown in Fig. 29 the rate of uptake of ^{3}H-estradiol under initial velocity conditions was linear over a wide range of steroid concentration (more than sufficient to saturate receptor). In addition, this rate was the same for both target and nontarget tissues in the presence or absence of *N*-ethylmaleimide or excess diethylstilbestrol. Any of these procedures would interfere with carrier-mediated transport and/or processes employing cytoplasmic receptor as a carrier. The entry of steroid is not dependent on metabolic energy as demonstrated by preincubation with 2,4-dinitrophenol for 30 s at 37°C prior to the study of uptake.

An argument often raised in favor of a pump or carrier-mediated uptake process is the temperature dependence of steroid entry into target or nontarget cells. This was demonstrated by Terenius (1968), Williams and Gorski (1971), and Milgrom et al. (1973b). However, while this finding is *consistent* with a carrier-mediated transport process, it is not *sufficient* to establish the same. In fact, as pointed out in elementary physiology texts (Giese, 1973), activated diffusion (as opposed to facilitated diffusion) may result from the chemical properties of solutes and their affinities for the lipid matrix of the cellular membrane. In activated diffusion, transport rate is linearly related to solute concentration and shows a temperature dependence indicative of activated states (kinetic properties similar to carrier-mediated diffusion or transport). The diffusion of steroids is expected to be "activated" (see Giese, 1973) since they are lipophilic and likely to encounter a diffusional barrier before entering the aqueous environment of the cytoplasm from the hydrophobic environ of the membrane matrix. In addition, the state of lipid bilayers is temperature dependent, undergoing transition from a liquid to more solid state with decreasing temperature (Overath and Trauble, 1973; for review see Singer, 1974). Finally, the mobility within bilayers of optical probes such as 8-anilino-1-napthalenesulfonate is restricted by these phase transitions (Tsong, 1975a, b). Thus, transitions of membrane state may impose a temperature dependence on estrogen entry. To summarize a number of studies, the temperature dependence of estrogen entry is as likely to substantiate a diffusion process as to refute it.

The reader is reminded, however, of the low-density lipoprotein (LDL) pathway, which functions to regulate the uptake, storage, and synthesis of cholesterol (Brown and Goldstein, 1976a; Goldstein and Brown, 1976; for review see Goldstein and Brown, 1977). In this system plasma LDL, consisting primarily of esterified cholesterol surrounded by phospholipids, free cholesterol, and protein, is recognized by and binds to a specific receptor on the cell surface. The receptor LDL complex is subsequently internalized by a temperature-dependent endocytotic process (Goldstein et al., 1976; Brown and Goldstein, 1976b). The internalization step is inhibited by paraformaldehyde (Brown et al., 1976) and *N*-ethylmaleimide (Brown and Goldstein, 1976c). Metabolic inhibitors also prevent the endocytosis of LDL (Goldstein and Brown, 1977).

Is it possible that a similar system functions in the uptake of estrogens? Our data demonstrating no effects of *N*-ethylmaleimide and 2,4-dinitrophenol on estrogen uptake certainly argue against such a system. In addition, the absence of any "saturable" nature of the uptake process and lack of competition between ^{3}H-estradiol and unlabeled diethylstilbestrol argue against such a model. However, Pietras and Szego (1977) have described specific binding sites for estrogens on the outer surfaces of isolated endometrial cells. Furthermore, there are additional binding sites for estrogens (such as Type II, see Chap. IV), and these vary in number within target tissues as a function of estrogen treatment. In addition, they release steroid more readily than the cytoplasmic receptor itself, a requirement if they are to be utilized for transmembrane transport of estrogens. At the present time such a transport system remains unidentified. The availability of estrogen target cells in culture may afford evidence for or against its existence.

To reiterate our previous conclusions, estradiol appears to diffuse through the membranes of target and nontarget cells by a process not requiring a carrier. In addition, *free* estradiol is available to the nuclear compartment of target and nontarget cells (Peck et al., 1973a). Assuming that the cell membrane acts in a semipermeable manner with respect to steroids and thus that the concentration of *free* estradiol outside equals that of *free* inside, we demonstrate that translocated cytoplasmic receptor (i.e., R_nE) retains estrogen within the nuclear compartment by a simple noncooperative and noncovalent interaction. To test the model, serum albumins were employed to lower the concentration of free estradiol outside, and as expected initial rates of uptake of ^{3}H-estradiol as well as levels of retention of R_nE were reduced in a manner predicted by the concentration of serum albumin employed and the association constant of serum albumin for estradiol (Peck et al., 1973a). From studies such as this we conclude that at physiologic temperature and at physiologic concentrations of steroid, no requirement exists for an active pump or facilitated diffusion mechanism for uterine tissue.

As stated above, steroid hormones may enter nuclei without assistance from cytoplasmic receptors. Indeed, in systems such as chicken liver it has not been possible to demonstrate R_c; rather it appears that estrogen receptors exist predominantly in the nuclear fraction (see Chap. V for references and discussion). Whether or not target tissues of a given species possess R_c may depend on the levels of steroid found in their blood. Mammals have vanishingly small quantities of *free* steroid in the blood, i.e., 10^{-10}–10^{-12} *M*, depending on the species. These are extremely low concentrations that may demand of mammalian cells some sort of concentrating and retention mechanism, a function that could be served by cytoplasmic receptors. In addition, other steroid-binding sites have been described which might serve this purpose (see Sect. 4 below).

Birds have higher blood levels of estrogen than mammals, and hence higher levels of *free* estrogen are available to their target organs (Van Tienhoven, 1968). Chickens, for instance, have blood levels of estrogen in the nM range, and other birds have relatively high levels of estrogen during the breeding season (Bahr and Nalbandov, 1977). Therefore the advantage provided by cytoplasmic receptors in mammals may be superfluous in birds. This situation may vary with the hormone. The blood levels of progesterone in birds are very low (1–10 nM), whereas in mammals they are relatively high (10–100 nM) (Moghissi et al., 1972; Butcher et al., 1974; Stabenfeldt and Shille, 1977; Robertson, 1977). Birds have readily demonstrable cytoplasmic receptors for progesterone (Schrader et al., 1972), whereas mammalian target tissues possess lower numbers of progesterone receptors. Again, these soluble receptors may permit the detection and accumulation of hormone in birds despite the relatively low levels of free hormone in avian blood.

Thus, the concentration of free estrogens in mammals is so low that a soluble form of cytoplasmic receptor may be required in large quantities to accumulate and retain estrogen against a concentration gradient. In birds, on the other hand, the concentration of free estrogen is much higher and cytoplasmic receptors may not be required. In other classes of vertebrates, such as reptiles and amphibians, the level of estrogen in the blood is relatively high during the breeding season. From this it may be anticipated that these animals also lack cytoplasmic receptors for estrogen. This aspect of comparative endocrinology should be explored in the future since it may provide insight into the requirement for compartmental states of hormone receptors.

4. Mechanisms for Steroid Accumulation not Involving Receptors

Approximately 50% of the ^{3}H-estradiol found in the uterus after an injection is not bound to receptor (Peck et al., 1973a). However, it is retained by the uterus against a declining concentration gradient. Therefore other mechanisms than high-affinity receptors must exist for retaining the hormone.

One of the mechanisms that may account for this observation was first described by Keller et al. (1969). Steroid hormones, especially estrogens, have the capacity to cause the dilation and fenestration of capillaries in certain target tissues. When this occurs, serum components, such as albumin and leukocytes, accumulate in the extracellular spaces of target organs (see Chap. V for further discussion). Spaziani (1963) demonstrated that albumin accumulation was particularly striking in the uterus after estrogen treatment. Serum albumin binds estradiol with a relatively low affinity ($K_d \sim 10^{-5}M$); however, when present in large quantities, it serves as an important binding site for estrogens and could account, at least in part, for the retention of that ^{3}H-estradiol not bound to the receptor. Cytoplasmic Type II binding sites (discussed in Chap. IV) may also fall into this category. The capacity of sex steroids to promote the accumulation of serum binders would result in an elevation of steroid levels in these tissues. Furthermore, complexes of these serum-binding sites with estrogens could function as a large reservoir of steroid, which could be available through dissociation for binding to receptors.

Chapter IV

Characteristics of Cytoplasmic and Nuclear Receptor Forms

Although this monograph is primarily concerned with the relationship between hormone-induced responses and nuclear binding, it is necessary to relate nuclear receptors to their cytoplasmic counterparts.

1. Cytoplasmic Forms of Estrogen and Progesterone Receptors

a) Sucrose Density Gradient Analysis

The original demonstrations of specific estrogen-binding macromolecules were the sucrose density gradient analyses of Toft and Gorski (1966) and the gel filtration studies of Talwar et al. (1964). Toft and Gorski (1966) showed that cytosol of immature rats contained an estradiol-binding protein or receptor, which sedimented at approximately 8–9*S* on low ionic strength sucrose gradients. This 8*S* receptor is apparently an aggregate of smaller subunits, which sediment as 4*S* entities in gradients containing 0.3 *M* KCl (Korenman and Rao, 1968). The experiments of Vonderhaar et al. (1970) also suggested that the receptor was made up of subunits. The determination of these *S* values has been the subject of some confusion because they are influenced by the concentration of cytosol protein on the gradient (Stancel et al., 1973a); however, the 8*S* $\rightleftarrows$ 4*S* conversion generally holds for concentrations of cytosol protein of 1–5 mg/ml (Korenman and Rao, 1968; Stancel et al., 1973a). Hence this interrelationship forms a frame of reference upon which many studies are based.

Jensen et al. (1967, 1968) first showed that a 5*S* complex could be extracted by 0.4 *M* KCl from rat uterine nuclei after estradiol exposure. In the original two-step hypothesis of Jensen, the receptor was translocated to the nucleus as a 4*S* entity and subsequently associated with some unknown molecule, coverting it to a 5*S* form. This view is no longer held by the Jensen group because they could demonstrate that 4→5*S* conversion occurred in cytosol and was not dependent on the presence of nuclei. In contrast, however, the work of Siiteri et al. (1973) and Linkie and Siiteri (1978) shows clearly that the 4→5*S* conversion takes place in pure nuclei as a function of time after exposure to estradiol.

Work from the laboratory of Yamamoto (1974) and Yamamoto and Alberts (1974) shows that the 4→5*S* conversion is accelerated by the presence of DNA. Nielsen and Notides (1975), Notides and Nielson (1975), and Notides et al. (1975) have examined this conversion step extensively. Their results indicate that the in vitro transformation involves the association of a 4*S* subunit, which has a molecular weight of 80,000 daltons, with a second component of molecular weight 50,000 daltons. This dimerization results in the formation of a 5*S* form of receptor with a molecular weight of 130,000. For a recent review of these studies see Notides (1978).

It has also been suggested that receptor transformation involves a Ca^{2+}-activated transforming factor, the so-called RTF or receptor transforming factor (Puca et al., 1972). Puca et al. (1977) have suggested that RTF is an autoproteolytic factor that causes a reduction in receptor size from 5.3*S* to 4.5*S* as well as a change in shape which enables the R_cE complex to enter the nucleus. Mild trypsin treatment of the receptor also produces smaller fragments that retain their ability to bind estrogen; however, these appear different from the form described by Puca (Erdos and Fries, 1974, Rat et al., 1974). The molecular weight of trypsin-treated receptor in 6 *M* guanidine HCl is 28,000, whereas that of RTF-treated receptor is 35,000. Receptor exposed to RTF will bind to chromatin, whereas trypsinized receptor does not bind to chromatin or DNA (Sala-Trepat and Vallet-Strouve, 1974; Andre and Rochefort, 1975).

This disparity in results of different laboratories could result from artifacts that are produced by several variables. Various sedimentation values, ranging from 4 to 8, for the cytosol estrogen receptor can be obtained by changing the salt concentration in the gradients (Chamness and McGuire, 1972; Stancel et al., 1973a). Physiologic salt concentrations of 0.15–0.2 *M* KCl yield a 6S form (Reti and Erdos, 1971; Chamness and McGuire, 1972). In addition, Chamness et al. (1972) and Stancel et al. (1973a) have shown that various *S* values can be obtained by varying the quantity of protein and by including heparin in the gradient. Conformational changes in the receptor without effect on the molecular weight of the molecule could result in changes in the sedimentation coefficient (Notides and Nielson, 1975). The addition of EDTA, proteolysis inhibitors, sulfhydryl reagents, etc., by some investigators and not by others has naturally resulted in distinct differences. In addition, as discussed below, binding sites that are probably not receptors have undoubtedly contributed to this confusion.

Sucrose density gradient analysis of cytoplasmic progesterone receptors has yielded variable results, probably for the same reasons as for the estrogen receptor. Sedimentation values ranging from 3.5 to 8*S* have been reported depending on conditions and species (Milgrom et al., 1970a, b; Sherman et al., 1970; Baulieu et al., 1971; Reel et al., 1971; Schrader et al., 1972; O'Malley and Schrader, 1972; Faber et al., 1972, 1973; Atger et al., 1974; Feil and Bardin, 1975; Saffran et al., 1973, 1976). Schrader et al. (1975) have shown that the progesterone receptor in the chick oviduct exists as a 6*S* dimer composed of two nonidentical 4*S* subunits. Other investigators have proposed a similar subunit relationship in the guinea pig uterus (Milgrom et al., 1970; Faber et al., 1972) and in the rabbit uterus (Faber et al., 1973).

Although sucrose density gradient analysis is useful, it has not provided a clear picture of the various forms of the receptor in cytosol. Other methods discussed below help to clarify some of these problems.

b) Chromatographic and Electrophoretic Analysis

DEAE chromatography of chick oviduct cytosol reveals two peaks (*A* and *B*) that bind progesterone in a specific manner (Schrader et al., 1972b) and have 4*S* sedimentation coefficients on sucrose density gradients (Schrader et al., 1975). Each peak is thought to represent a subunit of the progesterone receptor and is composed of a single polypeptide chain containing one hormone binding site (Kuhn et al., 1975). Association of subunits *A* and *B* results in the formation of the 6*S* form,

Table 2. Physical properties of the progesterone receptor[a]

Parameter	Monomeric forms		Dimer
	A	B	A–B
Sedimentation coefficient (*S*)	4.2	4.2	6
Stokes radius (Å)	55	63	
f/f_0	1.7	1.9	
Axial ratio	14	18	
Isoelectric point	4.5	4.0	
Molecular weight	110,000	117,000	227,000
Dissociation constant	$10^{-10}M$	$10^{-10}M$	

[a] Schrader and O'Malley (1972); Schrader et al. (1975).

which is observed on sucrose density gradients. A summary of the various physical characteristics of these progesterone receptor forms is shown in Table 2. The binding of these forms to nuclei will be discussed later in this chapter.

Sherman et al. (1976) have used preparative ion exchange filtration and electrophoresis in highly cross-linked polyacrylamide gels to resolve five forms of progesterone binding site. Two of these appear similar to the *A* and *B* subunits discussed above, and two others may be products of the proteolytic cleavage of *A* and/or *B*. One of these has been called the mero-receptor because it is the smallest fragment of the receptor containing a steroid binding site. The formation of mero-receptor is dependent on Ca^{2+} and may be related to Ca^{2+}-activated protease activity (Sherman et al., 1974). The remaining form is heterodisperse and may represent aggregates of the other forms. Thus, the principal forms of the progesterone receptor are two 4*S* subunits with molecular weights of ~100,000. Proteolysis of either or both of these produces fragments that maintain the ability to bind progesterone. This limited proteolysis may be analogous to the proteolysis of the estrogen receptor discussed above. The role, if any, of proteolysis in vivo is unknown.

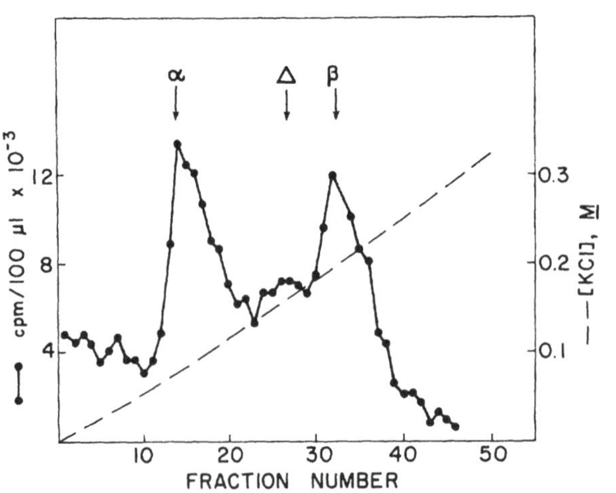

Fig. 30. Various forms of estrogen-binding sites in uterine cytosol. Cytosol was incubated with 20 n*M* ^{3}H-estradiol and applied to a 5-ml DEAE column that had been equilibrated with buffer (Tris-EDTA-mercaptoethanol). The column was extensively washed before elution with a linear KCl gradient

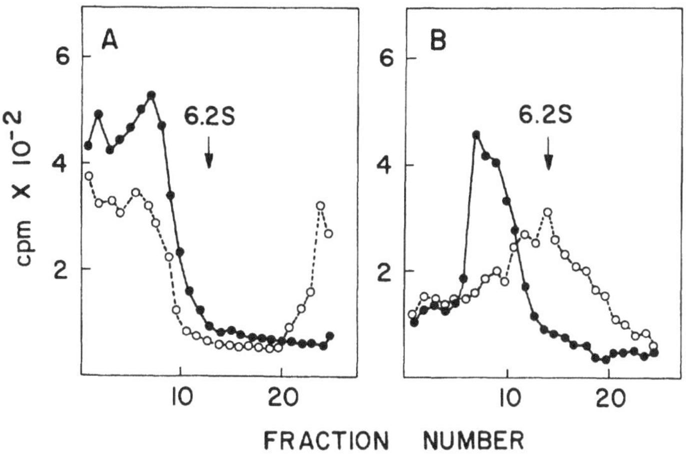

Fig. 31 A and B. Sucrose density gradient analysis of α and β peaks from DEAE column chromatography. Fractions containing either α or β forms of the receptor were collected from a DEAE column as in Fig. 30. Sucrose density gradients containing 10 percent glycerol and either no KCl (○) or 0.3 *M* KCl (●) were run on the α (**A**) and β (**B**) fractions. Bacterial alkaline phosphatase was used as a marker for the sedimentation coefficient. Top of gradients is on the left

Two or more forms of the estrogen receptor also exist in uterine cytosol (Metzger et al., 1978; Fig. 30). In these experiments uterine cytosol was labeled with ^{3}H-estradiol and eluted from DEAE cellulose by a linear KCl gradient. Two major peaks of estrogen-binding activity (α and β) were observed to correspond roughly to those of Schrader et al., (1975) for the progesterone system. Binding of labeled estrogen to both peaks was inhibited by excess unlabeled DES or estradiol. In addition, bound estrogen was observed in the area between these peaks which may constitute a third form (Δ) of the receptor. This form will be discussed below. A similar pattern is obtained when unlabeled cytosol is chromatographed and ^{3}H-estradiol is added after elution from the column. Thus, the α and β peaks do not result from an interaction of receptor with estradiol to produce two peaks. Pollow et al. (1975) have observed three similar peaks by DEAE chromatography of cytosol from human endometrium.

Sucrose density gradient analysis also demonstrates the dissimilarity of α and β (Fig. 31). In low salt gradients the α form sediments primarily as a 4*S* molecule with some aggregation. In high salt gradients the aggregates disappear, but the α form remains a 4*S* entity (Fig. 31A). On the other hand, β shows a salt dependence of sedimentation coefficient. Thus, in low salt gradients it sediments as a broad peak in the 6*S* region (Fig. 31B), whereas in high salt gradients only a 4*S* peak of radioactivity is present.

These receptor forms could result from different conformations of the same molecule, which in turn might have different ion exchange properties. To examine whether these are different molecules, we used gel filtration on agarose A 1.5*M*, which permits good resolution of large molecular weight species. Gel filtration of labeled cytosol demonstrated two major peaks of bound radioactivity corresponding to 40 Å and 26 Å species, with a minor peak at 35 Å (Fig. 32). Peak fractions were pooled, concentrated by dialysis against solid sucrose, and studied with both density gradient centrifugation and DEAE cellulose chromatography. Both major peaks

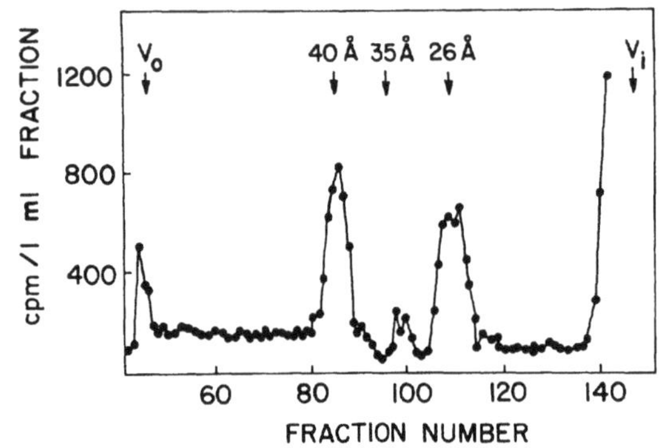

Fig. 32. Various forms of the cytosol estrogen receptor from gel chromatography. Uterine cytosol was applied to a 2.5 × 100 cm agarose A 1.5 *M* gel column, which was equilibrated with buffer containing 0.3 *M* KCl. Stokes radii were determined by calibration with ovalbumin, chymotrypsin, ribonuclease A, and thyroglobulin. Exclusion volume, V_o inclusion volume V_i

sediment in low and high salt gradients as 4*S* species. Using the method of Siegel and Monty (1966), the molecular weights of the 26 Å and 40 Å species were calculated to be ~ 50,000 and 80,000 daltons, respectively. On DEAE the 26 Å form coelutes with α (0.1 *M* KCl), and the 40 Å species coelutes with β (0.22 *M* KCl).

Heat transformation of labeled cytosol results in a shift in sedimentation coefficient of the estrogen receptor from 4*S* to 5*S*. As discussed earlier, Notides and Nielson (1975) and Yamamoto and Alberts (1974) have suggested that this is a bimolecular reaction, an association of a 4*S* receptor with a second component to form the 5*S* nuclear form. We have shown (Fig. 32) that uterine cytosol contains 50,000 and 80,000 dalton species, both of which display characteristics of receptors. If both are subunits of a 5*S* dimeric form, we would expect that α and β would participate in the transformation (activation) reaction. When cytosol is warmed (activated) the properties of these forms of the receptor depend on the presence or absence of EDTA. In the presence of EDTA, the various species aggregate and appear in the void volume on gel filtration. On DEAE cellulose, a considerable number of bound counts are found in the void volume. When EDTA is omitted from the homogenization buffer, less aggregation is observed. However, only when 1*M* urea is included prior to warming is aggregation eliminated as assessed by gel filtration or sucrose density gradient analysis. Therefore, in the following experiments cytosol was prepared in the absence of EDTA and in the presence of 1*M* urea. All three forms of the receptor (α, β, and Δ) are present under these conditions in nontransformed cytosol (Fig. 33). An additional peak that elutes at 0.05 *M* KCl may represent a secondary estrogen binding site, which will be discussed later in this chapter. When cytosol is warmed, both α and β decrease dramatically, and the quantity of the Δ peak, which elutes at 0.14 *M*, increases significantly. The broadness of the Δ peak may be due to unreacted α and β forms of receptor. When the Δ peak is pooled and rechromatographed on DEAE, no change is observed in the elution profile of Δ. When an aliquot of this peak is examined on sucrose density gradients containing 0.3*M* KCl, only a 5*S* peak is observed. The disappearance of α and β with the concomitant rise in Δ is not the result of changes occurring solely in α or β. When α or β is incubated at 30°C for 30 min prior to rechromatography, no change is observed in the individual elution patterns. Thus, α and β may participate in a dimerization reaction to produce the 5*S* form of the estrogen receptor.

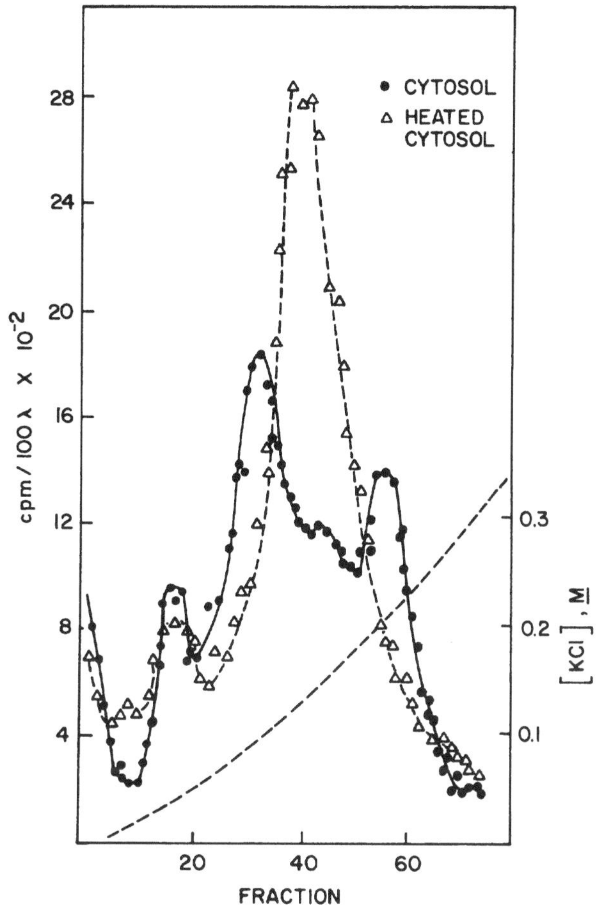

Fig. 33. Effect of temperature on the forms of the estrogen receptor. Nonactivated cytosol (●) was kept at 4°C, and DEAE chromatography was performed without extensive washing of the column as in Fig. 30. Activated cytosol (△) was prepared in buffer containing Tris, DTT, 0.1 *M* KCl, and 1.0 *M* urea. Prior to chromatography the cytosol was incubated at 30°C for 30 min. Both cytosols were exposed to 1 percent dextran-coated charcoal to adsorb free steroid

These experiments provide evidence for the existence of heterogeneity in the estrogen-receptor population of the rat uterus. Both α and β forms may represent subunits, which are present in approximately equimolar amounts. Each form sediments as a 4*S* species on sucrose density gradients, yet they are dissimilar and not interconvertible. Beta exhibits many of the properties of the 80,000 dalton, 4*S*, classic estrogen receptor. Alpha may be a form of the estrogen receptor which has not been previously identified or characterized or may result from the proteolysis of β. It has a molecular weight of 50,000 daltons and may correspond to the "Subunit X" proposed by Yamamoto and Alberts (1974) and Notides and Nielson (1974). An association of α and β may result in the translocation of the resultant complex to the nucleus. Support for this conclusion will be provided in subsequent parts of this chapter in which both α and β forms dissappear from the cytosol after estradiol injection while α, Δ, and an additional form, δ, can be extracted from nuclei.

c) Type II Binding Sites in Uterine Cytosol

In this chapter we have dealt primarily with high-affinity binding sites for estrogen and progesterone which meet the criteria for receptors as defined in Chap. II. The suggestion that other estrogen binders — perhaps not directly involved with receptor function — are present in the uterus has been made by several investigators. Early work indicated that dissociation of the R_cE complex exhibited two first-order phases. Dissociation of estrogen from a single binding site could not easily account for this multiplicity and so two or more sites were proposed (Sanborn et al., 1971; Best-Belpomme et al., 1970; Erdos et al., 1969; Ellis and Ringold, 1971; Puca et al., 1971). In addition, Rochefort and Baulieu (1969) had previously noted the presence of a secondary site in the uterus which bound estradiol with low affinity but very high capacity.

As pointed out in Chap. III, only 50 percent of the ^{3}H-estradiol present in the uterus after in vivo injection or in vitro exposure is bound to the estrogen receptor (Peck et al., 1973a). We have suggested that the remaining 50 percent is bound to secondary sites which could include lipids, serum albumin, and other proteins. Such secondary sites are usually ignored and/or considered of no physiologic significance. However, the proper evaluation of such secondary binding sites is necessary for the valid measurement of estrogen receptors. In addition, an understanding of the characteristics of these secondary sites should provide insight into their possible function.

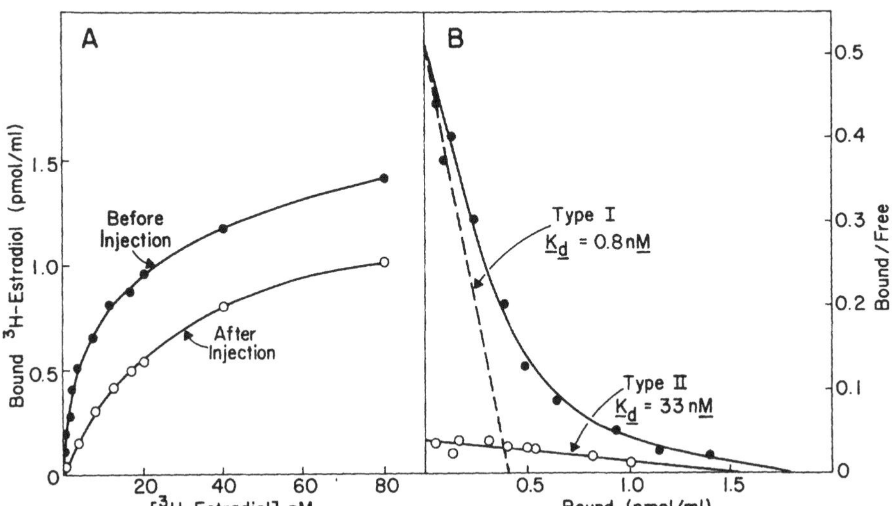

Fig. 34 A and B. Saturation analysis of estrogen binding in rat uterine cytosol. **A** The quantity of specifically bound ^{3}H-estradiol was determined in uterine cytosols from noninjected rats (●) and rats injected with 2.5 μg estradiol 60 min prior to being killed (○). Uteri were homogenized in TE-buffer (0.01 *M* Tris-HCl, 0.0015 *M* Na_2EDTA, pH 7.9 at 0°C) with a motor-driven glass-glass Duall homogenizer. The homogenate was centrifuged at 800 x *g* for 10 min in a Beckman JS 7.5 rotor. The supernatant was recentrifuged at 180,000 x g for 30 min and diluted with TE-buffer to 25 mg/ml. Cytosol was incubated with ^{3}H-estradiol or ^{3}H-estradiol plus excess DES for 18 h at 4°C and hydroxylapatite used to separate free from bound steroid. **B** Scatchard analysis of the data in **A**. The amount of ^{3}H-estradiol bound to uterine cytosols from estrogen-treated animals (○-Type II) was subtracted from the total binding in the system (●) to yield the *dashed line* labeled Type I (Clark et al., 1978b)

Saturation analysis of uterine cytosol over a wide range of ^{3}H-estradiol concentrations produces the data shown in Fig. 34A (Clark et al., 1978b; Eriksson et al., 1978). Although this curve appears to consist of only one binding component, it is actually composed of two that can be resolved by Scatchard analysis, using the Rosenthal-Feldman method for resolution of curved Scatchard plots (Scatchard, 1949; Rosenthal, 1967; Feldman, 1972; see Chap. II for discussion of these methods). One of these sites, Type I (Fig. 34B), conforms to the characteristics expected of the classic estrogen receptor, having a K_d of 0.8 nM with 0.6 pmol of sites per uterus (30–35 mg wet wt.). In addition, this site is depleted from the cytoplasm after an injection of estradiol (Fig. 34A). The other site, Type II (Fig. 34B), has a lower affinity for estradiol ($K_d \sim 30$ nM) but a higher binding capacity (> 2.0 pmol/uterus). Note that Type II sites do not disappear from the cytosol after an estradiol injection (Fig. 34).

To further characterize these sites, we have used sucrose density gradient analysis of uterine cytosol (Clark et al., 1978b). This method has been used extensively for

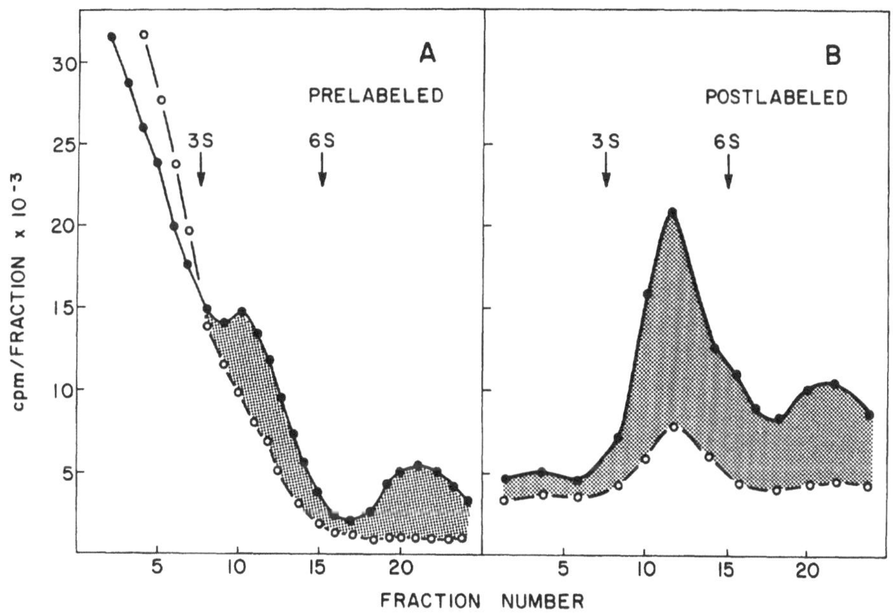

Fig. 35 A and B. Sucrose density gradient analysis of Type I and II estradiol-binding sites by prelabeled and postlabeled methods. **A** Prelabeled gradients. Linear 5–20 percent sucrose gradients (4.9 ml) were prepared with a Beckmann density gradient former. Sucrose solutions were prepared in TE-buffer plus 1 mM dithiothreitol (DTT). Uterine cytosol (250 µl) adjusted to 1 mM DTT was incubated at 4°C for 60 min with nM ^{3}H-estradiol (●) or the same concentration of ^{3}H-estradiol plus a 100-fold molar excess of diethylstilbestrol (DES, ○). The cytosol was layered on the gradients and centrifuged at 189,000 × g for 16 h. **B** Postlabeled gradients. Sucrose density gradients were prepared as described above. The cytosol, either labeled or unlabeled with ^{3}H-estradiol, was centrifuged for 16 h as described above. After centrifugation either 0.2 or 0.4 ml fractions were collected in tubes that contained ^{3}H-estradiol (20 nM, final concentration). An identical gradient was fractionated into tubes that contained the same concentration of ^{3}H-estradiol plus a 100-fold molar excess of DES. The tubes were incubated for 60 min at 20°C, and the measurement of specific estrogen binding was performed by HAP assays (Clark et al., 1978b)

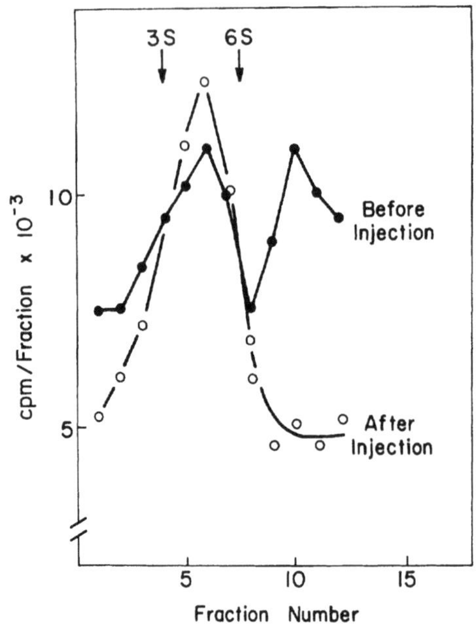

Fig. 36. Effect of estradiol injection on Type I and II binding sites in uterine cytosol. Uterine cytosol was examined by postlabeled sucrose density gradient analysis as described in Fig. 31 in noninjected rats (●) and in rats that had been injected 60 min prior to being killed (Clark et al., 1978b)

qualitative and quantitative studies of estrogen receptors in both normal and abnormal tissues (see Sect. 1.a above). In this method cytosol is usually prelabeled with ^{3}H-estradiol and subsequently centrifuged at high speed for 12–16 h. We anticipated that ^{3}H-estradiol would dissociate rapidly from Type II sites during this 12–16 h period and hence that very little bound hormone could be observed after centrifugation. On the other hand, dissociation from *Type I* sites is quite slow ($T_{1/2} \sim 20$ h), and hence at least 50 percent of the bound hormone should be observed after gradient analysis. To avoid this problem and to allow the demonstation of Type II, we have added ^{3}H-estradiol to gradient fractions after centrifugation and employed the HAP assay to separate free from bound steroid (see Chap. II). In Fig. 35 the prelabeling and postlabeling of receptors for sucrose gradient analysis is compared. As shown in Fig. 35A, the quantity of ^{3}H-estradiol bound in a specific manner, i.e., that amount which is displaced by an excess of DES, is significant in both the 4*S* and the 8*S* regions of prelabeled gradients. However, neither peak accounts for the predicted quantity of Type II binding. Postlabeling of gradient fractions followed by a HAP assay of each fraction reveals the presence of large quantities of an estrogen-binding molecule in the 4*S* region of the gradient. The relative quantities of bound ^{3}H-estradiol in the 4*S* and 8*S* region are compatible with the assumption that the 8*S* region is made up of Type I sites while Type II sites are in the 4*S* region. Type II sites also exist in the uterine cytosol from castrate mature rats and human mammary cancer (Clark et al., 1978b; Panko et al., in press).

Additional evidence that the 4*S* and 8*S* regions of gradient contain Type I and II sites, respectively, was obtained by postlabeling gradients of cytosol from rats that either had or had not been injected with estradiol. Injection of 2.5 μg estradiol into immature rats will deplete cytosol estrogen receptors (Type I), but Type II sites should remain in the cytosol. It is clear from Fig. 36 that the 8*S* region is depleted of

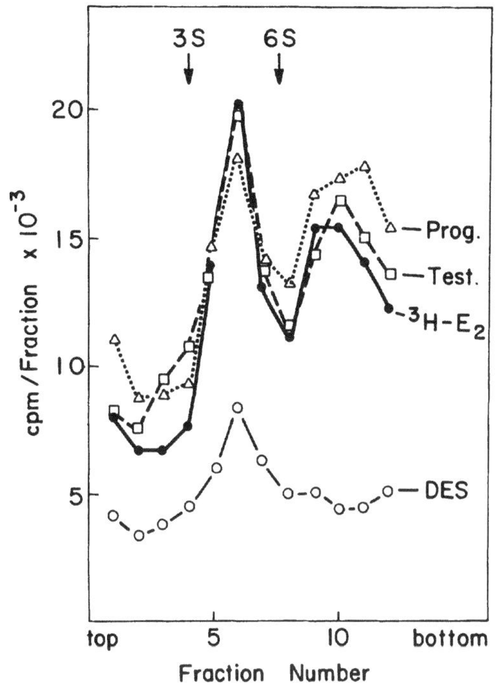

Fig. 37. Hormone specificity of Type I and II estradiol binding sites. Uterine cytosol from immature rats was examined by the postlabeled sucrose density gradient technique as described in Fig. 35. HAP assays were performed with ^{3}H-estradiol (●) or ^{3}H-estradiol plus progesterone (△), testosterone (□) and DES (○). The concentration of each added steroid was 2 μ*M*. (Clark et al., 1978b)

specific binding sites while the quantity of bound ^{3}H-estradiol in the 4*S* region does not change. This agrees with the observation that cytosol from estrogen-injected rats contains only Type II sites when analyzed by HAP assay (Fig. 34).

Hormone specificity of Type I and II binding sites was examined by postlabeling gradients as well as by direct HAP assay of high-speed cytosols. The HAP assay of

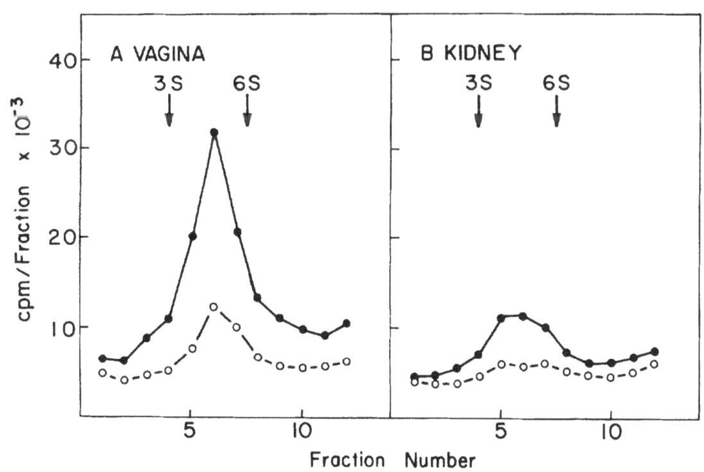

Fig. 38 A and B. Tissue specificity of Type I and II estradiol-binding sites. Cytosols were prepared from vagina (**A**) and kidney (**B**) and postlabeled sucrose density gradient analysis was performed as described in Fig. 35. (Clark et al., 1978b)

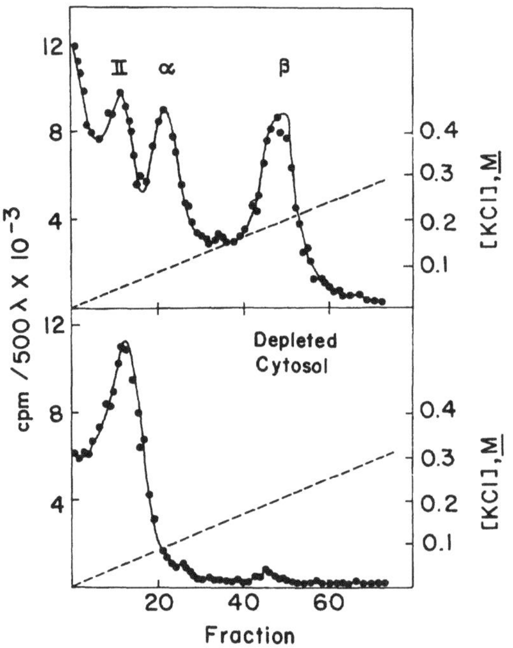

Fig. 39. Effect of estradiol on the depletion of various forms of the estrogen receptor from uterine cytosol. In the *upper panel* uterine cytosol from noninjected animals was labeled with ^{3}H-estradiol, and DEAE cellulose chromatography was performed as in Fig. 30. The *lower panel* shows the profile from cytosol prepared from rats injected with 10 μg estradiol 1.5 h before being killed

high-speed cytosol showed that estradiol and DES inhibited the binding of ^{3}H-estradiol to both types of sites, whereas progesterone, testosterone, and cortisol did not. The specificity of both sites was also shown by the postlabeled gradient technique (Fig. 37).

The presence of multiple estrogen binding sites in other tissues was examined by the postlabeled gradient method as well as by direct HAP assay of high-speed cytosols. The vagina contains large quantities of the Type II binding site (Fig. 38A) whereas the kidney contains significant but much lower amounts (Fig. 38B). This is also true for the spleen (data not shown). In these experiments the animals were injected with 2.5 μg estradiol 60 min before being killed; hence very little Type I receptor is present in the vagina; however, Jensen and DeSombre, (1972) have demonstrated the presence of a Type I binding in this tissue from animals not treated with estrogen. Thus, the absence of Type I receptor in Fig. 38 A indicates that translocation has occurred. Saturation analyses of ^{3}H-estradiol binding to serum and cytosols from spleen and kidney by direct HAP assay do not detect significant levels of Type II binding.

Type II sites can also be separated from the classic receptor by DEAE column chromatography as seen previously in this chapter (Fig. 33). These secondary sites elute from DEAE at 0.04 *M* KCl and are readily separable from α and β forms (Fig. 39). Both α and β forms appear to represent subunits of the Type I receptor as explained in the previous section and are depleted from uterine cytosol by estrogen injection, whereas Type II sites are unaffected (Fig. 39).

Thus, there exist at least two general types of macromolecules that bind estrogen in a stereospecific fashion: Type I, which has the properties of the classic cytosol estrogen receptor and Type II, which has a lower affinity and a higher capacity for estradiol than Type I. Type II sites also differ from Type I in that they do not undergo translocation from the cytoplasm to the nucleus after an estrogen injection. The implications of the existence of Type II sites are far-reaching. Their presence interferes with the measurement of Type I sites, producing an overestimate of Type I and/or an incorrect identification of Type II as Type I.

That Type II is a macromolecule with a sedimentation coefficient of 4*S* also has important implications. Sucrose gradient analyses usually employ the labeling of cytosol before centrifugation. During the 12–16 h centrifugation period, hormone dissociation from Type I and Type II may take place, and both types of binding sites will be underestimated. Depending on their rates of dissociation a multitude of estimates of their number can be obtained. The HAP assay combined with postlabeling of gradient fractions eliminates this problem and results in a reproducible estimate of both Type I and II binding sites.

Bound ^{3}H-estradiol in the 4*S* region of gradients is usually attributed to the presence of α-fetoprotein in the immature rat uterus. Certainly this serum protein makes some contribution (Michel et al., 1974). However, in these experiments DES was used as a competitive inhibitor, and since α-fetoprotein has a very low affinity for DES, the Type II binding observed cannot be due to α-fetoprotein (Soloff et al., 1971). In addition, Type II sites are readily observed in adult rat uteri (Figs. 33 and 40), which are devoid of α-fetoprotein. Type II sites have also been observed in mammary tumors from mice and humans (Watson and Clark, submitted; Panko et al., submitted) as well as in chick oviduct and human uterus (Smith et al., 1978).

The presence of estrogen receptor in human breast tissue is used by several laboratories to determine whether endocrine ablation should be employed in the treatment of breast cancer (McGuire et al., 1975; Jensen et al., 1976). Since breast tissue cytosols are assayed by the prelabeled sucrose gradient technique, it is obvious from the results shown in Fig. 35 that much of the estrogen-binding capacity of the tissue is underestimated. The method of postlabeled sucrose density gradients coupled with the HAP assay should provide a tool for a more thorough and valid estimate of estrogen binding sites in breast tissue and thus increase the accuracy and predictability of estrogen receptor assays in breast cancer therapy.

It is possible that Type II sites are extracellular binding proteins that help accumulate estrogens. If concentrated in the extracellular spaces of the uterus, they could act as an effective estrogen concentrating agent, maintaining local or organ levels of free estrogen at high concentrations relative to the blood. As discussed above, approximately half of the ^{3}H-estradiol in the uterus after in vivo injection or in vitro incubation is bound to the estrogen receptor (Peck et al., 1973a). The remaining estradiol may be distributed between Type II sites and other sites such as serum albumin. Such a mixture of binding sites with variable affinities and capacities may create an extracellular estrogen buffer. Alternatively, Type II sites may represent a precursor of Type I sites. Thus, one could envision a cytoplasmic reserve of low-affinity macromolecules that serve to replace Type I sites that have been translocated to the nucleus.

In summary, uterine cytosol contains two types of specific estrogen-binding macromolecules. Type I conforms to the characteristics of the high affinity-low capacity estrogen receptor, which undergoes translocation to the nucleus. Type I sites exist as α, β, and Δ forms, which may represent subunits and/or different states of the receptor. Type II sites have a lower affinity and greater capacity for estrogen than Type I sites and do not undergo translocation. Whether or not Type II sites exist in various forms is unknown. At the present time the physiologic significance of Type II sites is also unknown.

2. Nuclear Binding of Estrogen and Progesterone Receptors

a) Receptor Forms and Nuclear Binding

As explained in Chap. III and Sect. 1 above, receptor-steroid complexes undergo translocation from the cytoplasm to the nucleus of target cells. The majority of the receptor-hormone complexes appear to be bound to components of chromatin (King et al., 1965b; Maurer and Chalkley, 1967; Teng and Hamilton, 1968; Shyamala and Gorski, 1967; O'Malley et al., 1971; Schrader et al., 1972b). As discussed below, the binding of receptor-hormone complexes to DNA, chromosomal proteins, nuclear membranes, and ribonucleoprotein particles has been reported. Hence, there is no universal agreement on the precise binding interactions between R_nS complexes and nuclear components. The resolution of this problem is confounded by the relative ineffectiveness of nuclear extraction procedures. Many investigators have shown that a portion of receptor sites is not readily extracted by 0.3–0.4 *M* salt solutions (Puca and Bresciani, 1968; DeHertogh et al., 1973; Mester and Baulieu, 1975). As an example, in the chicken liver only 10–20 percent of the estrogen receptor is extracted by 0.4 *M* KCl (Lebeau et al., 1973; see Chap. V for further discussions of this point).

High salt extracts of uterine nuclei or chromatin contain 4*S* and 5*S* forms of the estrogen receptor after an injection of estradiol. Most investigators have reported that the 5*S* form predominates; however, as discussed in Sect. 1, the 4*S* form is present (Jensen et al., 1968; Shyamala and Gorski, 1967; Giannopoulos and Gorski, 1971b; McGuire et al., 1973; Siiteri et al., 1973). The relative distribution between 4*S* and 5*S* species in nuclear compartments can be quite variable. For instance, it is reported that 5*S* species form very slowly in targets of the central nervous system (Linkie, 1975, 1977), and thus the ratio of 4*S* to 5*S* is time dependent. The 5*S* nuclear form is converted to a 3.6*S* form by exposure to 4 *M* urea and 1 *M* KCl. This is also true for the cytoplasmic forms of the estrogen receptor. Stancel et al. (1973a, b) have suggested that the 3.6*S* form represents a common subunit of the cytoplasmic and nuclear forms of the estrogen receptor. Thus, this basic subunit may combine with various cytoplasmic and/or nuclear components to form larger molecular complexes as originally suggested by Jensen et al. (1968).

From observations such as these it is possible to visualize a nuclear mechanism that involves the transformation of cytoplasmic 4*S* receptor to a larger 5*S* dimer that binds to chromatin. This may involve a dimerization of α and β forms to produce the Δ-transformed receptor as discussed previously in this chapter and in Metzger et al. (submitted). When salt extracts of uterine nuclei from estrogen-treated rats are analyzed by DEAE chromatography, all three forms of the receptor are observed. The Δ form appears to predominate; however, no clear separation is possible. Thus, the transformed 5*S* or Δ form of the receptor may result from a dimerization of α and β subunits to form the larger complex, which can bind to nuclear components.

Table 3. Receptor-estradiol binding to isolated chromatin from rat uterus[a]

Fraction	cpm/uterus	% of total chromatin-bound activity
Total chromatin-bound activity	12,800	100
KCl-extractable intact receptor-estradiol complexes	2,200	17
KCl-extractable free hormone	1,200	9
KCl-resistant receptor-estradiol complexes (activity extracted with ethanol + solubilized with NCS)	8,400	66
Losses	1,000	8

[a] Immature rat uteri were exposed in vitro to 20 n*M* ^{3}H-estradiol, plus or minus a 100-fold excess of DES, at 37°C for 60 min; chromatin was prepared and then extracted with 0.4 *M* KCl.

This scheme differs somewhat from that proposed by Schrader et al. (1972a, 1975). They propose that the cytoplasmic progesterone receptor exists as a 6*S* dimer composed of nonidentical subunits. Activation by warming, which changes the sedimentation values and chromatographic characteristics of the estrogen receptor, causes the 6*S* progesterone receptor to dissociate into 4*S* subunits (Buller et al., 1975a, b). This dissociation may occur in vivo when the 6*S* dimer binds to nuclear sites. That is, the 6*S* dimer may bind to nuclear sites and separate into subunits: one

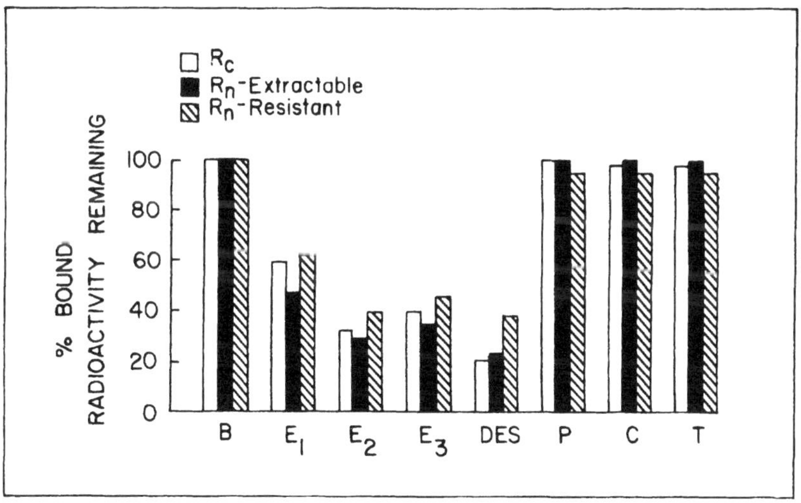

Fig. 40. Hormone specificity of the KCl-extractable and KCl-resistant forms of the nuclear-bound receptor-estrogen complex. Cytoplasmic receptor proteins were prepared from uteri of untreated immature rats (□); KCl-extractable (■) and KCl-resistant (▧) nuclear receptor-steroid complexes were prepared from uteri of immature rats treated with 2.5 μg unlabeled estradiol 1 prior to being killed. The receptor preparations were incubated with 20 n*M* ^{3}H-estradiol + 100-fold excess of the listed unlabeled steroids. Cytoplasmic receptor was incubated at 4°C for 60 min and extractable and resistant sites were incubated at 37°C for 30 min. After incubation the amount of labeled estradiol bound as receptor-steroid complex was determined

binds to nonhistone chromosomal proteins and the other, to DNA (Schrader et al., 1972b; Spelsberg et al., 1971a, b, 1972). As a consequence of this reaction, nuclear extraction yields only subunits and not the dimer (Schrader et al., 1972a).

These suggestions must be viewed with caution because only receptor forms that are extractable can be analyzed. Thus, if the functional or active form(s) of nuclear receptor-steroid complex remain bound to chromatin, that which has been extracted and analyzed by sucrose gradient analysis may be of little physiologic signficance.

Receptor-steroid complexes that remain bound to nuclear components after salt extraction are primarily found associated with chromatin. The quantity of salt-insoluble or unextractable receptor varies with experimental conditions and as a function of time (Clark et al., 1978c). When chromatin is isolated from immature rat uteri previously incubated in vitro with ^{3}H-estradiol at 37°C and is subjected to extraction with 0.4*M* KCl, 66 percent of the nuclear-bound radioactivity remains bound to the chromatin pellet (Table 3). Similar observations have been made for the pituitary (unpublished observations). These salt-insoluble complexes have the same hormone specificity as cytoplasmic and salt-extracted nuclear forms (Fig. 40).

The rate of dissociation of estradiol from the salt-insoluble form of the R_nE complex is slower than that of extracted complexes. Sala-Trepat and Rete (1974) and Jaffe et al. (1975) observed a slower rate of dissociation of estradiol from receptor-hormone complexes when bound to chromatin. Additionally, DeHertogh et al. (1973) observed a pool of nuclear-bound receptor in vivo which exhibits a slow rate of exchange when compared to the total nuclear pool of receptor. These results support the concept that salt-insoluble forms may represent receptor-hormone complexes, which are bound to acceptor sites on chromatin. This binding interaction may cause conformational changes in the receptor-hormone complex which result in a slower rate of dissociation of the hormone. If the maintenance of intact receptor-hormone complexes is a requirement for chromatin binding, this may explain why some complexes are retained for longer periods than others (see Chaps. II, IV, and X for further discussions of acceptor site binding and salt-insoluble complexes).

Table 4. Effects of various treatments on the release of KCl-resistant estradiol-receptor complexes from uterine nuclei

Treatment	% of KCl-resistant solubilized	% of solubilized recovered as intact complex
DNase (Mg^{2+}) (4°C/60 min)	3	67
Pronase (20°C/60 min)	56	18
Proteinase K (20°C/60 min)	54	22
Trypsin (20°C/60 min)	51	33
SDS (1% in buffer)	77	5
Dioxane (1% in buffer)	27	26
$CaCl_2$-extraction (0.1 *M*)	15	33
LiCl (2 *M*) - Urea (4 *M*)	70	46
Acid-extract (pH 1)	58	31
Alkaline-extract (pH 11)	30	33

Attempts to dissociate these salt-insoluble receptor-estrogen complexes by other means have also met with limited success. Treatment of nuclei with DNase did not result in the release of any receptor-estradiol complexes nor did it result in the destruction of any of the bound complexes (Table 4). Treatment with the proteolytic enzymes, pronase, proteinase K, and trypsin resulted in the release of approximately 50 percent of the bound radioactivity. Of the material released, between 18 percent and 33 percent was recovered as protein-hormone complexes. Attempts to solubilize salt-insoluble complexes with sodium dodecylsulfate resulted in the release of more than 75 percent of the bound activity. However, 95 percent of this material was in the form of free hormone. A one percent solution of dioxane solublized a smaller portion of the nuclear-bound radioactivity, most of it also in the form of free hormone. Salt extraction with $CaCl_2$ removed only a very small amount of the bound activity. The most successful treatment was a combination of lithium chloride (2*M*) and urea (4*M*). By this treatment, approximately 70 percent of the KCl-resistant complexes were released, and 46 percent of these were still present as receptor-hormone complexes as measured by the HAP assay. Finally, varying the pH resulted in the release of radioactivity (10%–58%), but most of it was in the form of free hormone.

b) Two Types of Estrogen Binding Sites in Uterine Nuclei

As discussed previously, an injection of estradiol will cause the depletion of Type I sites from uterine cytosol (Figs. 34 and 39). This depletion is accompanied by the accumulation of these sites in the nucleus and represents the well-known cytoplasmic

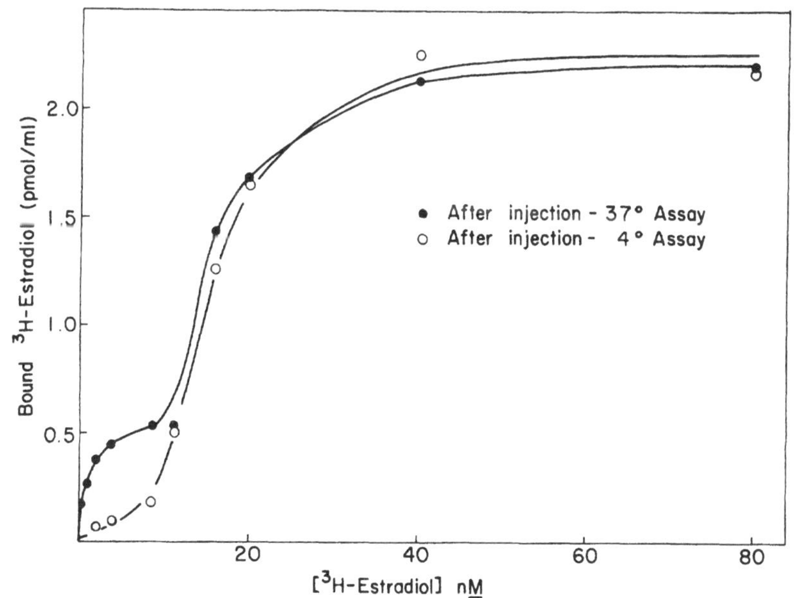

Fig. 41. Saturation analysis of estrogen binding in the nuclear fraction of the rat uterus. The quantity of specifically bound ³H-estradiol was determined by incubating uterine nuclear fractions at 4° (○) and 37°C (●) for 30 min. Immature rats ware injected with 2.5 µg estradiol 60 min before being killed. (Eriksson et al., 1978)

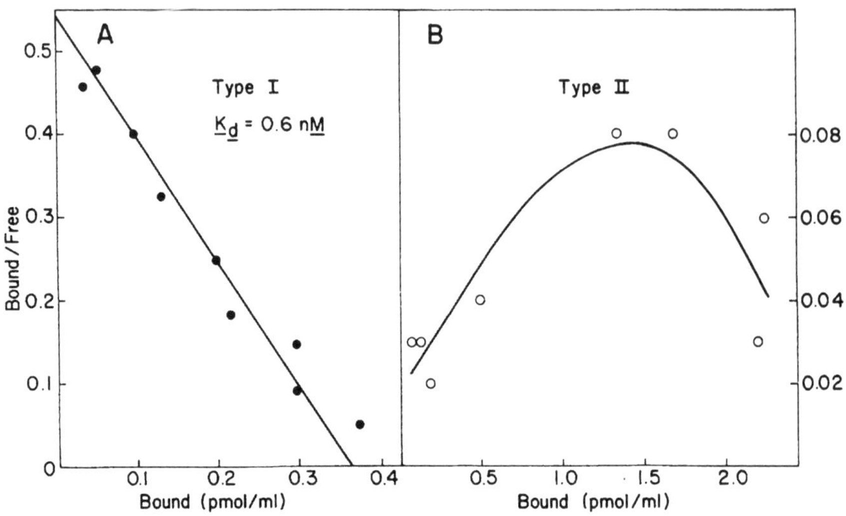

Fig. 42 A and B. Scatchard analysis of Type I and II estrogen-binding sites in the rat uterine nuclear fraction. **A** Scatchard analysis for type I nuclear sites. The amount of ^{3}H-estradiol bound to type I sites was obtained by subtracting the amount of ^{3}H-estradiol bound at 4°C from that observed at 37°C in Fig. 41 (Eriksson et al., 1978). **B** Scatchard analysis of Type II nuclear sites. The quantity of bound ^{3}H-estradiol observed at 4°C in Fig. 41 was used

to nuclear translocation phenomenon. As shown in Fig. 41, analysis of nuclear fraction for estrogen-binding sites by the ^{3}H-estradiol exchange assay reveals a complex picture that also involves at least two sites. One conforms to the Type I site, which was depleted from the cytosol and is undoubtedly identical to the classically described estrogen receptor. When the quantity of the second site, which we will call nuclear Type II, is subtracted from the total quantity of nuclear-bound hormone as measured by exchange, one obtains the amount bound to Type I. Scatchard analysis of the nuclear Type I reveals a K_d of 0.60 nM and a maximal number of sites of 0.36 pmol/ml (Fig. 42A). These values do not differ significantly from those of the cytosol receptor, Type I, which is depleted by estrogen treatment, and conform to the usually accepted properties of the estrogen receptor.

The nuclear Type II sites do not appear identical to the cytosol Type II sites and display cooperative binding behavior with a Hill coefficient of approximately 2 (Fig. 42B). These measurements were made with crude nuclear pellets; therefore, nuclear Type II sites might result from cytoplasmic contamination. However, this does not appear to be the case since purified nuclear preparations also contain these sites (Fig. 43). In this experiment immature rats were injected with 2.5 μg estradiol 1 h before assay, and nuclei were isolated as described in Chap. II. Specific estrogen binding was measured by [^{3}H]-estradiol exchange. Both types of sites are present in nuclei after estradiol treatment; however, neither of these sites appear in noninjected animals (Fig. 43). Since the cytosol Type II site does not undergo depletion, it appears unlikely that the nuclear Type II sites are derived from them. At this time, the relationship between these two classes of Type II sites is unknown.

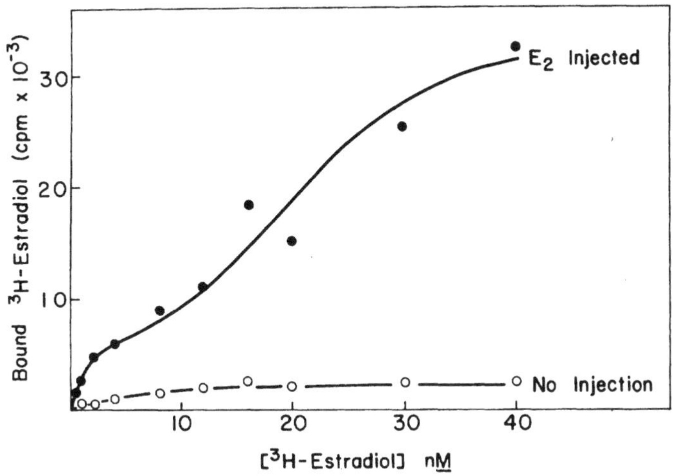

Fig. 43. Saturation analysis of Type I and II binding sites in purified uterine nuclei. Purified uterine nuclei were prepared from noninjected (○) and estradiol-treated (●) immature rats (2.5 μg estradiol, 60 min prior to being killed. The quantity of specifically bound ^{3}H-estradiol was determined by incubating the nuclei with ^{3}H-estradiol or ^{3}H-estradiol plus excess DES

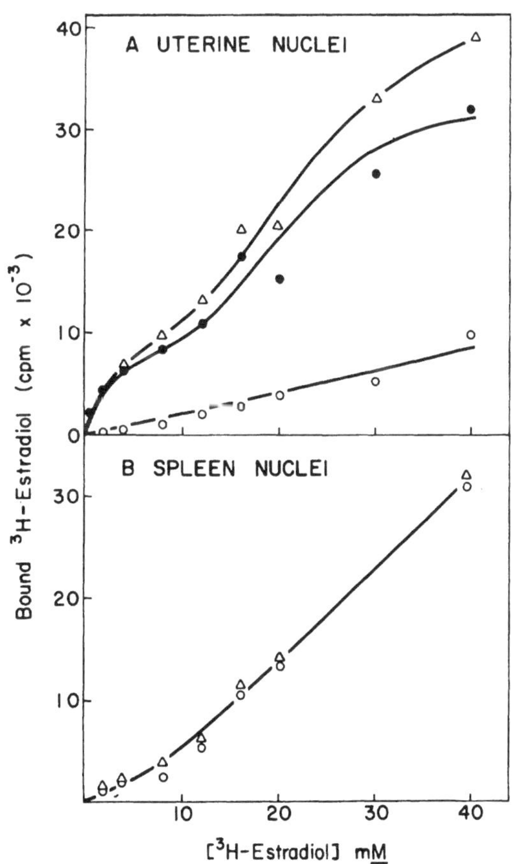

Fig. 44 A and B. Estradiol binding in purfied nuclei from the uterus and spleen. **A** Uterine nuclei were prepared as described in Chap. II. Non-specific binding (○) was subtracted from total binding (△) to yield the quantity of specifically bound estradiol (●). **B** Nuclei from spleen were prepared and analyzed as described in **A**

Nuclear Type II sites are hormone and tissue specific. Organs such as spleen and kidney, not generally considered estrogen targets, do not have any measurable quantities of this binding site. The data shown in Fig. 44 were obtained with purified nuclear preparations from the uterus and spleen of estradiol-treated rats. From these data, spleen nuclei will clearly bind large quantities of [^{3}H]-estradiol; however, all of this binding is of a nonspecific type. Hormone specificity was examined in purified nuclear preparations by exposing them to ^{3}H-estradiol alone or ^{3}H-estradiol plus competitor under exchange conditions. DES inhibits binding to both uterine sites, whereas nonestrogenic hormones have no inhibitory effect (data not shown). Thus, the nuclear Type II sites also display a binding specificity for estrogens which is similar to Type I estrogen receptors.

These results indicate that at least two types of specific estrogen-binding site can be found in estrogen-target cell nuclei. One of these, Type I, corresponds to the classic estrogen receptor and is probably derived from the cytosol Type I site. A second site, nuclear Type II, does not appear to be derived from the cytosol Type II site, but it is found in the nucleus after estrogen injection. The physiologic significance of the second nuclear site is unknown; however, as with studies of cytosol, the presence of these sites in nuclear fraction has important implications with respect to the validity of receptor measurement. Valid estimates of Type I binding must take into account the contribution made by the presence of nuclear Type II. Assays to differentiate the two sites are readily accomplished by saturation analyses of nuclear fractions via exchange at both 37°C and 4°C. At 37°C ^{3}H-estradiol exchanges with steroid bound to Type I sites and binds to Type II sites; hence, at 37°C both sites are measured. Since exchange of occupied Type I sites occurs very slowly at 4°C, only Type II sites are measured at this temperature. Subtraction of Type II sites as measured at 4°C from those sites measured via exchange at elevated temperature (Types I plus II) yields the contribution made by Type I alone. This evaluation is a necessity if accurate assessments of individual receptive sites are required. Previous reports from our laboratory and others have probably overestimated the quantities of Type I site in nuclear fractions. As pointed out above, this overestimate is also true for assays of cytosol receptor employing hydroxylapatite or other protein-adsorbant procedures. Fortunately, the error introduced by the presence of Type II sites makes an insignificant contribution to an estimate of Type I sites when the concentration of ^{3}H-estradiol employed for exchange determinations is below 10 n*M*. The vast majority of laboratories have used this or lower concentrations to assess Type I, and thus the quantity of Type I sites has not been significantly overestimated.

We do not understand the functional significance of these various forms of estrogen-binding sites; however, suggestions concerning their function may be made. As indicated earlier, the cytosol Type II site may be involved in the retention of estrogens within the uterus, thus creating an estrogen-rich environment for the Type I site, which translocates estrogen to the nucleus; alternatively, cytosol Type II sites may represent a precursor form of Type I sites. The ability of estrogen to elevate the level of nuclear Type II sites may represent a phase in the replenishment cycle for cytosol sites. Alternatively nuclear Type II sites may be a component of the machinery for "processing" the Type I nuclear complex. We have suggested that receptor-estrogen complexes bind to a small number of nuclear sites and undergo "nuclear processing" (see Chap. X for discussion). This processing may represent the

Table 5. Comparison of Type I and II sites

Cytosol	I	II
Quantity	~ 1.0 pmole	> 4.0 pmole
K_d	n*M*	~ 30 n*M*
Translocation	+	–
Forms	α and β	?
Nuclear	**I**	**II**
Quantity	Dependent on amount translocated from cytoplasm	Dependent on dose and time after hormone injection (not translocated)
K_d	n*M*	~ 30 n*M*
Forms	Δ plus some α and β	?

mechanism by which RNA synthesis is stimulated, and the receptor-estrogen complex is eventually detached from nuclear sites.

In conclusion, Type I estrogen-binding sites are probably identical to the classic estrogen receptor, which exists in the cytosol as α and β subunits. These subunits may undergo dimerization to form an activated complex, Δ, which binds to nuclear sites. Cytosol Type II sites do not appear to be involved with the translocation process; however, studies of the cytosol Type II sites are incomplete, and it is not possible at this time to determine their relationship to cytosol Type I sites. Nuclear Type I sites can be detected in salt extracts and as insoluble complexes. In addition, a second type of site, nuclear Type II, can be observed. These sites do not appear to be derived from the cytoplasmic Type II sites and constitute a separate component of the nuclear machinery. The exact relationship between these various sites remains to be elucidated; however, their current status is summarized in Table 5.

3. Nuclear Acceptor Sites and Nuclear Binding

The demonstration of acceptor sites in the nucleus has been the subject of considerable controversy (Chamness et al., 1973, 1974; Buller et al., 1975a, b). Several investigators have shown nuclear binding of receptor-steroid complexes is a saturable phenomenon, thus suggesting that specific acceptor sites exist (Alberga et al., 1971; Fang and Liao, 1971; Mainwaring and Peterken, 1971; King and Gordon, 1972; O'Malley et al., 1973; Higgins et al., 1973; Kalimi et al., 1973; Leclercq et al., 1973; Buller et al., 1975; Spelsberg, 1976); others claim that limited numbers of specific nuclear sites do not exist and that nuclear saturation is an artifact (Chamness et al., 1973, 1974; Williams and Gorski, 1974; Andre and Rochefort, 1975). Much of this conflict probably stems from the difficulties inherent in the detection of a low number of specific binding sites in the presence of a large number of nonspecific sites (Yamamoto and Alberts, 1975). This is especially true in view of the very "sticky"

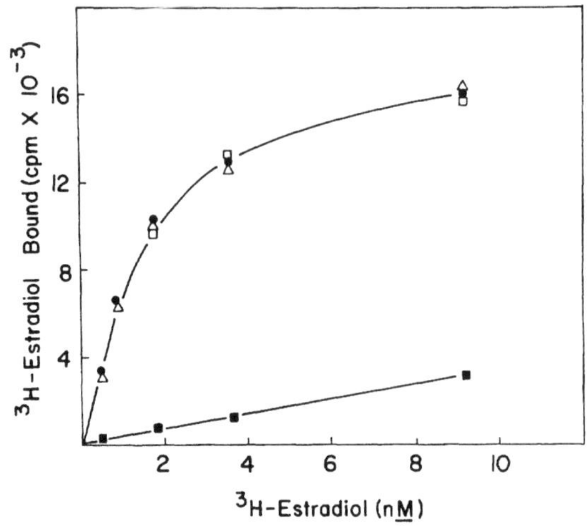

Fig. 45. Binding of receptor ^{3}H-estradiol complexes to biologic and nonbiologic surfaces. Uterine cytosol or buffer (■) containing ^{3}H-estradiol was incubated with nuclear fractions from the uterus (●) or kidney (△) or with ground-glass pellets (□) for 30 min at 25°C (from Clark and Gorski, 1969)

nature of the receptor-estrogen complex after it has been exposed to elevated temperatures (20°–30°C) at physiologic salt concentrations (Fig. 45) (Clark and Gorski, 1969). This process has been observed by many investigators and is generally called "activation" (Brecher et al., 1967; Musliner et al., 1970; Gschwendt and Hamilton, 1972; Jensen and DeSombre, 1972; Buller et al., 1975a, b; Kalimi et al., 1975; Atger and Milgrom 1976). Thus, studies of the binding of the receptor-estrogen complex to nuclei, chromatin, and/or DNA in cell-free systems are susceptible to the error introduced by the masking effect of nonspecific binding to many surfaces (Conn, 1976).

Acceptor binding of labeled receptor-steroid complexes should be a specific interaction, i.e., it should be inhibited in a competitive fashion by nonlabeled complexes. We have used this principle to measure acceptor sites in uterine nuclei. In these experiments purified nuclei are incubated with various concentrations of tritiated receptor-estradiol complex (RE*) and in parallel experiments with RE* complexes as well as receptor-DES complexes (RDES) at 10 × the concentration of RE* complexes. The assay tubes containing only RE* complexes are used to measure the total binding of such complexes to nuclei, i.e., acceptor-site binding, plus nonspecific interactions. The other set of assay tubes, which contain both RE* and RDES complexes, is used to measure the level of these nonspecific interactions of complexes with various surfaces. Subtraction of nonspecific binding from total binding yields the amount of acceptor site binding. This system is analogous to the one described in Chap. II for the determination of specific binding of steroids to receptors, but is here applied to the determination of specific nuclear-acceptor sites.

When this method is used to detect nuclear-acceptor sites in uterine nuclei, the results obtained depend on the type of buffer used to prepare the cytosol. If buffer containing 0.01 *M* Tris and 0.1 *M* KCl is employed, the quantity of nonspecific binding is very high and a very small quantity of specific (acceptor) binding is observed. As discussed in Chap. II, these conditions virtually preclude the accurate measurement of specific sites. However, when 0.0015 *M* EDTA and 0.001 *M* monothioglycerol are included in the Tris-KCl buffer, the amount of nonspecific binding decreases dramatically and the quantity of specific binding can be measured (Fig.

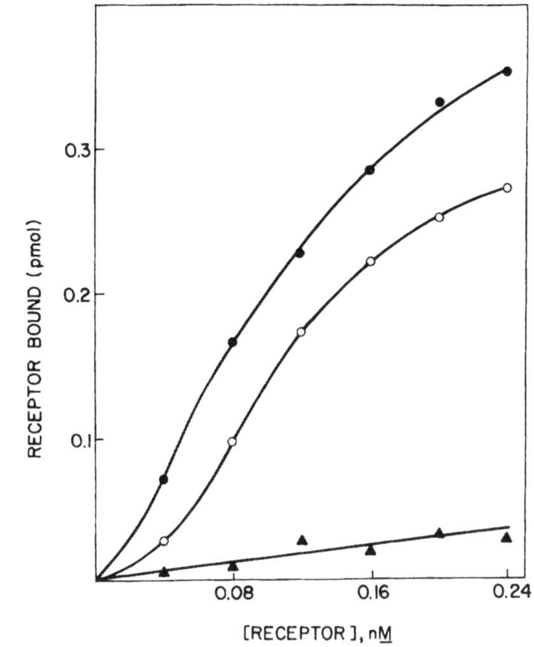

Fig. 46. Binding of receptor-^{3}H-estradiol complex to purified nuclei. Increasing concentrations of cytosol labeled with ^{3}H-estradiol were incubated with 30 μg DNA equivalents of purified nuclei in the presence or absence of cytosol saturated with unlabeled DES. Total binding, ● (RE* alone); nonspecific binding, ▲ (RE* plus excess RDES); specific binding, ○ (total minus nonspecific). The concentration of free RE* was determined by subtracting bound RE* from total added

46). Thus, EDTA seems to reduce the amount of nonspecific "stickiness" of the RE* complex but does not block the binding of the complex to acceptor sites. Monothioglycerol probably acts to protect the binding components from inactivation.

Before these data can be used to estimate the binding parameters of RE*-acceptor site complexes, another factor must be considered. As mentioned previously, RE* complexes will bind to many surfaces including the glass walls of test tubes. Since a significant portion of this binding exhibits noncompetitive inhibition by RDES com-

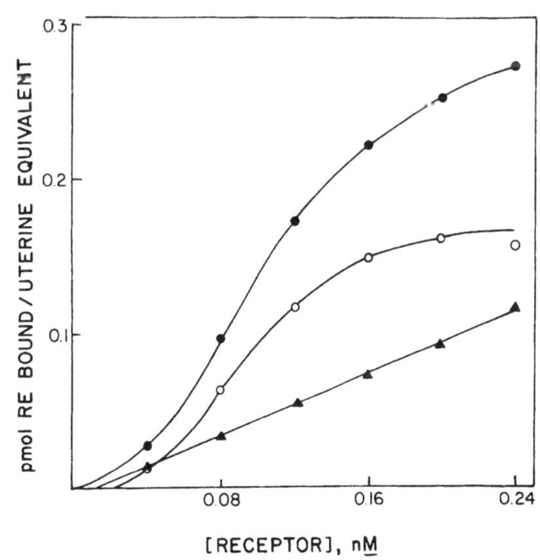

Fig. 47. The effect of binding of receptor-^{3}H-estradiol complexes to glass on the determination of their association with nuclear acceptor sites. Specific nuclear binding from Fig. 46, ●; displaceable complexes bound to glass as measured by displacement of RE* by RDES in tubes without nuclei, ▲; calculated specific binding to nuclear sites, ○

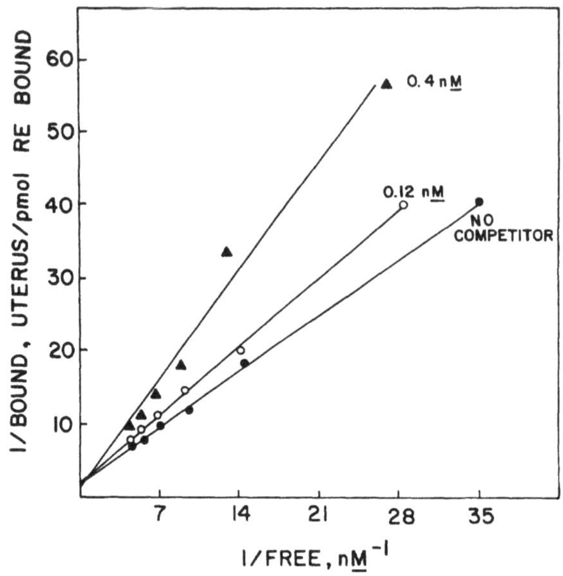

Fig. 48. Analysis of receptor ^{3}H-estradiol binding in the presence of varying concentrations of RDES. The lines of best fit were drawn by linear regression analysis. No competitor, ●; 0.12 n*M* RDES, ○; 0.4 n*M* RDES, ▲

plexes, it must be subtracted from the specific binding observed in Fig. 46. Data obtained from this subtraction are shown in Fig. 47. The apparent sigmoid nature of this curve makes precise determinations of the dissociation constant and number of acceptor sites difficult; however, it is possible to obtain rough estimates from the data shown in Fig. 48. This is done by assuming that the K_d for the interaction of RE* with acceptor sites is equivalent to the K_d for RDES and thus that the K_d of the RE*-acceptor interaction is equivalent to the K_i of the RDES interaction. Such analysis (Fig. 48) yields a K_d (or K_i) of 0.3 n*M* and a number of acceptor sites of 0.27 pmol/uterus or 3500 sites/nucleus. These numbers are in reasonable agreement with those obtained by salt extraction (Clark and Peck, 1976a) and those of other investigators (see Chap. V).

If these acceptor sites are the point of interaction between RE* complexes and specific gene sites, target tissues should contain more sites than nontarget tissues. This is the case when nuclei from uterus and spleen are compared. The number of acceptor sites in the uterus is four times as great as in the spleen, and the affinity of RE* complexes for uterine nuclear sites is much greater.

These results indicate that specific, acceptor-like binding may occur between RE* complex and nuclear sites. The number of sites obtained by this method agrees well with estimates made by other methods (see Chap. V). However, the binding system presented here is not a simple one, and caution should be used in drawing conclusions from it. The sigmoid nature of the saturation curve is suggestive of positive cooperativity. This may be more apparent than real for the following reasons: (a) true equilibrium may not have been reached under the conditions employed; (b) degradation or inactivation of receptors or acceptor sites may be occurring, especially at low concentrations of RE*; and (c) the apparent positive cooperativity may be a function of interactions taking place between receptor molecules, i.e., subunit interactions or interactions between nuclear receptors. Thus, many complex interactions may mimic positive cooperativity (Rodbard, 1973; Boeynaems and Dumont,

1975). Therefore any conclusions concerning receptor binding to acceptor sites must be tempered in light of these possibilities.

As explained earlier, activated receptor-steroid complexes are very "sticky" and hence will bind to many surfaces. This property probably explains why the literature is replete with divergence of opinion concerning specific binding to cellular components. It has been suggested that specific binding interactions occur with DNA (Harris, 1971; Schrader et al., 1972b; Toft, 1972; King and Gordon, 1972; Yamamoto and Alberts, 1974), with acidic proteins of chromatin (Steggles et al., 1971; Spelsberg et al., 1971a, b, 1972; Spelsberg, 1976; O'Malley et al., 1972), with basic nuclear proteins (Puca et al., 1974, 1975), with nuclear membranes (Jackson and Chalkley, 1974) and with ribonucleoproteins (Liao et al., 1973; Liang and Liao, 1974). None of the above studies employed a competitive binding analysis for acceptor (specific) binding; hence, it is not possible to assess their validity. The competitive assay outlined in this chapter should provide a tool for such determination.

Chapter V

Nuclear Binding and Biologic Response

The mechanism by which female sex steroids stimulate cellular functions has been the subject of several recent reviews (Jensen and DeSombre, 1973; Jensen et al., 1974; Baulieu et al., 1975; Katzenellenbogen and Gorski, 1975; O'Malley and Means, 1974; Gorski and Gannon, 1976; Yamamoto and Alberts, 1976; Clark et al., 1978c). In this chapter we will not attempt to cover all of the topics discussed in these reviews, but instead we will explain the experiments and concepts that we have developed.

The binding of receptor-steroid complexes in the nucleus is presumed to be involved in the mechanism of action of steroids. This does not imply that extranuclear actions of steroids are unimportant and, in fact, these will be examined in Sect. 6 later in this chapter. However, the primary purpose of this chapter will be to discuss the relationships between the quantity of estrogen receptor in the nuclear fraction of the rat uterus and uterotropic responses.

The relationship between steroid binding and response is a complex interaction. It involves many steps including (1) translocation of the R_cS complex to the nucleus;

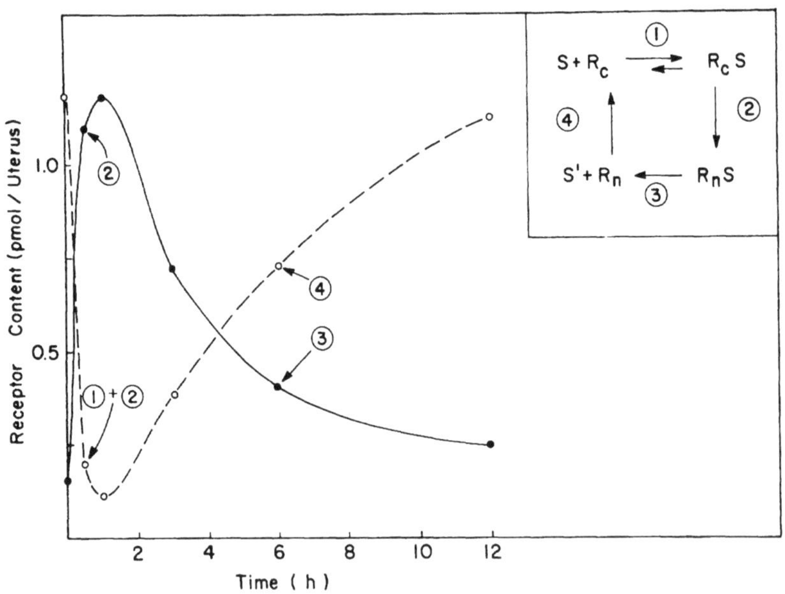

Fig. 49. Relationship between the translocation of receptor-steroid complexes to the nucleus and replenishment of the cytoplasmic receptor. Cytoplasmic (○) and nuclear (●) estrogen receptors were measured by ligand binding and/or exchange at various times after the injection of estradiol. Various numbers in the inset refer to the sequence of events that are involved in receptor-steroid binding and relate to the time functions shown in the main Fig. See text for details

(2) interactions of the R_nS complex with nuclear sites, with possible processing and metabolism of R and S; (3) reactivation and replenishment of R_c; and (4) rebinding of R_c with S (Fig. 49). Our in vivo experiments to examine these steps have utilized injections of estrogens dissolved in saline, not oil, so that the hormones would bind to receptors, promote their response, and be rapidly cleared by the body before significant receptor replenishment could take place. This is important because replenishment of R_c is stimulated by the hormone, and "new" R_cS complexes could complicate the interpretation of receptor-response relationships. In this way, we have examined steps 1, 2, and 3 with little interference from replenishment, step 4 (Fig. 49).

1. Nuclear Occupancy and Response

a) Nuclear Retention of the Estrogen Receptor

The quantity of receptor-estradiol complex (R_nE_2) which accumulates in the nuclear fraction of immature rat uteri immediately after estradiol treatment is proportional to the quantity of estradiol injected, an observation first made for the binding of labeled estradiol to the whole uterus by Jensen and Jacobsen (1962). In the experiment shown in Fig. 50, immature rats were injected with estradiol (0.01–2.5 μg), and the quantity of R_nE_2 was determined at various times after injection by the ^{3}H-estradiol exchange assay (Anderson et al., 1972a; Clark et al., 1973a). The concentration of R_nE_2 is at a maximum 1 h after estradiol injection; however, at 1 h physiologic doses (0.05–0.1 μg) of the hormone do not result in the saturation of nuclei with R_nE_2. The decline in nuclear receptor-estrogen complex between the 1st and 6th h after treatment is dependent on the quantity of estradiol administered. During this period 70–80 percent of the R_nE_2 complexes are lost from the nuclear fraction following pharmacologic doses of the hormone (1.0–2.5 μg). The differential rates of disappearance of R_nE_2 during this 5 h period result in approximately equivalent quantities of nuclear complex at 6 h, whether R_nE_2 is produced by physiologic (0.1 μg) or pharmacologic (0.4–2.5 μg) doses of the hormone. Therefore, the *retention* of R_nE_2 by the nucleus is not proportional to the dose of estrogen. Instead, a rapid loss of R_nE_2 is observed

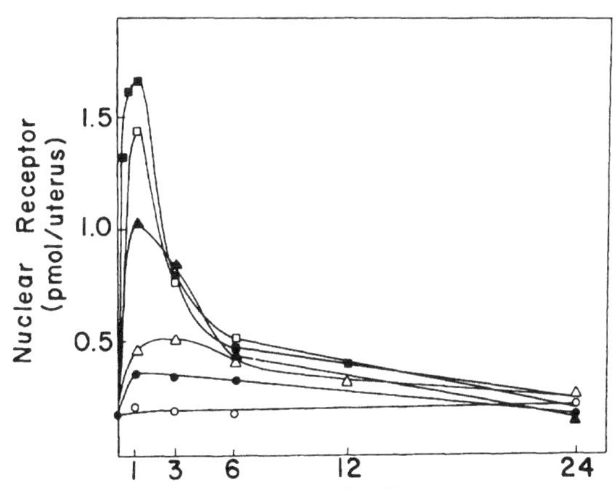

Fig. 50. Effects of estradiol receptor-steroid complexes as a function of time and dose. Rats were injected subcutaneously with saline or estradiol (0.01 μg, ○, 0.05 μg, ●, 0.1 μg △, 0.4 μg, ▲, 1.0 μg □, or 2.5 μg, ■). At the designated times after treatment the animals were killed, and the quantity of specific nuclear sites was determined by [^{3}H]-estradiol exchange (Anderson et al., 1973)

after administration of high dose levels, whereas a gradual loss is seen during the first 6 h after injection of low doses. The maximum number of R_nE_2 complexes which can be translocated to the nuclear fraction during early periods after treatment (30 min–1 h) is approximately equal to the total number of cytoplasmic sites, 15–20,000 sites/cell. Therefore, large quantities of estradiol can saturate all available receptors, and these can accumulate in the nucleus; however, 80 percent are lost within 6 h and do not undergo long-term retention. On the other hand, low doses of estradiol (~ 0.1μg) cause the translocation of about 2–3,000 sites, which, regardless of dose or the number of R_nE complexes present in the nucleus immediately after estrogen treatment, is the number of sites retained by the nucleus for periods longer than 4–6 h.

b) Early Uterotropic Responses and the Retention of Nuclear Estrogen Receptor

Early uterotropic responses include a vast array of events (for details see Sect. 6 in this chapter) which occur within the first 3–4 h after an estrogen injection and involve the generalized initiation of metabolic and biosynthetic mechanisms of the uterus. We have chosen water imbibition and glucose oxidation as two parameters representative of this early time period. These responses are examined as a function of nuclear-bound estrogen receptor at 1, 3, and 6 h after estrogen injection (dose range 0.01–2.5 μg) in Fig. 51. This figure compares the dose-response relationships between R_nE_2 complexes and uterine responses. The dose of estradiol required to increase uterine wet weight and R_nE content above control levels is 0.05 μg at 3 and 6 h after estradiol treatment. The minimal dose that results in both maximal weight and R_nE content at 3 h is 0.4 μg. Similarly the estradiol dosage resulting in maximal weight and nuclear R_nE concentration 6 h after treatment is 0.1 μg. Such relationships between response and R_nE suggest that these complexes in the nuclear fraction are involved in the early weight response. On the other hand, the conversion of glucose to CO_2 at 1, 3, and 6 h after injection is maximal with estradiol doses that are lower than those required to elicit maximal nuclear R_nE content. Similar relationships are observed between glucose oxidation and R_nE levels 3 h after estrone or estriol injection (Anderson et al., 1973a; Clark et al., 1973a). In addition, 0.01 μg estradiol maximized glucose metabolism at 6 h but did not significantly increase levels of R_nE over that of controls. Therefore either the R_nE complex is not involved in the glucose oxidation response or the quantity of R_nE required for this response is so small that it cannot be measured by the exchange assay.

If the R_nE complex is involved in the initiation of the fluid imbibition, or wet weight response, then the extent of the early weight response should be proportional to the number of R_nE complexes. We have observed that a linear correlation exists between the level of R_nE and uterine weight 3 h after an injection of estrone, estradiol, or estriol (Clark et al., 1973a; Anderson et al., 1975). The identity of the estrogen occupying the receptor site was not evaluated; however, Jensen et al. (1966) showed that estradiol and estriol remained unchanged and were retained by the whole uterus. Ruh et al. (1975) have shown that estrone will bind to the estrogen receptor and stimulate the synthesis of the induced protein (IP) in the uterus. Therefore, it seems likely that the identity of the estrogen in the nucleus is not a critical

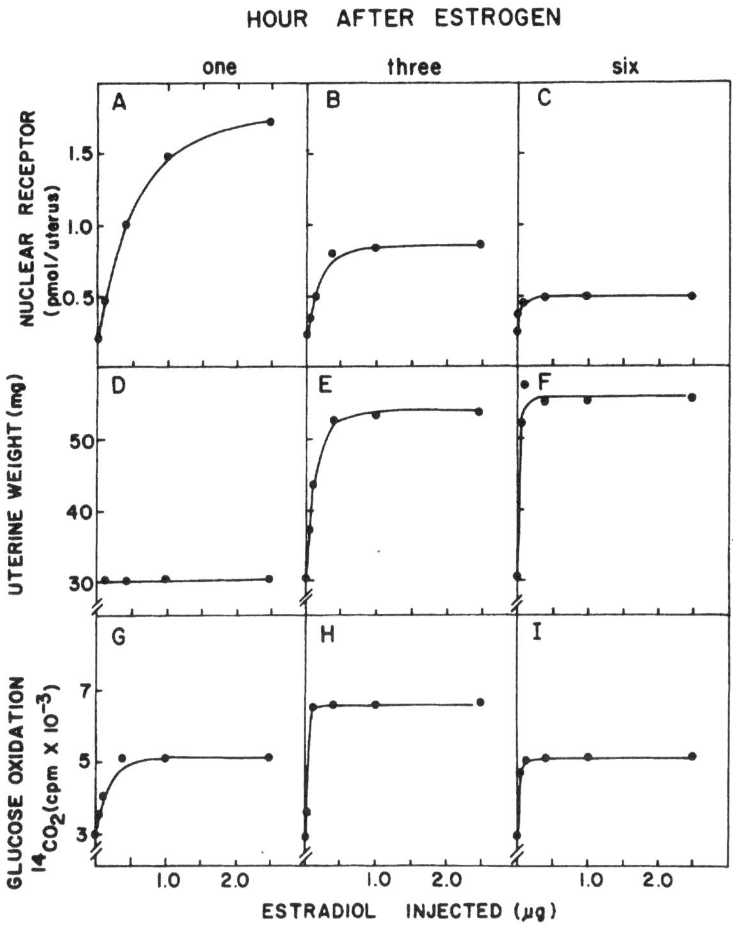

Fig. 51. Comparison of nuclear estrogen receptor (**A–C**), uterine wet weight (**D–F**), and glucose oxidation (**G–I**) at 1, 3, and 6 h after estradiol injection (dose range 0.01–2.5 μg). (Anderson et al., 1973)

factor in the fluid imbibition response, but rather that the quantity of receptor-estrogen complex determines the degree of response.

Although the correlation between uterine weight and R_nE content suggests that the R_nE complex plays a role in the initiation of uterine fluid imbibition, the receptor might serve to transport estrogen to the nucleus where the steroid could influence nuclear events directly. A nuclear event in the early fluid imbibition response has been suggested by studies demonstrating that the increase in weight evoked by estradiol is inhibited by actinomycin-D (Noteboom and Gorski, 1965). However, it should be noted that actinomycin-D causes increased adrenal activity due to stress, and this results in elevated blood levels of corticosterone (Lippe and Szego, 1965). Since glucocorticoids are known to inhibit the water imbibition response, these relationships are not clear. The suggestion that estrogen stimulates the accumulation of eosinophils in the uterus and that these cells in turn mediate the water imbibition response further confuses the issue (Tchernitchin, 1972; Tchernitchin et al., 1975, 1976; Tchernitchin and Tchernitchin, 1976). Regardless of its mechanism, water imbibition is not obligatory for uterotropic responses since it can be blocked without inhibiting uterine DNA, RNA, or protein synthesis (Gorski, 1964; Tchernitchin, 1976).

In conclusion, the involvement of nuclear binding and the stimulation of early uterotropic events is complex. No clear picture has emerged at this time. The initial interactions of estrogen with the uterus are manifold, involving vascular components as well as intracellular receptors. This multiplicity will be discussed in more detail later in this chapter.

c) Late Uterotropic Responses and the Retention of R_nE Complexes

Late uterotropic responses are associated with cellular hypertrophy and hyperplasia of the uterus and are considered to be true growth responses. These responses represent the culmination of biosynthetic events that are maximal by 24–36 h after an estradiol injection. At this time both estrogen level (Jensen and Jacobsen, 1962) and the quantity of R_nE (Fig. 50) have declined to control values. However, at this time maximal or near maximal rates of protein synthesis are observed; in addition both nucleic acid synthesis and content are elevated (Mueller, 1953; Aizawa and Mueller, 1961; Hamilton, 1964, 1968; Hamilton et al., 1968a and b). These elevated rates of macromolecular synthesis result in increased uterine dry weight and thus in hyper-

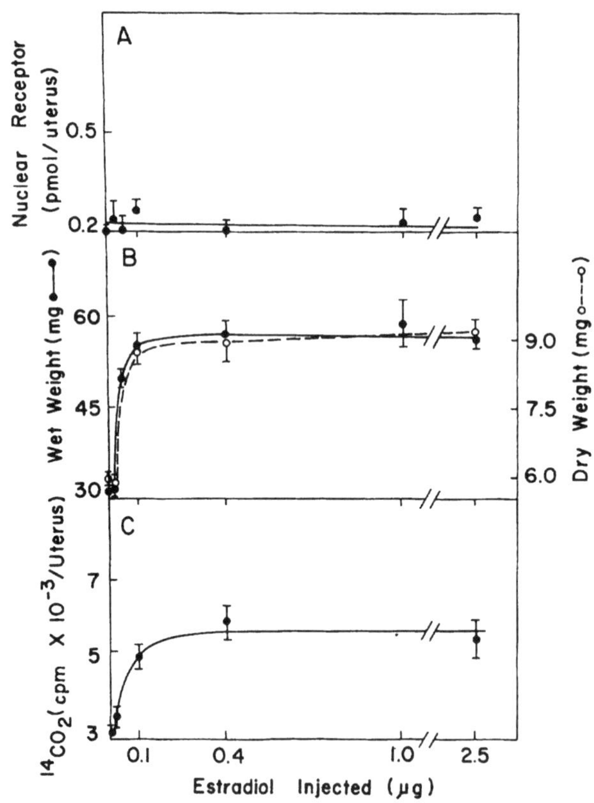

Fig. 52. Comparison of nuclear estrogen receptor (**A**), uterine wet weight (**B**), and glucose oxidation (**C**) 24 h after an estradiol injection (dose range 0.01–2.5 μg). (Anderson et al., 1973)

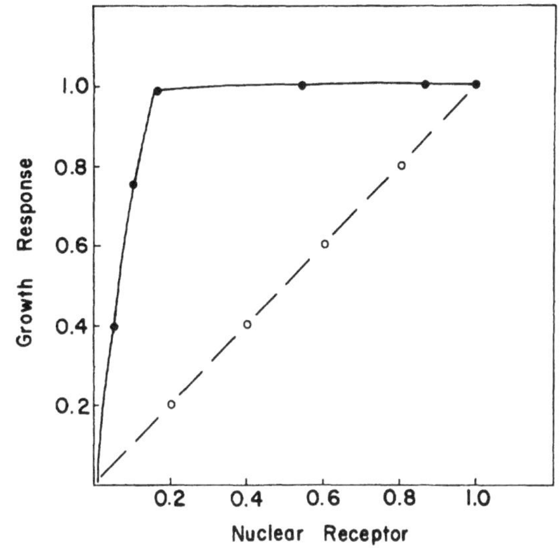

Fig. 53. Relationship between the quantity of nuclear receptor and uterine growth response. The fractional increase in uterine growth 24 h after estradiol treatment (●) is plotted as a function of the quantity of estrogen receptor in the nuclear fraction 1 h after treatment (dose range 0.01–2.5 μg). Hypothetical data (○) for a direct relationship between response and receptor occupancy, as seen for R_nE and early water imbibition, is also given

trophy and hyperplasia of the uterus (Astwood, 1939; Hisaw, 1959; see Segal et al., 1977 for review). To account for this disparity between low levels of R_nE within the uterus and maximal uterotropic responses at 24 h, one must suggest that the hormone altered uterine cells prior to that time.

We have examined this relationship between R_nE complexes and late uterotropic responses by injecting rats with estradiol at several dose levels and measuring glucose oxidation and weight of the uterus 24 h after treatment. Figure 52 demonstrates that 0.1 μg estradiol elicits maximal uterotropic responses 24 h after treatment. Thus, the elevated levels of R_nE at 1 and 3 h after hyperphysiologic levels of estradiol (0.4–2.5 μg) are not required for the true uterine growth (Figs. 50–52). The quantity of nuclear R_nE at 1–6 h after injection of 0.1 μg estradiol (~ 0.2 pmol/uterus) is approximately 20 percent of the total number of estrogen receptors in the rat uterus. This number, which corresponds to 2,000–3,000 sites/nucleus, may be limited by the number of nuclear acceptor sites present in the nucleus. The number of high-affinity sites on chromatin that bind the R_nE complex may determine the degree to which R_nE is retained in the nucleus. The estimate of 2,000–3,000 acceptor sites agrees well with the values that were determined in the cell-free system described in Chap. IV.

These results can be summarized with the data shown in Fig. 53. Our findings indicate that about 10–20 percent of the total number of available receptors are required for full uterotropic response. This is in contrast to the assumption that response should be directly proportional to the amount of bound receptor. These concepts will be discussed more fully in Chap. X.

The uterotropic responses used in the preceeding sections of this chapter, although good for the evaluation of overall responses, are very complex and involve multiple steps. To examine a response more closely related to direct nuclear mediation, we studied the ability of estradiol to stimulate endogenous DNA-dependent RNA polymerases I and II in intact nuclei. In these experiments immature rats were injected with 0.1 or 1.0 μg estradiol, and the activities of RNA polymerases I and II were

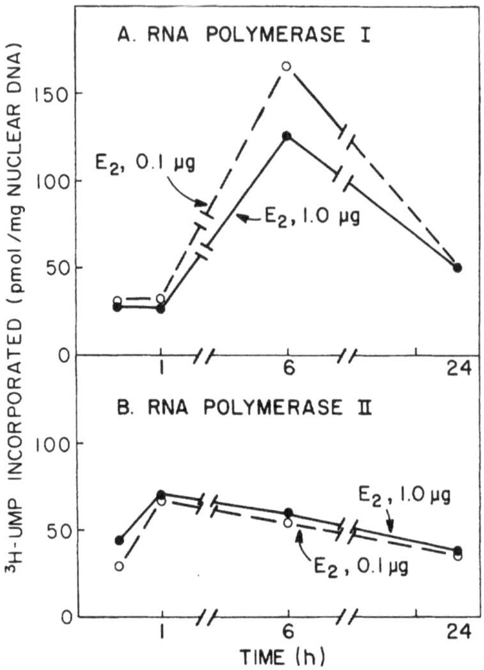

Fig. 54 A and B. The effect of low and high doses of estradiol (*E*) on RNA polymerase activity. Immature rats were injected with 0.1 (○) or 1.0 (●) μg estradiol, and the activity of RNA polymerase I (A) and II (B) were assayed as a function of time

determined in purified nuclei (Hardin et al., 1976). It can be seen in Fig. 54 that 0.1 μg estradiol is capable of stimulating RNA polymerases I and II to an extent equal to that produced by 1.0 μg estradiol. This has also been observed by Stormshak et al. (1976) for DNA synthesis in the uterus. Since 0.1 μg estradiol causes 2–3,000 R_nE_2 sites to accumulate in the nucleus whereas 1.0 μg estradiol will result

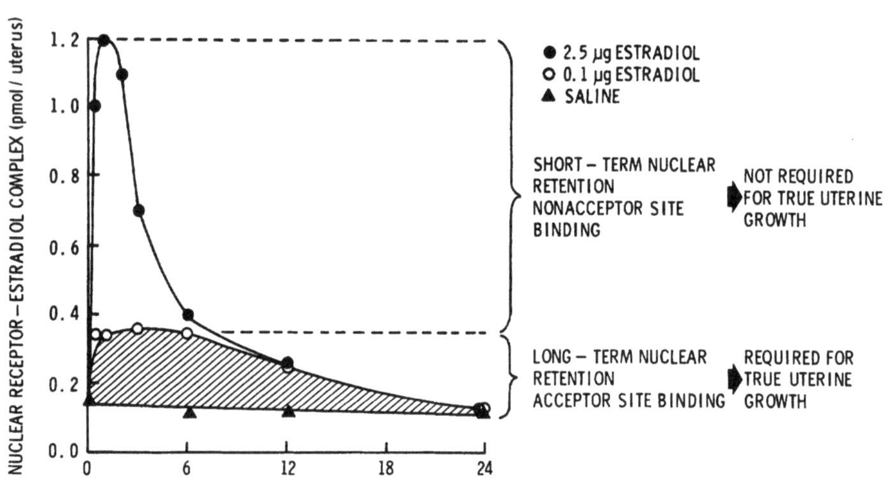

Fig. 55. Relationship between nuclear retention of the estrogen receptor and uterine growth. The quantity of nuclear-bound estrogen receptor was determined by ^{3}H-estradiol exchange in immature rats injected with 0.1 (○) and 2.5 μg estradiol (●). See text for discussion of this figure. (Clark and Peck, 1976a)

in the binding of ~ 10,000 sites, it is apparent that only a small percentage of the total receptors available to the nuclear compartment are required to stimulate estrogen-directed responses maximally. We have suggested that the binding of R_nE complexes involves at least two kinds of nuclear sites (Fig. 55), (a) nonacceptor sites, which exist in large numbers and serve to maximize the probability of nuclear translocation of R_nE complexes, and (b) acceptor sites, which exist in small numbers and are involved in the retention of R_nE complexes for long periods of time. The interaction between these acceptor sites and R_nE complexes results in the stimulation of true uterine growth.

2. Nuclear Acceptors and Long-Term Nuclear Retention

The acceptor hypothesis mentioned above was tested by differential salt extraction of R_nE complexes from uterine nuclei. The rationale for the use of this technique was based on the observation that extraction of nuclei with 0.3–0.4 *M* KCl does not remove all the nuclear-bound estrogen (Puca and Breciani, 1968; De Hertogh et al., 1973; Mester and Baulieu, 1975). The phenomenon of KCl-resistance of nuclear binding has also been reported for glucocorticoid receptors in fibroblasts and thymocytes (Middlebrook et al., 1975; Cidlowski and Munck, 1977). Immature rats were injected with either a small (0.1 μg) or large (2.5 μg) dose of estradiol. The animals were killed and their uteri removed 1 or 6 h later. Uterine nuclear pellets were prepared and extracted with various concentrations of KCl, and the quantity of estrogen receptor (R_n + R_nE) remaining in the nucleus was measured by the ^{3}H-estradiol exchange assay.

The uterine nuclear fraction from rats that received 2.5 μg estradiol 1 h before being killed contained a large number of receptor sites extractable with concentrations of KCl less than 0.4 *M* (Clark and Peck, 1976a, b). The quantity of receptor remaining in the nuclear fraction after exposure to KCl concentrations of 0.4 *M* or greater was ~ 0.1 pmol per uterus or 1,400 sites per cell. This small number of KCl-resistant receptor sites could represent those receptor sites that exhibit long-term nuclear retention. If this were true, the number of KCl-resistant sites in animals receiving the large dose of hormone should equal that of those receiving the small dose, since both doses result in long-term retention of the same number of sites and in the same maximal response (Anderson et al., 1973a, 1975; Clark et al., 1973a). As Fig. 56 shows, this seems to be the case. In addition, the number of KCl-resistant sites should be equivalent 6 h after injection of either dose of estrogen and approximately the same as that number 1 h after the small dose. This is the case as shown in Fig. 56. Therefore, we conclude that those receptor sites remaining in the nucleus after high-salt extraction are correlated with those exhibiting long-term nuclear retention. The differential extraction of the receptor-estrogen complex by KCl may reflect the binding of these complexes to different types of nuclear acceptor sites. Most (80%–90%) of these nuclear acceptors are of low affinity, and R_nE complexes bound to these are extractable with KCl. However, a minority (10%–20%) of these acceptor R_nE complexes are of high affinity and are nonextractable. Recently Ruh and Baudendistel (1977) have confirmed these findings and have shown that DES treatment also initiates long-term nuclear retention and KCl-resistant nuclear sites.

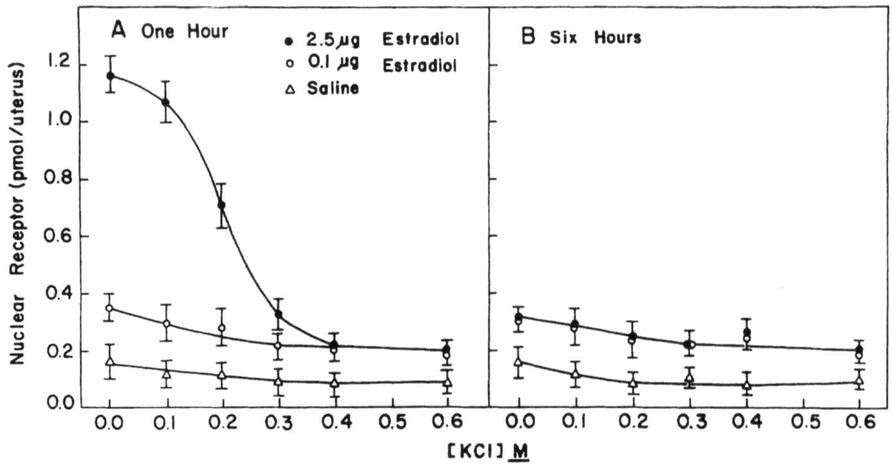

Fig. 56 A and B. Differential salt extraction of nuclear-bound estrogen receptor. Uterine fractions were prepared 1 h (A) and 6 h (B) after injection, and differential KCl extraction was performed as described in Chap. II. Following extraction, the quantity of receptor remaining in the nuclear fraction was measured by the ^{3}H-estradiol exchange assay (Clark and Peck, 1969)

It has also been possible to demonstrate that KCl-resistant sites are formed in a cell-free system (Metzger et al., in press). When uterine cytosol is incubated with uterine nuclei as described in Chap. IV salt-insoluble complexes are formed between RE and nuclear sites. This reaction occurs at 22°C but not at 4°C even though nuclear binding occurs equally at both temperatures. Thus, the binding interaction between RE complexes and acceptor sites which results in salt-insoluble nuclear complexes is a temperature-dependent phenomenon and represents sites that are similar to those observed in vivo. The topic of salt-insoluble sites will be discussed in more detail in Chap. X.

Several observations suggest that receptor-hormone complex may manifest different physical characteristics following translocation to the nuclear compartment. The rate of dissociation of the receptor-estrogen complexes is much slower when bound to chromatin than when free (Sala-Trepat and Rete, 1974; Jaffee et al., 1975). DeHertogh et al. (1973) has observed a small pool of nuclear-bound receptor in vivo which exhibits a relatively slow rate of estradiol exchange when compared with the majority of receptor. We have also observed two rates of exchange of the receptor-estrogen complex when it is bound to the nucleus (unpublished). The relationship of these observations to salt-extractable or inextractable sites is not clear. However, these changes in rate of dissociation may reflect conformation changes in receptor-estrogen complexes attendant to their association with specific nuclear acceptor sites. Such conformational alterations might result in associations of great affinity between receptor-hormone complexes and acceptor sites, thereby accounting for their resistance to KCl extraction. Since the number of these KCl-resistant nuclear sites is identical to that number of sites required for maximal uterine growth, we propose that these binding sites in the nucleus represent specific acceptor sites.

3. Uterotropic Response and Nuclear Retention of the Estrogen Receptor

To test the concept that long-term nuclear retention of the R_nE complex is necessary for the stimulation of uterine growth, we have used the differential effects of various estrogens on nuclear binding and uterine growth. It has been known for many years that estrogens vary in their ability to stimulate true uterine growth, i.e., to stimulate cellular hypertrophy as well as hyperplasia. Hisaw (1959) demonstrated that estriol (E_3) was as potent as estradiol (E_2) in stimulating early uterotropic responses, i.e., those responses that occur within 1–6 h after hormone treatment, but was essentially ineffective in causing significant increases in uterine weight at 24–48 h. We have used this differential effect of estradiol and estriol on uterine growth to examine the relationships between nuclear retention and growth responses. In these experiments animals were injected with equal quantities of estradiol and estriol, and nuclear retention and uterine growth responses were measured at various intervals thereafter (Fig. 57 and 58).

The concentrations of nuclear-receptor estrogen complex are equivalent at 3 h after an injection of either estradiol or estriol (Fig. 57). Uterine responses are also equivalent at 3 h (Fig. 58). However, by 6 h the concentration of nuclear RE elicited by estriol has declined to near control levels, while that by estradiol remains well above control (Fig. 57 B). This rapid decline in the R_nE_3 complex when compared to the R_nE_2 complex is paralleled by a corresponding inability of estriol to maintain

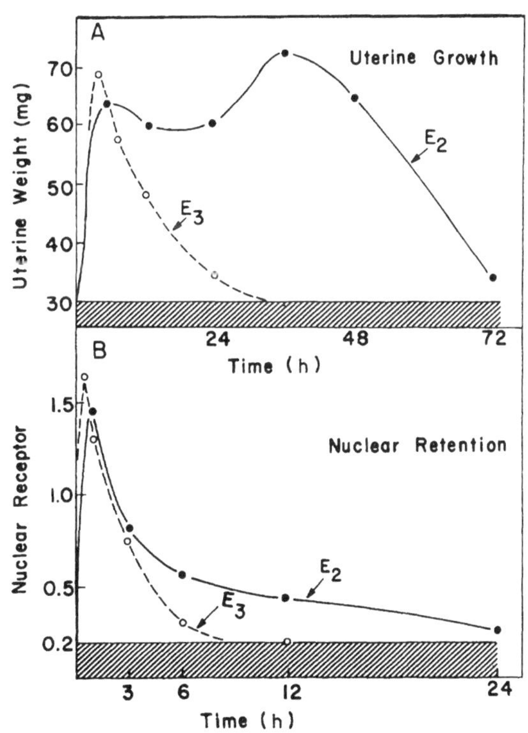

Fig. 57 A and B. Effect of estradiol and estriol on nuclear retention of the estrogen receptor and uterine growth. Immature rats were injected with estradiol (●) or estriol (○), and the uterine wet weight (A) and nuclear-bound estrogen receptor (B) were determined as a function of time (Anderson et al., 1975)

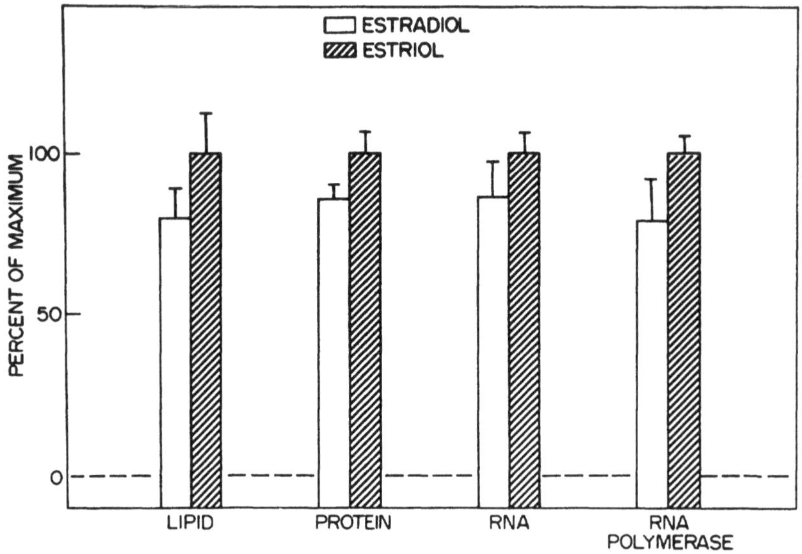

Fig. 58. Effects of estradiol and estriol on short-term uterine responses. Immature rats were killed 3 h after an injection of 1 μg either estradiol or estriol. The incorporation of ^{14}C from ^{14}C-glucose into lipid, protein, and RNA as well as RNA polymerase I activity was measured and normalized to the maximum response

uterine wet weight and to elicit long-term growth responses (Fig. 57 A). Other investigators have reported that estradiol and estriol are equally effective in stimulating early uterotropic responses (Szego and Roberts, 1953; Hisaw, 1959; Hamilton, 1963; Gorski and Baker, 1973) and that estriol fails to stimulate late uterotropic responses (Hisaw, 1959; Gorski and Raker, 1974; Katzenellenbogen and Gorski, 1975; Lan and Katzenellenbogen, 1976). These experiments support the idea that retention of the R_nE complex for at least 6 h is required for the stimulation of nuclear events resulting in long-term uterine growth responses. Estriol is a weak estrogen in this regard because it does not promote long-term retention of the receptor within the nuclear compartment. Thus, R_nE_2 complexes remain within the nucleus for longer periods of time than R_nE_3 complexes, perhaps at a limited number of nuclear acceptor sites. This long-term retention may be necessary for the sustained stimulation of nuclear events important for the production of true uterine growth.

The suggestion that long-term retention of hormones in tissues is necessary for full response has been made by other investigators (Martin, 1969; Miller and Emmens, 1969; Emmens and Miller, 1969; Terenius and Ljungkvist, 1972). Dimethylstilbestrol, DMS, and 16 oxoestradiol are "weak" estrogens when injected in saline and do not cause vaginal cornification after a single injection in the rat. However, repeated injections of these compounds do cause full vaginal cornification (Martin, 1969). We have shown that injection of estriol or pellet implants every 3 h for 15 h result in uterine growth equivalent to that produced by one injection of estradiol (Anderson et al., 1975; Clark et al., 1977c). Recently Lan and Katzenellenbogen (1976) have confirmed these findings and have used estriol derivatives to demonstrate that nuclear occupancy is correlated with the biologic half-life of the estrogen. Estriol administered in oil to mice causes significant true uterine growth (Martin et al., 1976). This probably results from the prolonged exposure of the uterus to estriol due to its slow release from oil. Continuous exposure of estriol has also been shown to facilitate mammary tumor formation equally as well as estradiol (Rudali et al., 1975). Thus, continuous occupancy of receptors by an estrogen will cause full biologic response. In the past estriol has been called a "weak" or an "impeded" estrogen (Huggins and Jensen, 1955). It is clear from the above observations that this nomenclature needs revision. Instead of "weak", *short acting* might be a more valid term for estrogens like DMS, estriol, mesestrol, and 16 oxoestradiol. The concept of differential estrogen potencies and their relationship to hormone antagonism will be discussed further in Chap. VI.

Other early metabolic and biosynthetic events are equally stimulated by estradiol and estriol (Fig. 58). These include an almost endless array of responses involving the vascularity of the uterus, expansion of the extracellular space, substrate availability and changes in uterine permeability as discussed previously in Sect. 1 of this chapter. These early changes probably make up the necessary supportive and preparative events maximizing the ability of the uterus to grow. All of these early responses set the stage for uterine growth; however, it is clear that the mere stimulation of early uterotropic events does not lead to true uterine growth.

These findings with estradiol and estriol have important implications with respect to some of the theories proposed to explain estrogen-induced uterine growth. Early transport and metabolic responses have been suggested to constitute the primary stimulatory action of estrogen by providing substrates that are rate limiting for growth (Szego and Roberts, 1953; Szego, 1971). Alternatively several authors have indicated that early changes in RNA and protein synthesis were the primary events that result in the production of true uterine growth (De Angelo and Gorski, 1970; Mayol and Thayer, 1970; Baulieu et al., 1972a; Talwar et al., 1973). The above concepts depend on a cascade of events ultimately culminating in uterine hypertrophy and hyperplasia. Our results make either of these hypotheses unlikely since E_3 has the capacity to stimulate all of these early events yet does not cause significant uterine growth when injected. Instead of a cascade phenomenon, we propose that R_nE complexes must remain in the nucleus for a period of time greater than 4–6 h in order to stimulate and sustain all of the biosynthetic events that are obligatory for true uterine growth. Therefore the time of nuclear residency is of prime importance among those factors responsible for the growth of the uterus.

In contrast to the short nuclear residency of the R_nE_3 complex, Nafoxidine and other triphenylethylene derivatives cause nuclear retention of the receptor for long periods of time (Fig. 59). For references and further discussion of this point see Chap. VI. This long-term retention of the R_nN complex correlates with the extended stimulation of true uterine growth by Nafoxidine. This stimulation is equal in magnitude to that after a single injection of estradiol and superior to that of estradiol in the duration of growth stimulation. Thus, the R_nN complex undergoes long-term nuclear retention, which is correlated with sustained uterotropic stimulation. In the immature rat this phenomenon of prolonged nuclear retention and the attendant uterotropic stimulation can last for weeks (Clark et al., 1973b, 1974a). The prolonged nuclear retention that is induced by triphenylethylene derivatives does not appear to be due to the formation of large numbers of KCl-resistant nuclear sites (Ruh and Baudendistel, 1977; Jordan et al., 1977). This point will be discussed further in Chap. X. Regardless of the nuclear binding mechanism, Nafoxidine is representative of a class

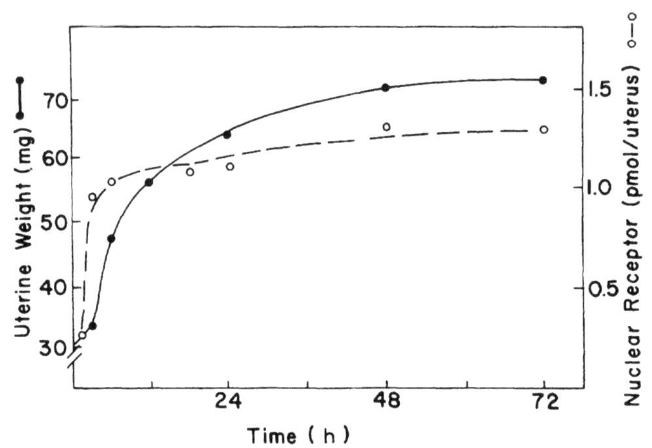

Fig. 59. Relationship between uterine growth and nuclear-bound estrogen receptor in Nafoxidine-treated rats. Immature rats were injected with 100 µg Nafoxidine, and uterine weight and nuclear-bound estrogen receptor were determined as a function of time

of compounds causing anomalous behavior with respect to both nuclear retention and uterine growth stimulation. As a consequence Nafoxidine is very useful in comparative studies with estradiol and estriol for the detection of those response profiles that are associated with the stimulation of true growth.

4. Nuclear Receptor Binding and the Control of Transcriptional Events in the Uterus

a) RNA Polymerase Activity

As pointed out above, estrogens stimulate many early uterotropic responses; however, the most important of these probably involve the augmentation of RNA synthesis. Estrogen administration to either immature or ovariectomized animals is known to increase RNA synthesis in the uterus (for reviews see Mueller et al., 1958; Hamilton, 1968; O'Malley and Means, 1974; Katzenellenbogen and Gorski, 1975; Segal et al., 1977). Noteboom and Gorski (1965) and Gorski (1964) demonstrated that estradiol stimulated RNA polymerase activity within 60 min after treatment. Quantitative changes in RNA synthesis are most marked in rRNA and are measurable within 4–6 h following hormone administration (Hamilton, 1968; Hamilton et al., 1965, 1968 a, b; Billing et al., 1968). Elevated levels of specific mRNA are assumed to occur and have been studied indirectly through the use of RNA synthesis inhibitors (DeAngelo and Gorski, 1970; Keran and Barker, 1976). Several groups have reported an early increase in very high molecular weight RNA (DNA-like) and have suggested that the marked increase in total RNA which follows may be dependent on this early appearance of mRNA (Knowler and Smellie, 1971; Luck and Hamilton, 1972; Borthwick and Smellie, 1975). Complimentary to these results on high molecular weight RNA, estrogen treatment also results in an early increase in endogenous nuclear RNA polymerase II activity. This initial surge of polymerase II activity is followed at later times (2–4 h) by an increase in the activity of RNA

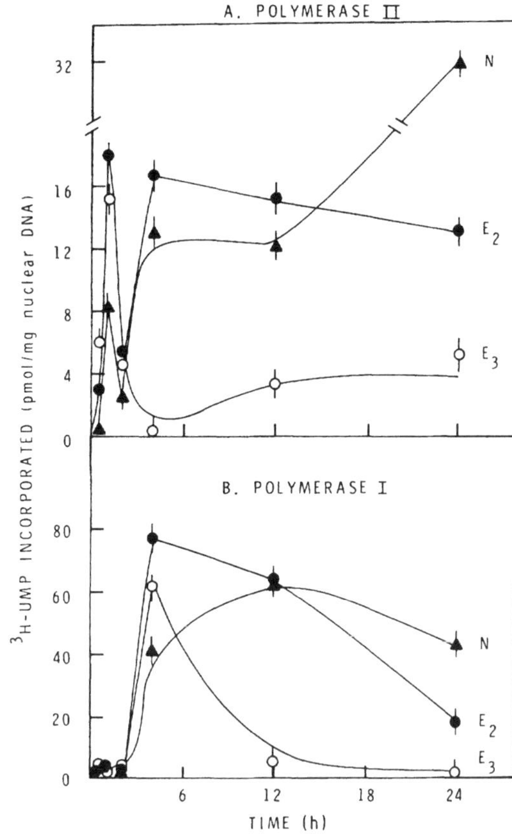

Fig. 60. Nuclear RNA polymerase II (**A**) and I (**B**) activities as a function of time after hormone injection. Animals were injected with either 1 μg estradiol (E_2, ●), estriol (E_3, ○) or 50 μg Nafoxidine (N, ▲). (Hardin et al., 1976)

polymerase I and a second rise in RNA polymerase II activity (Glasser et al., 1972; Borthwick and Smellie, 1975). Similar observations have been made in the chick oviduct system by Spelsberg and Cox (1976). These findings suggest that RNA polymerase activity could be used as one endpoint for the detection of important differences between receptor binding and uterotropic stimulation by estradiol, estriol, and Nafoxidine. Therefore we have examined the RNA polymerase response profiles induced by these compounds and compared them to their ability to cause long-term retention of the R_n complex by the nucleus (Hardin et al., 1976).

A single injection of estradiol, estriol, or Nafoxidine causes a transient rise in endogenous nuclear RNA polymerase II activity which reaches a peak within 1 h after hormone treatment (Fig. 60). The greatest response at this time is elicited by estradiol and estriol, both of which cause a significant increase in enzymatic activity as early as 30 min after hormone treatment. Nafoxidine has little, if any, effect at this very early time and evokes less response at 1 h than do the other two compounds. This decreased response is correlated with the slower rate of nuclear accumulation of the receptor-Nafoxidine complex.

The activity of polymerase II declines dramatically by 2 h after injection in all three groups. This decline is followed by a second elevation in activity in estradiol-treated animals which reaches a maximum by 4 h (Fig. 60). The early elevation of polymerase II activity in Nafoxidine-treated animals is followed by a second large increase occuring between 12 and 24 h after the injection. A second elevation in enzyme activity was not observed in estriol-treated animals; rather the activity of polymerase II remained low between 4 and 24 h.

Following the early transient increase in uterine nuclear RNA polymerase II activity, an increase in RNA polymerase I activity occurred in all hormone groups by 4 h. Estriol caused a transient rise in this activity, which was characterized by a regression phase between 4 and 12 h. Estradiol treatment results in a marked elevation in RNA polymerase I activity by 4 h. This rise is followed by a slow decline in activity; however, values remain significantly above those of the control at 24 h. Nafoxidine treatment results in a similar elevation of endogenous RNA polymerase I activity by 4 h which is maintained at levels as high or higher than those induced by estradiol.

The correlations made earlier in this chapter between nuclear retention of the receptor and estrogenic potency are also apparent in these polymerase experiments. The greatest increase in early polymerase II activity occurred in response to estradiol and estriol, the compounds that cause the most rapid accumulation of receptor by the nucleus. On the other hand, Nafoxidine, which accumulates more slowly as does R_nN, produces a peak of polymerase II activity which is approximately 50% less than that produced by either estradiol or estriol at early times after hormone administration. The second elevation of polymerase II activity, observed by 4 h in estradiol- and Nafoxidine-treated animals, but not in estriol-treated animals, may be dependent upon long-term retention of the receptor by the nucleus. The failure of estriol to stimulate this second rise in activity may be one of the factors involved in the inability of this estrogen to stimulate true uterine growth.

The elevation of polymerase I activity, which was stimulated by all three compounds, was maintained for much longer periods of time by treatment with estradiol or Nafoxidine. The rapid decline in polymerase I activity in the estriol-treated animals may result from the short-term residency of the R_nE_3 complex in the nucleus. These data demonstrate that the secondary rise in polymerase II activity, as well as the duration and magnitude of the increase in polymerase I activity, correlate with the nuclear retention of receptor-estrogen complexes.

The early (30 min) estrogen-dependent rise in uterine endogenous nuclear RNA polymerase II activity was first measured directly by Glasser et al. (1972) in the mature ovariectomized rat and subsequently confirmed in the immature rat (these data) and immature rabbit (Borthwick and Smellie, 1975). It has been suggested that this early stimulation of putative mRNA synthesis (Knowler and Smellie, 1971; Glasser et al., 1972; Borthwick and Smellie, 1975) may play a role in the synthesis of uterine proteins involved in later increases of both polymerases. A similar suggestion was made by Raynaud-Jammet et al. (1972) and Baulieu et al. (1972a, b) who found an RNA product sensitive to α-amanitine apparently necessary for estrogen-induced RNA polymerase I activity. A similar cascade mechanism has been proposed by DeAngelo and Gorski (1970) for the estrogen-induced protein (IP). Baulieu et al. (1972b) have suggested that IP plays a key intermediary role in the mechanism of action of estrogen and hence is a primary event in a cascade reaction ultimately culminating in uterine growth. Our results do not support such a mechanism; instead they indicate that the action of estrogen on RNA synthesis is more complex and difficult to interpret than simply relating an early effect in RNA polymerase II activity to the synthesis of a product that later modulates increases in the activities of both polymerases. Thus, an injection of estriol to immature rats, which evokes as great an increase in polymerase II activity at 30–60 min as does estradiol (Fig. 60), fails to produce the secondary increase in polymerase II activity and evokes only a transient increase in RNA polymerase I activity. The failure of estriol to continue to stimulate RNA synthesis is reflected in both the disappearance of the receptor-estriol complex from the nucleus and the failure of the hormone to stimulate true uterine growth.

In comparison Nafoxidine causes only a small early increase in polymerase II activity but produces a substantial and sustained increase in polymerase I and a large secondary stimulation in polymerase II activity as well as true uterine growth. The elevated activity of RNA polymerase II in Nafoxidine-treated animals may mean that many genes are "turned-on" and remain in the "on" position. This seems even more likely when one considers the large number of R_nN complexes that are present in the nucleus. Approximately 10,000 R_nN complexes remain in the nucleus for long periods of time. This quantity exceeds the number of R_nE complexes required to maximize uterine growth by a factor of 5 and leads to the speculation that R_nN complexes are binding to secondary and tertiary nuclear sites that control RNA synthesis. If this is the case, these binding interactions might lead to the stimulation of many uterine genes that would not normally be "turned-on" and thus could account for the marked elevation in RNA polymerase II activity. This anomalous nuclear-binding behavior may also be linked to the anti-estrogenicity of these compounds. One can imagine that either inappropriate genes are "turned-on" or they are activated in an incorrect temporal sequence. The activation of such "right and wrong" genes might result in the inhibition of growth. This is particularly relevant when one considers that estrogens act in concert with other hormones to produce the biochemical changes that occur during the estrous cycle of the rat. These changes involve both synthetic and degradative enzymatic activities and account for the fluctuations in uterine size and morphology during the cycle. Thus, the uterus is programmed for both anabolic and catabolic events, and any compound altering the proper sequence of these events could result in the anomalous growth state observed with Nafoxidine. The mechanism of action of estrogen antagonists will be discussed in more detail in Chap. VI.

In summary, the stimulation of true uterine growth is not characteristic of every estrogen when administered by injection. Of the varied multiple effects characterizing the biologic action of each estrogen only certain correlated sets of events stimulate true uterine growth. Of prime importance, the estrogen receptor must be retained in the nucleus for some critical period to effect true uterine growth. Retention of the receptor correlates with a secondary increase and sustained activation in RNA polymerase II activity and a sustained elevation in RNA polymerase I activity. These effects on RNA polymerase I and II appear necessary for true uterine growth.

b) RNA Polymerase Initiation Sites in the Uterus

The changes in RNA polymerase activity discussed above do not appear to be due to an increase in the number of polymerase molecules (Courralin et al., 1976; Weil, et al., 1977). Rather these changes may result from either alterations in chromatin template activity or increases in "factor(s)" that directly modulate RNA polymerase activity. Among the first demonstrations that estrogens have an effect on uterine biochemistry at the level of transcription was the observation that estrogen induces changes in chromatin template activity (Barker and Warren, 1966; Hamilton, 1968; Church and McCarthy, 1970; Glasser et al., 1972). However, due to the complexity of the transcription process, these studies provide no information as to which components of the chromatin are altered or which steps in transcription are affected by the hormone.

Procedures from prokaryotic systems have been recently adapted for measuring the specific initiation of RNA chains in eukaryotes (Tsai et al., 1975). These techniques take advantage of the specific inhibition by the antibiotic, rifampin, of bacterial RNA polymerase not bound in a highly stable initiation complex (Chamberlin, 1974). When a mixture of chromatin and bacterial RNA polymerase is challenged with a mixture of ribonucleoside triphosphates and rifampin, only those RNA polymerase molecules bound at true initiation sites are capable of synthesizing an RNA chain. This facilitates the accurate measurement of the number of RNA chains initiated on a given quantity of chromatin DNA. We have used this technique to examine the effect of various hormones on R_nE complexes and the promotion of RNA polymerase initiation sites.

Figure 61 illustrates typical titration curves of rat uterine chromatin incubated with increasing amounts of *Escherichia coli* RNA polymerase in the presence of rifampin and heparin. The latter two components are included to inhibit RNA chain reinitiation and RNase activity, respectively (Tsai et al., 1975). Several points are evident from this plot. The amount of RNA being synthesized increases linearly with increasing enzyme concentration until a transition point is reached. Tsasi et al. (1975) have shown that the coordinates of this transition point can be used to calculate the number of high-affinity binding sites for RNA polymerase as well as the number of RNA chains initiated.

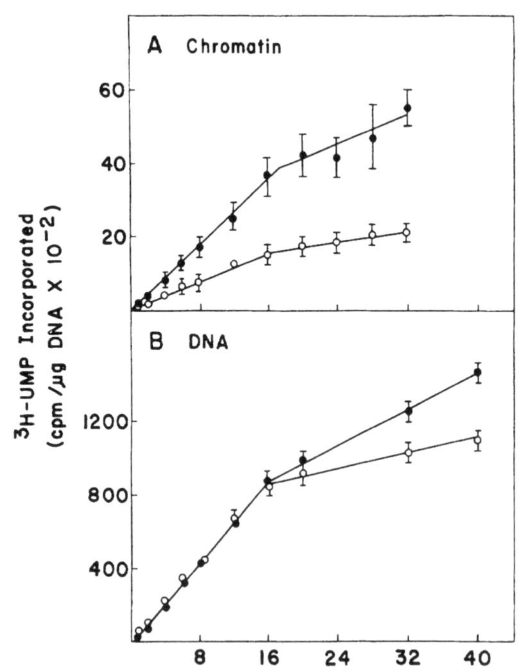

Fig. 61. Titration of chromatin (**A**) and DNA (**B**) with *E. coli* RNA polymerase. Immature rats were injected with saline (○) or 1.0 µg estradiol (●) and killed at 24 h. Assays were performed as described in the text or Markaverich et al., 1978

The data presented in Fig. 61 also demonstrate that treatment of immature rats 24 h prior to preparation of chromatin significantly increases the number of rifampin-resistant RNA chains initiated. Chromatin isolated from both control and estradiol-treated animals saturates at 16 μg RNA polymerase. However, chromatin from estradiol-treated animals allows the initiation of more than twice as many RNA chains as reflected by the increased incorporation of ^{3}H-UMP into RNA. These results indicate that estradiol treatment results in a marked increase in the number of RNA chain initiation sites on chromatin. These results agree with previous reports on template activity in the uterus and support the concept that estrogens act at the level of chromatin transcription.

The increased incorporation of ^{3}H-UMP into RNA using chromatin from estradiol-treated animals as compared to controls could result from either an increased number of RNA chains initiated or an increased length of the RNA synthesized. This dichotomy results from the nature of the assay in which only RNA polymerase molecules that are bound to chromatin in a stable preinitiation complex are able to synthesize RNA.

The number of average chain length of the RNA synthesized in all portions of this study ranged from 340 to 515 nucleotides. This variation in size was not sufficient to account for any treatment effects observed in the rifampin assay. In fact most of the RNA preparations fell into a more narrow size range (390–470 nucleotides). These findings indicate that estradiol-induced increases in ^{3}H-UMP incorporation into RNA were the result of increased number of specific RNA chain initiation rather than increased length of the RNA synthesized as has been suggested by Barry and Gorski (1971).

The role of chromatin structure in steroid-induced RNA synthesis in target tissues was also examined. One approach to this problem is illustrated in Fig. 61 B. DNA isolated from the chromatin used as template in Fig. 61 was employed in titration analyses in the presence of rifampin and heparin. Both estradiol and control DNAs had essentially the same transition point. This is in sharp contrast to the result obtained when chromatin was used as a template (Fig. 61 A). It should also be noted that purified DNAs are much more efficient templates for RNA synthesis than chromatin. These data indicate that the structure of chromatin plays an important role in the expression of steroid hormone effects on RNA synthesis.

We have previously discussed the correlation existing between the patterns of endogenous nuclear RNA polymerase activities and the retention of estrogen-receptor complexes in nuclei of the immature rat uterus (Sect. 5 this chapter and Hardin et al., 1976). All classes of estrogenic compounds examined resulted in altered RNA polymerase activities. Only the estrogen-receptor complexes, which were retained in nuclei for periods greater than 6 h, produced a secondary rise in RNA polymerase II activity and the sustained elevation in RNA polymerase I activity required for true uterine growth. When we examined the effects of estradiol, estriol, and Nafoxidine on the number of RNA initiation sites on uterine chromatin, a similar general pattern was observed. As shown in Fig. 62 estradiol treatment gradually elevates the number of initiation sites to very high levels by 12 h, and this is followed by a slow decline in the number. In contrast, estriol causes a slight elevation in the number of initiation sites which fades away. These observations mirror the effect that estradiol and estriol have on the long-term retention of the receptor in the nucleus. Likewise the long-term retention of the receptor induced by Nafoxidine is correlated with a high level of sus-

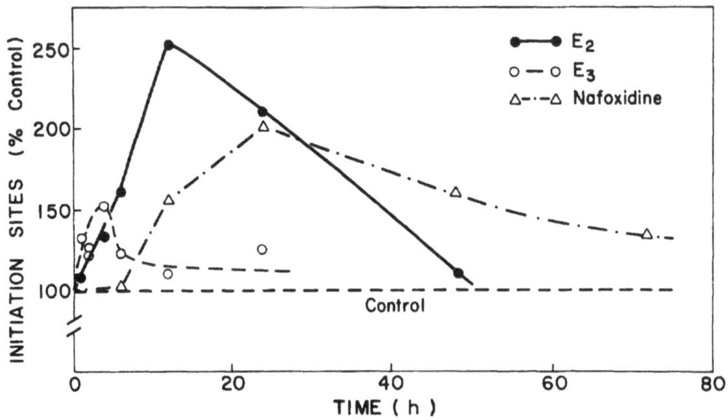

Fig. 62. Initiation sites as a function of time after hormone injection. Animals were injected with either 1 μg estradiol (●), 1 μg estriol (○), or 50 μg Nafoxidine (△). Uteri were removed, chromatin, isolated, and chromatin RNA polymerase initiation sites were determined by the rifampin-challenge assay. Control levels represent 42,000 initiation sites /pg DNA. (Markaverich et al., 1978)

tained numbers of initiation sites. As can be seen in Fig. 62 the number of initiation sites declines by 48 h in estradiol-treated animals. This is expected since estradiol does not cause retention of the receptor for as long a period as Nafoxidine (see Fig. 59).

The marked elevations in RNA polymerase II activity observed within 60 min after estradiol, estriol, and Nafoxidine (Fig. 61) do not correlate well with the barely detectable increases in initiation site number at this time (Fig. 62). Since the amount of polymerase remains constant during this period, it seems likely that factors other than template availability are operating to bring about this elevation in enzyme activity. The secondary rise in RNA polymerase II and the primary elevation in RNA polymerase I activity, which are stimulated by estradiol and Nafoxidine, but not by estriol, are correlated with increases in the number of initiation sites. Thus, the later increases in both RNA polymerase activity and polymerase initiation sites are stimulated by hormones that cause long-term retention of the receptor in the nucleus. In addition, these increases in initiation sites closely correlate with the conversion of nuclear receptor-hormone complexes to a form which is resistant to 0.4 *M* KCl extraction (Clark and Peck, 1976a; Clark et al., 1978c). Such forms of the receptor molecule may be associated with acceptor sites on chromatin, resulting in their long-term retention in the nucleus of target cells.

5. Interrelationships Between Early and Late Uterotropic Responses

Although this chapter is concerned primarily with relationships between nuclear-bound receptor and response, we do not mean to imply that interactions at other cellular sites are unimportant. Smith and Ecker (1969) suggested that progesterone acts at the level of the membrane to cause oocyte maturation in frog eggs. Liang and Liao (1975) and Liao et al. (1973) have suggested that androgen-receptor complexes interact with ribosomes to initiate protein synthesis; in fact, post-transcriptional controls have been proposed by several investigators (Tomkins et al., 1970). In addition estrogen is known to increase many uterine functions that do not appear to involve nuclear-mediated mechanisms (Szego, 1971). The relationship between these extranuclear events and those mediated via nuclear mechanisms is unknown. However, in this section we will attempt to place all of the known uterotropic events in perspective and to describe possible relationships between them.

The mammalian uterus is a highly complex organ made up of many cell types, including connective tissue, nerves, and blood vessels (see Wynn, 1977, for anatomic details), and it responds to estrogen and progesterone in a complicated manner. The primary purpose of estrogen is to stimulate uterine cells to increase in size and number. This hypertrophy and hyperplasia sets the stage for subsequent modifications induced by progesterone which result in a receptive environment for blastocyst implantation.

Uterotropic responses to estrogen can be classified according to their time of appearance and functional relationship (Tables 6 and 7). Early responses include both

Table 6. Early uterotropic responses

A. Supportive or metabolic	
1. Hyperemia	Loeb and Kountz, 1928; Markee, 1932; MacLeod and Reynolds, 1938
2. Histamine mebolization	Shelesnyak, 1959; Szego, 1965
3. Eosinophil infiltration	Tchernitchin et al., 1975, 1976
4. Water imbibition	Astwood, 1938; Szego and Roberts, 1953
5. Albumin accumulation	Kalman and Lowenstein, 1958; Spaziani, 1963
6. Increased electrolytes	Kalman and Lowenstein, 1958
7. Lysosome labilization	Szego, 1971
8. Increased cyclic nucleotides, prostaglandins, and associated enzyme activation	Szego and Davis, 1967; Kuehl et al., 1974; Ham et al., 1975
9. Increased glucose metabolism and associated enzyme activity	Szego and Roberts, 1953; Moulton and Barker, 1971
10. Increased uptake of RNA and protein precursors	Herranen and Mueller, 1956; Means and Hamilton, 1966
11. Increased calcium influx	Pietras and Szego, 1976
B. Biosynthetic	
1. Increased lipid synthesis	Aizawa and Mueller, 1961
2. Increased activity of RNA polymerase I and II	Gorski, 1964; Glasser et al., 1972; Hardin et al., 1976
3. Synthesis of the induced protein (IP) and its mRNA	DeAngelo and Gorski, 1970; Barnea and Gorski, 1970; Katzenellenbogen and Gorski, 1972
4. Increased synthesis of glucose-6-phosphate dehydrogenase	Smith and Barker, 1974
5. Increased chromatin template activity and RNA polymerase initiation sites	Barker and Warren, 1966; Glasser et al., 1972; Markaverich et al., 1978
6. Increased synthesis of histone and nonhistone proteins	Barker, 1971; Glasser et al., 1972

Table 7. Late uterotropic responses

A. Supportive or metabolic	
Many of the functions that are listed in Table 6 continue for many hours after estrogen administration. For a comprehensive review see Segal et al., 1977.	
B. Biosynthetic	
1. Increased general and specific protein and RNA synthesis	Aizawa and Mueller, 1961; Smith and Barker, 1974
2. Continued stimulation of RNA polymerase activity	Glasser et al., 1972; Hardin et al., 1976
3. DNA synthesis and mitosis	Epifanova, 1966; Aizawa and Mueller, 1961
4. Cellular hypertrophy and hyperplasia	Allen et al., 1937; Hisaw, 1959
5. Increased synthesis or changes in histone and nonhistone proteins	Barker, 1971; Glasser et al., 1972

biosynthetic and metabolic activities. (For a recent comprehensive review of this topic see Segal et al., 1977). The relationship between these events can be visualized in the following way. Hyperemia, calcium influx, histamine release, eosinophil infiltration, increased RNA and protein precursor uptake and enhanced glucose oxidation are due to the ability of estrogen to mobilize every possible physiologic function in order to optimize biosynthetic activity. Early responses also include increased RNA and protein synthesis, components of the biosynthetic machinery that eventually causes the uterus to grow. However, as was discussed in Sect. 4 of this chapter, the stimulation of these biosynthetic events is not necessarily obligatory in the stimulation of uterine growth. Late responses, some of which are simply extensions of those which were begun during the early period, include increased and sustained RNA and protein synthesis. This biosynthetic activity results in cellular hypertrophy and eventual DNA synthesis and hyperplasia. These late responses are considered true growth responses of the uterus. Obviously true growth would occur most readily in an environment in which substrate availability has been optimized. This environment is provided by the increased blood flow and other supportive events that are listed in Tables 6 and 7.

It has been suggested that some of these early events, which we are considering as supportive, are involved in estrogen's primary mechanism of action. Spaziani and Szego (1958, 1959) and Szego and Lawson (1964) have suggested that uterine growth is mediated by estrogen-induced histamine release, which causes increased capillary permeability and hyperemia. Although these events undoubtly maximize the ability of the uterus to grow, most investigators do not consider them primary events. As indicated below, these events can be blocked in the estrogen-treated animal without the loss of ability of the uterus to respond to estrogen (Zor et al., 1973; Tchernitchin et al., 1976).

Very rapid elevations in cyclic AMP levels have been reported following estrogen treatment in the uterus, and it has been suggested that cyclic AMP is a mediator of estradiol effects (Szego and Davis, 1967, 1969). Other investigators have not confirmed this effect, and it remains to be shown that activation of adenylate cyclase is involved in estrogen action (Zor et al., 1973; Korenman et al., 1973). Szego has also suggested that estrogen causes nuclear accumulation of lysosomes and that enzymes released into the nucleus control gene activity (Szego, 1971). Because of the uncertain role of lysosomes in the mechanism of hormone action, they have been included in both categories of early responses. Tchernitchin and Tchernitchin (1976) have suggested that estrogen causes eosinophils to be attracted to uterine capillaries where they migrate into the extracellular spaces of the uterus. The eosinophils then release hydrolytic enzymes that depolymerize the uterine ground substance (Tchernichin et al., 1976). Eosinophils may also cause mast cells to release histamine, and this, coupled with the hydrolysis of mucopolysaccharides, increases vascular permeability and creates an osmotic environment favoring water imbibition and precursor uptake. These early events appear supportive and not obligatory for true uterine growth, since they may be blocked without changes in estrogen-induced DNA, RNA, and protein synthesis (Tchernitchin et al., 1976). This observation, together with the fact that estriol stimulates all of these events yet does not produce significant true growth when injected, mitigates against these early events as obligatory or primary in the sequence that produces true growth (see Sect. 4 this chapter and Anderson et al., 1975).

Therefore we conclude that two interlinked pathways exist by which estrogens elicit uterotropic action: (1) a supportive one, which provides increased substrates, oxygen, and energy sources and thus provides an optimal environment for growth and (2) an obligatory pathway regulated by mechanisms which control DNA, RNA, and protein synthesis. This latter pathway is assumed to involve receptor binding in the nucleus of uterine cells where it stimulates transcriptional events that result in true uterine growth.

6. Nuclear Binding and RNA Transcription in the Chick Oviduct

In addition to the rat uterine model, the effect of estrogens on RNA polymerase initiation sites has been examined in the chick oviduct (Kalimi et al., 1976). Estrogenic stimulation of the chick oviduct leads to growth and differentiation accompanied by a dramatic increase in the synthesis of ovalbumin, which constitutes approximately 60 percent of the total protein (O'Malley et al., 1969; Oka and Schimke, 1969; Chan et al., 1973; O'Malley and Means, 1974; Harris et al., 1975). Ovalbumin synthesis is directly correlated with de novo synthesis of ovalbumin mRNA (Chan et al., 1973; Harris et al., 1975). Prior to the increase in ovalbumin mRNA, there is an overall elevation of chromatin template activity which results in an increased number of binding sites for RNA polymerase and the generation of many new initiation sites for RNA synthesis on chromatin (O'Malley et al., 1969; Tsai et al., 1975; Schwartz et al., 1975). These transcription events should be related to receptor binding in the nucleus.

To study these relationships an exchange assay was developed to measure the quantity of nuclear-estrogen receptor in the oviduct (Kalami et al., 1976). In these experiments we demonstrated that chick oviduct nuclei contain an estrogen receptor. Incubation of oviduct nuclei from DES-treated chicks with ^{3}H-estradiol results in a saturable exchange process that is inhibited by estrogens but not by nonestrogenic compounds. The number of receptor sites was 4,000–5,000 per cell, and the apparent K_d of the receptor-estrogen complex was 2 n*M*. Tissue specificity of the estrogen receptor was also apparent. Target tissues for estrogen, such as the oviduct and liver, contained nuclear sites while nontarget tissues, such as kidney, spleen, and skeletal muscle, did not contain significant quantities of these estrogen-binding sites. Low but significant levels were found in the heart, the significance of which is not known; however, retention of steroid hormone by the heart has been observed (Stumpf, 1977). From the above, estrogen-binding sites of the chick oviduct manifest the basic characteristics that are expected of a receptor, i.e., finite capacity, high affinity, hormone specificity, and tissue specificity.

A detectable increase in the quantity of estrogen receptor in nuclei of the oviduct can be observed within 10 min after a single injection of 2.5 mg DES to chicks (Fig. 63). A maximum level is reached by 20 min, followed by a rapid decline. Nuclear R_nE reaches a steady-state level by 60 min. Between 1 to 4 h the amount of R_nE remains constant and is elevated significantly above control values. If the complex in the nucleus is directly involved in the activation of estrogen-inducible genes, one would expect an increase in the number of RNA chain initiation sites following saturation of nuclear acceptor sites with R_nE. Indeed, Fig. 63 shows that this is the case. Within 30 min after estrogen administration, there is a 100 percent increase in the number of RNA chain initiation sites. These reach a maximum by 1 h and remain constant for up to 4 h. These levels of initiation sites are in the range of the predicted number of haploid genes (Ohta and Kimura, 1971). Nevertheless, we do not wish to defend the thesis that these initiations, as measured in vitro, are identical with the actual promoters. The evidence to support such a conclusion does not exist at this time.

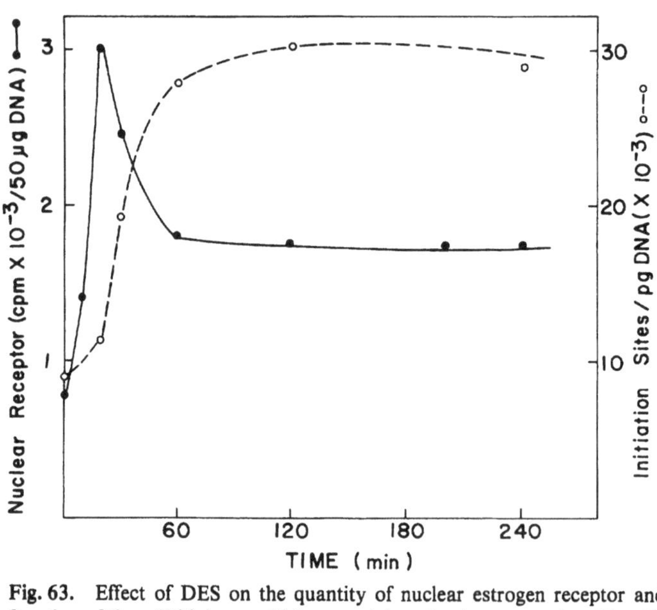

Fig. 63. Effect of DES on the quantity of nuclear estrogen receptor and RNA initiation sites as a function of time. Withdrawn chicks were injected subcutaneously with saline (0.9 % NaCl solution) or 2.5 mg DES. At the indicated times after treatment, the chicks were sacrificed, oviducts removed, and nuclei and chromatin prepared. The quantity of nuclear estradiol receptor (●) was determined by the [^{3}H]- estradiol exchange assay while RNA chain initiation sites (O) were measured by the rifampicin challenge assay.

Our results demonstrate that chick oviduct nuclei contain an estrogen-binding macromolecule with the characteristics of an estrogen receptor. These accumulate very rapidly following an injection of estrogen, and this accumulation precedes the stimulation of RNA chain initiation (Buller et al., 1975a, b; O'Malley et al., 1972). Nuclear receptor reaches a maximum by 20 min and declines rapidly by 1 h to a steady-state level that persists for at least 4 h. The maintenance of the receptor for as long as 4 h was observed in both withdrawn and unstimulated chicks. A similar pattern of nuclear binding in the rat uterus was discussed earlier and suggests a rapid accumulation of cytoplasmic receptors by the nucleus in excess of the number of nuclear acceptor sites. Those receptor-hormone complexes not bound to acceptor sites are then lost because of degradation or inactivation, while complexes that are bound to acceptor sites are retained for longer periods in the nucleus.

The receptor hormone complexes that remain at a constant level in the nucleus between 1 and 4 h are probably responsible for maintenance of the elevated and constant number of RNA initiation sites observed during this time (Fig. 63). Thus, the binding of R_nE complexes to acceptor sites in the nucleus may be directly related to the availability of DNA sequences for RNA synthesis. This proposal is further substantiated by the observations shown in Fig. 64. When hormone was withdrawn from DES-stimulated chicks, the number of RNA initiation sites on chromatin decreased in parallel with the decrease in the number of R_nE complexes. A similar direct relationship was observed when the number of R_nE complexes and the number of RNA initiation sites were measured following various dose levels of DES. The number of chromatin initiation sites were found to be similar when either *E. coli* or homologous hen RNA polymerase II was employed; this is consistent with recent

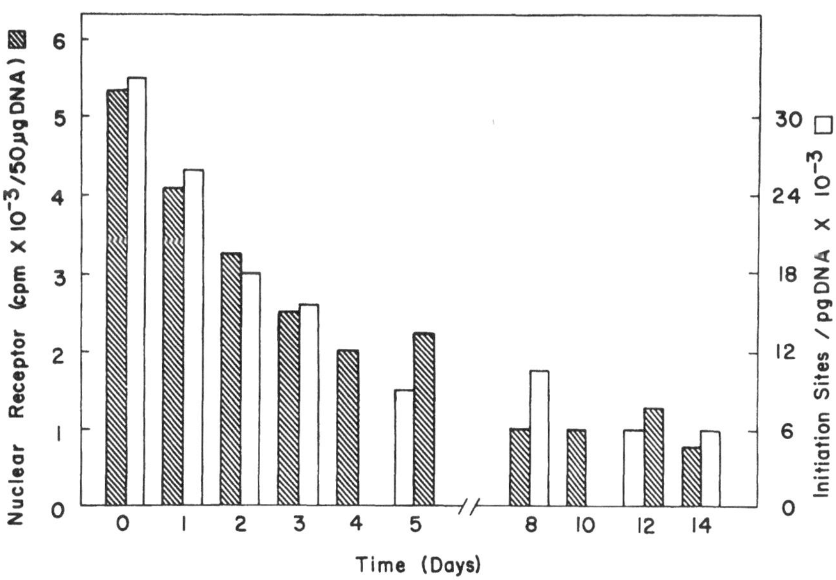

Fig. 64. Relationship between the quantity of nuclear estrogen receptor and RNA initiation sites following DES withdrawal. The quantity of estrogen receptor (*cross-hatched bars*) and the number of RNA initiation sites (*open bars*) were measured at various times following withdrawal from DES stimulation (14-day stimulation) (Kalimi et al., 1976)

observations that these enzymes appear to use the same sites on chromatin for initiation of RNA synthesis (Tsai et al., 1975). The presence of significant but low levels of R_nE in the unstimulated and withdrawn chicks may be due to small amounts of endogenous estrogens.

During the early stages of oviduct growth and differentiation, the quantity of both R_nE and RNA initiation sites increases at a similar rate and reaches a maximum on day 6. This time corresponds to the period of maximum gene expression (Kohler et al., 1969), protein synthesis (Means et al., 1971, 1972), and cell differentiatiǒn (O'Malley et al., 1972; Socher and O'Malley 1973; Spelsberg et al., 1972, 1976; Spelsberg and Cox, 1976). Following this period, intact oviduct cells show a decrease in RNA synthesis prior to reaching the mature differentiated state at days 10–15. Nevertheless, mature oviduct cells still require estrogen for synthesis of the majority of their intracellular proteins and control of metabolic functions (Spelsberg et al., 1972; Palmiter, 1972). It is of interest that following this differentiation of oviduct cells, the quantity of nuclear estrogen receptor does not decline, thus implying that a similar concentration of bound nuclear receptors is required for cell maintenance as is needed for induction of cell replication and growth. This is not surprising in view of previous conclusions that estrogen regulates and maintains the transcription of many thousands of unique gene sequences in the chick oviduct (Liarakos et al., 1973).

7. Nuclear Binding and the Control of RNA and Protein Synthesis in the Chicken Liver

The liver in birds and amphibians synthesizes large quantities of protein that are deposited in developing oocytes as yolk. Hormones appear to regulate this process by stimulating the synthesis of specific mRNAs for vitellogenin and very low density lipoproteins (VLDL) (Hillyard et al., 1956; Bergink et al., 1973; Luskey et al., 1974; Jost et al., 1975; Wittliff and Kenney, 1972; Wittliff et al., 1972c). Enhanced secretion of VLDL by chick liver is correlated with increased levels of translatable VLDL mRNA (Chan et al., 1976). The induction of vitellogenin synthesis also involves enhanced transcription, which is accompanied by the accumulation of cytoplasmic mRNA (Wetekam et al., 1975; Deeley et al., 1975; Mullinix et al., 1976; Deeley et al., 1977a, b; Tata, 1976; Skipper and Hamilton, 1977).

It is likely that the biosynthetic events described above are under the control of estrogen receptor interactions. Therefore, it might be anticipated that the usual cytoplasmic binding of estrogen by the receptor and subsequent nuclear translocation of an R_nE complex would occur; however, this has not been demonstrated. Clear evidence has been presented by a number of laboratories to show that estrogen administration to chickens and frogs causes increased binding of estrogen by receptors in liver nuclei (Mester and Baulieu, 1972; Lebeau et al., 1973, 1974; Gschendt and Kittstein, 1974; Gschwendt, 1976; Lazier, 1976). However, the origin of these receptors is uncertain since these investigators have been unable to demonstrate the presence of a cytosol receptor for estrogen and since the receptors do not preexist in the nuclear compartment (see Fig. 49 and discussion below). At the present time one cannot say with certainty whether or not a cytosol receptor exists because the assays currently in use may be inadequate for their detection. Indeed cytosol receptors for estrogens have been demonstrated in mammalian liver by Chamness et al. (1975b), Viladiu et al. (1975), Powell-Jones et al. (1976), Eisenfeld et al. (1976, 1977), and Beers and Rosner (1977). Since cytoplasmic receptors can be demonstrated in mammalian liver, the absence of these in chick liver may be related to a peculiarity of the avian species, not a characteristic of liver tissue. That the evolutionary and physiologic status of the bird might be important in receptor compartmentalization will be discussed in Chap. X.

a) Binding of Estrogen by Chick Liver Nuclei

A single injection of DES in the immature chick causes a 7-fold increase in the quantity of nuclear estrogen receptor 4 h after the injection (Fig. 65). Scatchard analysis of these data indicate that this receptor has a K_d of about 2.0 nM and that there exist about 1,100 sites per cell. These results generally agree with those of

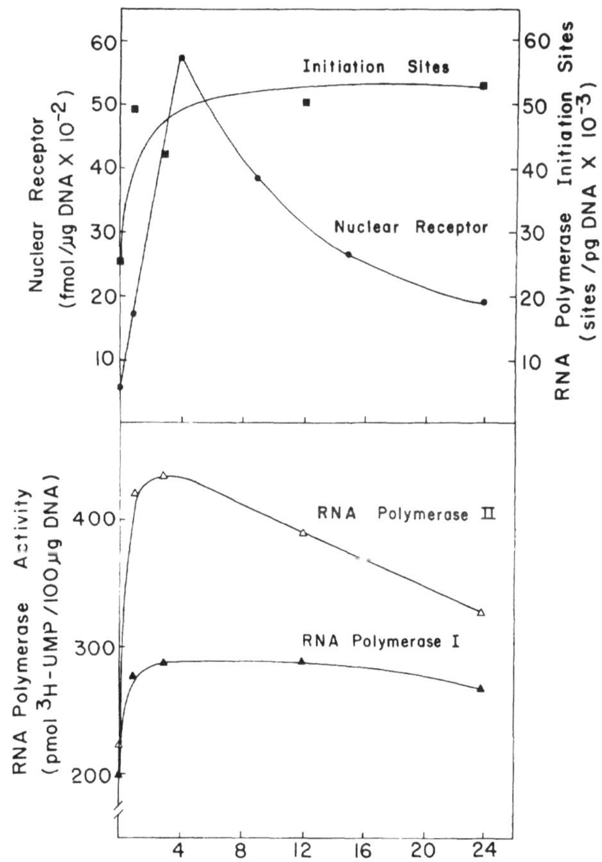

Fig. 65. Relationships between nuclear-bound receptor, RNA polymerase initiation sites, and RNA polymerase activity in chicken liver. Chicks were injected with 2.5 mg DES and the quantity of estrogen receptor in the nucleus (●), RNA polymerase initiation sites (■), and RNA polymerase I (▲) and II (△) activities were determined as a function of time (Snow et al., 1978)

others (Mester and Baulieu, 1972; Lebeau et al., 1973; Gschwendt and Kittstein, 1974; Gschwendt, 1976; Lazier, 1976) with respect to K_d; however, the quantity of receptor reported by our laboratory is greater. This disparity probably results from the assay conditions used in our studies. As described in Chap. II, the conditions for optimal measurement of liver nuclear receptors in the chick include incubation of nuclear fractions at 30°C for 30 min. Other investigators have used suboptimal conditions and thus have underestimated the quantity of nuclear receptor.

The binding of ^{3}H-estradiol to R_nE in the exchange assay is inhibited by estrogenic compounds and is not inhibited by other steroids (Snow et al., 1978). Thus, the receptor displays the stereospecificity expected of a specific estrogen binding site.

The nuclear receptor can be partially extracted by 0.4 *M* KCl; however, most of it remains bound to the nuclei (Lebeau et al., 1973; Snow et al., 1978). The dissociation constants do not differ between the KCl-extractable and KCl-resistant forms.

The receptor-DES complex extracted by 0.4 *M* KCl can be exposed to exchange conditions in the presence of ^{3}H-estradiol and then analyzed for its sedimentation characteristics by sucrose density centrifugation (Snow et al., 1978). Such studies indicate that the nuclear estrogen receptor in chick liver is a macromolecule that binds estrogen in a specific manner. It is partially extractable by KCl and hence similar to the estrogen receptor in the rat uterine nuclear fraction (Clark and Peck, 1976a).

Since the major portion of the nuclear receptor appeared to be resistant to KCl extraction, it was likely that the receptor was bound to chromatin as in the case of retained receptor in rat uterine nuclei (see Chap. II and above). This was examined in chromatin prepared from liver nuclei by a modification of the method of Spelsberg et al. (1971a). DES treatment substantially increased the quantity of R present in chromatin. These "new" R_nE complexes on chromatin account for most of the receptor in liver nuclei (Snow et al., 1978).

b) Relationship Between Nuclear Binding, RNA Polymerase Activity, and Polymerase Initiation Sites

Earlier in this chapter we demonstrated that nuclear binding of the estrogen receptor in the chick oviduct showed temporal and quantitative relationships to the opening of RNA polymerase initiation sites on chromatin. These relationships were also investigated in the chick liver system. The quantity of nuclear receptor increases rapidly within 1 h and reaches a maximum at 4 h, then declines gradually between 8 and 24 h after treatment to levels significantly above control (Fig. 65). By 1 h both RNA polymerase I and II activities are maximal, and elevated activities are observed over the 24-h period after the injection. A similar pattern of stimulation of polymerase initiation sites was also observed.

These results are in accord with the concept that the occupation of receptor by estrogen in the nucleus of the chick liver increases the number of sites available for initiation of RNA synthesis by RNA polymerase. Occupied nuclear estrogen binding sites approximately double by 1 h after injection (Fig. 65), and this doubling is correlated with a similar elevation of initiation sites by this time. Likewise RNA polymerase I and II activities are elevated by this time. The quantity of receptor in the nucleus continues to increase to very high levels as it does in the rat uterus and chick oviduct when hyperphysiologic doses of estrogen are used. This excess of nuclear receptor has been discussed earlier in this chapter and probably relates to the

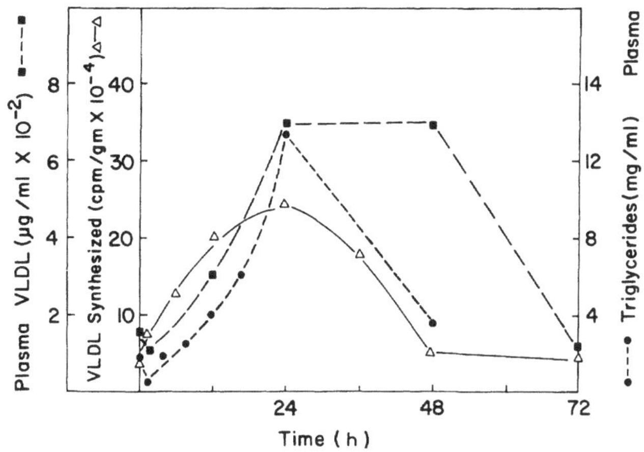

Fig. 66. Relationship between synthesis of *VLDL* and plasmic levels of *VLDL* and triglycerides after estrogen treatment. Immature chicks were injected with DES and plasmic level of *VLDL* and triglyceride were measured as a function of time. In addition, the synthesis of *VLDL* was determined by in vitro incubation of liver slices (Snow et al., 1978)

binding of estrogen to larger quantities of receptor than are required to maximize a given response. The relatively rapid decline in receptor levels by 24 h to approximately 20 percent of the total number of binding sites is accompanied by a very gradual decline in the activity of RNA polymerase I and II activities. Thus, RNA polymerase activities and initiation site quantities are maintained by only a small percentage of the total number of receptor sites. This has been shown to be the case in the rat uterus and chick oviduct system as discussed above.

c) Relationship Between Nuclear Estrogen Receptor and the Synthesis of VLDL and Plasma Triglycerides

The stimulation of VLDL and plasma triglycerides by estrogen administration in the liver is probably the result of increased RNA synthesis, which results from the binding interactions discussed above. To examine these relationships, the synthesis of VLDL and plasma levels of triglycerides were measured following DES injection in the immature chick (Fig. 66). The rate of VLDL synthesis, as measured in liver slices in vitro (Chan et al., 1976), increased 2-fold within 2 h after hormone treatment. The maximum increase was observed between 12 and 24 h. By 48 h the rate of synthesis had returned to control levels. Plasma levels of triglyceride declined at 1 h and then began a gradual elevation to 6-fold that of controls by 24 h. Subsequently triglycerides declined slowly to control levels.

The elevated synthetic rate for VLDL is probably due to an increased synthesis of mRNA for this protein complex. Chan et al. (1976) demonstrated that DES stimulates the accumulation of mRNA for a major VLDL protein (apo $VLDL_{II}$). Little is known about triglyceride synthesis, and the lag period for its appearance in the blood is probably related to complex secretory processes in the liver. Regardless of the reason, it follows that the synthesis of this lipid is under the control of nuclear events that are regulated by estrogen. This assumption is further substantiated by the R_nE

correlation observed between the quantity of nuclear estrogen receptor and levels of plasma triglycerides after DES administration (Snow et al., 1978).

In most respects estrogen acts in the avian liver through mechanisms quite similar to those observed in other classic target organs, such as the rat uterus and chick oviduct. Though the origin of the receptor is unknown, estrogens appear in liver nuclei bound to specific receptors within a short time of administration. The precise nature of the hormone receptors remains to be elucidated. Whether they exist in the cytoplasm as is the case for many other steroid hormone systems or are integral components of the chromosomal proteins as proposed for thyroid hormone is an open question (Spindler et al., 1975; Latham et al., 1976; Oppenheimer et al., 1976). Nevertheless, interaction of receptor-hormone complexes with liver chromatin results in stimulation of gene transcription. This increase in transcription activity is a consequence of an elevation of the number of RNA synthesis initiation sites on chromatin as well as an increase in RNA polymerase I and II activities. Subsequently, accumulation of VLDL mRNA occurs, which is closely correlated with an increased rate of VLDL synthesis (Chan et al., 1976). Newly synthesized VLDL is then secreted and appears in the circulation, resulting in elevated plasma VLDL and triglycerides.

Chapter VI

Control of Steroid Receptor Levels and Steroid Antagonism

In previous chapters of this book the action of each hormone was considered separately. However, it is clear that hormones interact to modify and sometimes antagonize each other (Clark et al., 1976b). These relationships have been recognized at the physiologic level for many years, but only recently have their mechanisms begun to be understood. These interactions will be discussed further as they relate to reproductive physiology in Chap. VIII.

1. Control of Estrogen Receptor Concentrations

Estrogen responsive cells of the uterus in a castrate rat maintain levels of receptor which enable it to respond to administered estrogen. This basal level of receptor is probably controlled by genetic mechanisms that are programed for the constitutive synthesis of the cytoplasmic receptor, R_c^E. In addition, as discussed later, endogenous steroids from the adrenal may play a role in this maintenance. Thus, estrogen target tissues can usually detect and respond to estrogens. This is also true of estrogen targets in the male animal which will respond readily to exogenous estrogen and which have estrogen receptor levels equal to those of the female (Clark et al., 1972b; Anderson et al., 1973a; Cidlowski and Muldoon, 1976).

Although estrogen target cells appear capable of maintaining a constitutive level of cytoplasmic receptor, this does not imply that sex hormones have no influence on the level of R_c^E. To the contrary, it is well known that steroid receptor levels are influenced by endogenous and exogenous steroids. As pointed out in Chap. I, an injection of estradiol causes a rapid depletion of R_c^E which appear in the nucleus as R_nE complexes. This is followed by a period during which R_c^E is replenished. The replenishment of R_c^E involves at least two processes, the reactivation of R_nE complexes and the de novo synthesis of R_c^E molecules. In tissues that do not grow in response to hormone, replenishment may involve only reactivation. This was first suggested by Munck et al. (1972) for glucocorticoid receptors in thymus cells. In fact, replenishment of cytoplasmic receptors for glucocorticoids does not depend on protein and RNA synthesis (Rousseau et al., 1973; Ishii and Aronow, 1973; Middlebrook et al., 1975), an observation that supports this suggestion.

In tissues that grow in response to hormone stimulation, such as uterus, vagina, and mammary gland, it is easy to envision the involvement of both reactivation and synthesis in R_c^E replenishment. It follows that synthesis of more R_c^E molecules is required in cells that will undergo cell division after hormone stimulation. In this manner a constant amount of R_c^E per cell is maintained. Cells that grow in size and do not divide may also require R_c^E synthesis to counteract the dilution effect brought on by cellular hypertrophy. The involvement of protein synthesis in the replenishment process has been suggested by Gorski et al. (1971) and Cidlowski and Muldoon (1974).

Recently Mester and Baulieu (1975) observed that the replenishment of R_c^E in uteri of immature rats involves two separate processes. The first process, 0–6 h after injection, could not be blocked by cycloheximide, whereas the second process, 6–11 h after injection, was dependent on protein synthesis. Thus, in cell types that grow in size or number in response to a steroid hormone, R_c replenishment may involve both recycling and synthesis of new receptor, whereas in cells that do not grow in response to their target steroid, R_c replenishment may involve only recycling.

The exception to the above proposal appears to be the effect of estrogen on R_c replenishment in the hypothalamus and pituitary (Cidlowski and Muldoon, 1974). In these tissues, which do not grow in response to estradiol administration, cycloheximide also blocks replenishment of R_c^E. Therefore it is difficult to make any clear distinctions based on growing or nongrowing tissues. Possibly all sex steroid hormones stimulate replenishment via a combination of reactivation and synthetic mechanisms. Regardless of the mechanisms involved, replenishment of R_c following steroid treatment appears to be requisite for subsequent stimulation and hence could serve as a control point by which other compounds antagonize or modify estrogen action.

2. The Relationship of Cytoplasmic Replenishment of the Estrogen Receptor to Biologic Response

The absence of cytoplasmic steroid receptors results in an inability of the tissue to respond to the appropriate hormone. This point was discussed in Chap. II and serves as a basic tenet in the receptor concept. Therefore, the replenishment of R_c that follows hormone-induced depletion should be an important factor in determining the ability of a cell to respond to subsequent hormonal stimuli.

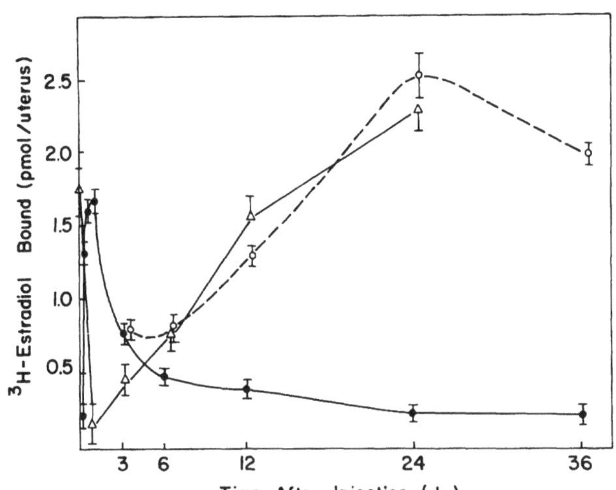

Fig. 67 Relationship between replenishment of the cytoplasmic estrogen receptor and the availability of that receptor for subsequent translocation. Immature rats were divided into two groups, both of which were injected with 2.5 μg estradiol. For one group the quantity of nuclear-bound (●) or cytoplasmic receptor (Δ) was determined by the ^{3}H-estradiol exchange assay as a function of time. A second group was reinjected with 2.5 μg estradiol at 3, 6, 12, 24 and 36 h after the initial injection and the quantity of nuclear-bound receptor was determined 30 min later (O)

To test the relationship between replenishment and the return of responsiveness, we first injected immature rats with estradiol and killed them 0, 3, 12, 24 or 36 h later to measure both the quantity of R_c^E that could be translocated to the nuclear fraction and the ability of the uterus to respond (Anderson et al., 1974b) (Fig. 67). This figure shows that as R_c^E is replenished after the first injection, subsequent injections are able to cause nuclear translocation of approximately the quantity of R_c^E that was replenished.

If the formation of R_nE complex is required for the induction of estrogenic responses, one would predict that when R_nE levels are refractory to estradiol reinjection as, for example, at 3 h (Fig. 67), uterine responses would also be refractory. Further, if the number of R_nE complexes is important to responsiveness, one would expect that the extent of response would be proportional to the levels of R_nE. The results in Fig. 68 A and B show the effect of estradiol reinjection on early uterine responses. In this study rats were injected with 2.5 µg estradiol and killed at various times after treatment. The uteri were removed, weighed, and the conversion of glucose to CO_2 was determined. Control rats received only a single injection (*open bars*) while a second group of rats (*hatched bars*) received an additional injection of estradiol (2.5 µg) at 3, 12, or 24 h after the first injection. Estradiol reinjection at 3 h does not significantly elevate the magnitude of uterine fluid imbibition (wet weight) or glucose oxidation. That is, when the second estradiol injection follows the first by a period of 3 h, early responses are equivalent to those elicited by a single estradiol

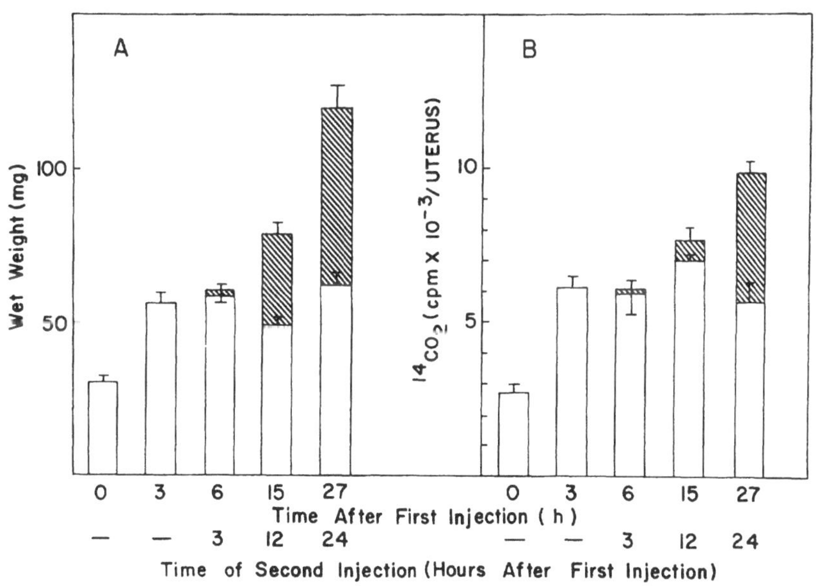

Fig. 68 A and B. Uterine sensitivity following an estradiol injection: short-term uterine responses. Two groups of rats were injected with 2.5 µg estradiol at time 0. The first group (*open bars*) were killed at various times after the first injection, their uteri were weighed (**A**), and the conversion of U-^{14}C-glucose to $^{14}CO_2$ determined (**B**). The second group (*cross-hatched bars*) received a second estradiol injection (2.5 µg) at 3, 12, or 24 h after the first injection. These animals were killed at 3 h following the second injection, their uteri were weighed (**A**), and the conversion of U-^{14}C-glucose to $^{14}CO_2$ determined (**B**). (Anderson et al., 1974b)

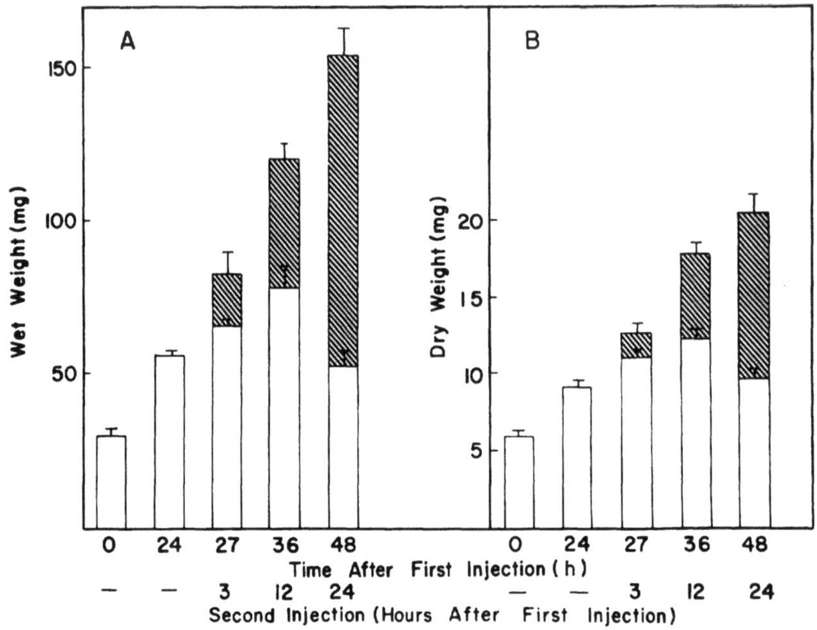

Fig. 69 A and B. Uterine sensitivity following an estradiol injection: long-term uterine responses. Two groups of rats were injected with 2.5 μg estradiol. The first group (*open bars*) were killed at various times after treatment, and uterine wet (**A**) and dry (**B**) weight determined. The second group (*cross-hatched bars*) received an additional estradiol injection (2.5 μg) at 3, 12, or 24 h after the first injection. These animals were killed at 24 h after the second injection and uterine wet (**A**) and dry (**B**) weight determined (Anderson et al., 1974b)

injection. However, estradiol reinjection at 12 or 24 h after the initial injection increases the extent of early responses.

The results in Fig. 69 A and B show that long-term growth responses are also dependent on the time interval between the two estradiol injections. The experimental procedure for this study was the same as in Fig. 68 A and B except that uterine dry weight and wet weight were examined 24 h after the second estradiol injection. Reinjection of estradiol at 3 h does not significantly alter uterine wet or dry weight ($p < 0.05$). Increases are observed when reinjection is performed at 12 h, and maximal response levels are obtained when the second estradiol injection follows the first by a period of 24 h.

This study demonstrates that the uterine capacity for the formation of R_nE complexes is closely correlated with the ability of the uterus to respond to estradiol. At 3 h after estradiol treatment, cytoplasmic binding capacity is low (Fig. 67; Jensen et al., 1968, 1969; Sarff and Gorski, 1971). At this time estradiol reinjection fails to increase nuclear receptor content or uterine response. However, at approximately 15–18 h after treatment, cytoplasmic receptor (Figs. 67–69; Sarff and Gorski, 1971), R_nE, and uterine sensitivity have returned to the levels before injection. Furthermore, at 24 h after estradiol injection, uterine cytoplasmic binding capacity has increased to levels greater than those of nontreated rats (Fig. 67; Sarff and Gorski, 1971). At this time estradiol reinjection elicits marked elevation in R_nE and uterine response (Fig. 68 A and B), a feature which may result from the large cytoplasmic binding capacity.

The replenishment of cytoplasmic receptor beginning about 6 h after estradiol is dependent on early protein and RNA synthesis (Jensen et al., 1969; Sarff and Gorski, 1971; Mester and Baulieu, 1975). Further, the increase in cytoplasmic sites after estradiol injection as well as the levels of R_nE induced by estradiol reinjection (Fig. 67) are paralleled by general uterine protein synthesis (Mueller, 1953). At 24 h after estradiol injection uterine protein content is near a maximum, whereas uterine DNA is equivalent to the noninjected value (Aizawa and Mueller, 1961; Schmidt et al., 1967). Thus, the elevated levels of cytoplasmic receptor (and hence R_nE after reinjection) at 24 h after estradiol injection are probably a reflection of uterine cellular hypertrophy.

At 3 h after estradiol treatment R_c^E is low, and the quantity of R_nE is approximately 35 percent of the level observed at 1 h (Fig. 67). In addition, estradiol reinjection at this time does not increase R_nE. These observations suggest that between the 1st and 3rd h after estradiol treatment, approximately 65 percent of the receptor has been degraded or inactivated. Similar observations have been described under in vitro conditions (Giannopoulos and Gorski, 1971b). These observations provide a possible mechanism to account for the significant loss of estradiol within the 1st h after administration (Jensen and De Sombre, 1972). Thus, degradation or inactivation of the receptor molecule may alter it such that it can no longer bind estradiol. The R_nE complexes, which are inactivated or degraded between the 1st and 3rd h after estradiol injection are not required for uterine growth responses (see Chap. V). These excess receptors may be those that are not bound to nuclear acceptor sites. One can speculate that physiologically inactive R_nE complexes not bound to nuclear acceptor sites are rapidly lost from the nuclear fraction because of a degradation or inactivation process. Active R_nE complexes, in contrast, may be protected from degradation by their association with acceptor sites in the nucleus. Similar observations concerning the replenishment of active R_c and the return of sensitivity have been made by Katzenellenbogen, 1975 and Katzenellenbogen et al. (1977). Their studies with nonsteroidal estrogen antagonists will be discussed later in this chapter.

3. Control of the Progesterone Receptor by Estrogen

The uterus is relatively insensitive to progesterone unless first exposed to estrogen. Thus, progesterone treatment in a nonestrogenized uterus will not produce a secretory uterine epithelium (Reynolds, 1951); however, with estrogen priming progesterone has dramatic effects on the production of secretory responses. These observations may be explained a priori by assuming that estrogen priming stimulates the synthesis of the progesterone receptor, thereby enhancing the ability of the uterus to respond to progesterone. Several investigators have shown that estrogen treatment increases the quantity of progesterone receptors (Corvol et al., 1972; Faber et al., 1972; Feil et al., 1972; Toft and O'Malley, 1972; Milgrom et al., 1973a; Rao et al., 1973; Freifeld et al., 1974; Thi et al., 1975; Reel and Shih, 1975; Hsueh et al., 1975, 1976; Horwitz and McGuire, 1978). These effects of estrogen occur in both the endometrium and myometrium of the guinea pig uterus, and hence estrogen does not appear to have any differential cell effect (Luu et al., 1975; Warembourg, 1974;

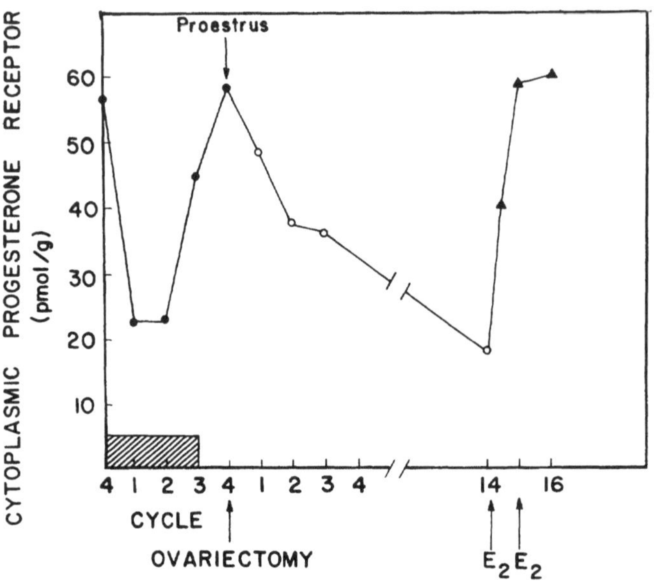

Fig. 70. Effect of ovariectomy at proestrus and subsequent estrogen treatment on uterine progesterone receptor. Results for the cycle (●) are provided for comparison. The concentration of binding sites declined more slowly after ovariectomy at proestrus (○) than between proestrus and estrus in intact animals. Estradiol treatment (15 μg/day) 14 days after ovariectomy restored the concentration of progesterone receptor to the proestrus level (▲) (from Leavitt et al., 1974)

Warembourg and Milgrom, 1977). The ability of estradiol to increase the level of the cytoplasmic progesterone receptor (R^P_c) had been suggested for neurons of the hypothalamus (Sar and Stumpf, 1973) and recently demonstrated by Kato and Onouchi (1977) and McEwen (1978).

Leavitt and Blaha (1972) and Leavitt et al. (1974) demonstrated estrogen's effect on R^P_c quite clearly in the hamster uterus. In these studies hamsters were ovariectomized at proestrus, and the quantity of cytosol progesterone receptor was measured as a function of time (see Fig. 70). The quantity of receptor fell dramatically and reached low levels by 2 weeks. Estrogen treatment at this time caused a marked elevation to proestrus levels within 24 h. A similar effect of estradiol on the level of cytoplasmic progesterone receptor (R^P_c) has been observed in the rat uterus (Walters and Clark, 1977b). In these studies either 22–23 day old immature rats or ovariectomized rats were injected for 2 days, and the quantity of R^P_c was determined by ^{3}H-progesterone exchange. During this 72-h period the quantity of R^P_c increased dramatically (2× on a weight basis and ~ 5× on a per uterus basis).

Other investigators have observed that not only does estrogen treatment increase the quantity of R^P_c but it also causes a shift in the sedimentation coefficient. Castrate animals contain primarily the 4*S* form of the receptor, and estrogen treatment causes a shift to the 7–8*S* form. This observation has been made in endometrium and myometrium of the monkey (Elsner et al., 1977; Illingworth et al., 1977), in the human uterus (Janne et al., 1975, 1976), in the guinea pig uterus (Freifeld et al., 1974), and the chick oviduct (Toft and O'Malley, 1972). From these observations it can be concluded that estrogen stimulates the uterus to produce qualitative, as well as quantitative changes in the cytoplasmic progesterone receptor, probably by de novo synthesis. Thus, estrogen sets the stage for the binding of progesterone, which is prerequisite for progesterone action.

Leavitt et al. (1978) have also shown this effect in vitro. Uterine strips from the hamster were incubated in vitro with estradiol, and the quantity of cytoplasmic progesterone receptor increased dramatically over a 24 h period of incubation. The increased quantity of R_c^P is blocked by the addition of cycloheximide and actinomycin-D, both in vitro in hamster uteri (Leavitt et al., 1978) and in vivo in the guinea pig (Milgrom et al., 1973a). Thus, estradiol stimulation of R_c^P probably involves RNA and protein synthesis.

4. Effects of Progesterone on the Control of Progesterone Receptors

Milgrom et al. (1973) and Freifeld et al. (1974) demonstrated that progesterone has the paradoxical effect of causing a rapid decline in the quantity of total progesterone receptor in the guinea pig uterus. This decline is not simply the result of receptor translocation to the nucleus, although this mechanism is clearly functioning immediately after an injection of progesterone (Walters and Clark, 1977a, b). The data in Fig. 71 show that nuclear accumulation and cytoplasmic depletion take place immediately after progesterone injection in estrogen-primed rats. In addition, some R_c^P replenishment takes place during the first 10–12 h after the injection. This period is followed, however, by a decline to low levels by 24–48 h similar to the observation of Milgrom et al. (1973a).

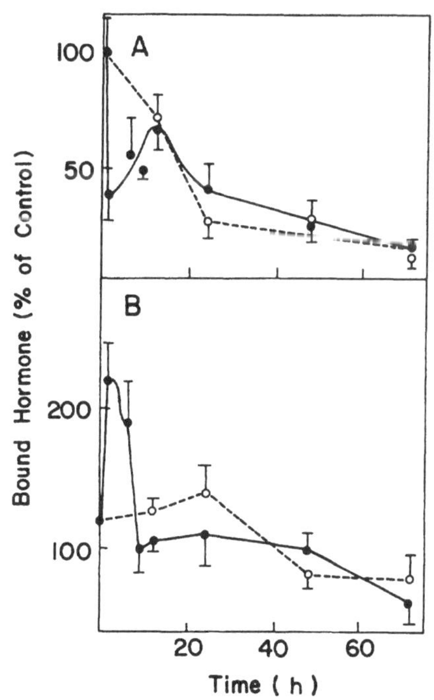

Fig. 71 A and B. Effect of progesterone on the distribution of cytoplasmic and nuclear progesterone receptor. Estrogen-primed castrate rats were injected with 0.5 mg progesterone (●) or saline (○) and the level of progesterone receptor was determined in the cytosol (A) and nuclear (B) fractions as a function of time. Note that the *open symbols* represent the change in receptor resulting from estrogen withdrawal, while *closed symbols* represent the change in receptor due to both progesterone and estrogen withdrawal (Walters and Clark, 1978)

Table 8. Effect of estradiol on the number of uterine cytosol progesterone receptors in progesterone-treated rats[a]

Treatment	Cytosol receptors (pmol/uterus)	Uterine wt. (mg)
Progesterone + estradiol	2.96 ± 0.77	217.5 ± 9.8
Progesterone	0.98[b] ± 0.39	174.9[c] ± 7.0
Vehicle + estradiol	3.59 ± 0.65	234.5 ± 8.1
Vehicle	1.02[b] ± 0.35	168.4[c] ± 11.5

[a] Mature castrate rats were petreated with estradiol (1.0 μg/rat, SC) 24 h before receiving either progesterone (0.5 μg/rat) or vehicle. 12 h later two groups of rats received a second dose of estradiol (1.0 μg/rat).
[b] $P < 0.05$.
[c] $P < 0.01$ when compared to estradiol-treated group; Student's one-tailed *t*-test; n = 4.

With the exception of initial translocation, the decline in R_c^P and R_nP levels after injection of progesterone appears due simply to estradiol withdrawal, since progesterone receptor levels depend on the presence of estradiol. It is clear from the results shown for noninjected animals that estrogen withdrawal results in a decline of R_c^P levels (Fig. 71) and, in addition, estrogen will maintain R_c^P when given subsequent to the administration of progesterone (Table 8). Estrogen withdrawal implies a lack of estrogen action which could result from either declining serum levels of estrogen or declining levels of R_c^E. Since progesterone suppresses the synthesis of R_c^E (see below) and since the synthesis of R_c^P is dependent on the action of estrogens via R_c^E, it is possible for progesterone to suppress the synthesis of its own receptor by serial antagonism. The results of Milgrom et al. (1973) and Freifeld et al. (1974) as well as a portion of the depletion observed during the estrous cycle may be explained by this mechanism.

5. Control of Estrogen Receptor Levels by Progesterone and Its Relationship to Estrogen Antagonism

Progesterone acts on the estrogen-primed uterus to alter cell function and reproductive competence. Often this ability of progesterone is considered to be antagonistic to estrogen; however, it probably should be referred to as a modifier of estrogen action rather than an antagonist. Nevertheless, progesterone will reduce the ability of estrogens to cause uterine growth and vaginal cornification (for reviews, see Lerner, 1964 and Huggins and Jensen, 1955). This ability of progesterone to modify or antagonize estrogen action is generally considered to involve receptor mechanisms.

a) Receptor Replenishment

Progesterone does not interfere with the initial binding of estradiol to the cytoplasmic receptor, R_c^E, or to the subsequent translocation of the R_cE complex to nuclear sites (Toft and Gorski, 1966; Anderson et al., 1972a; Clark et al., 1977b); hence, the

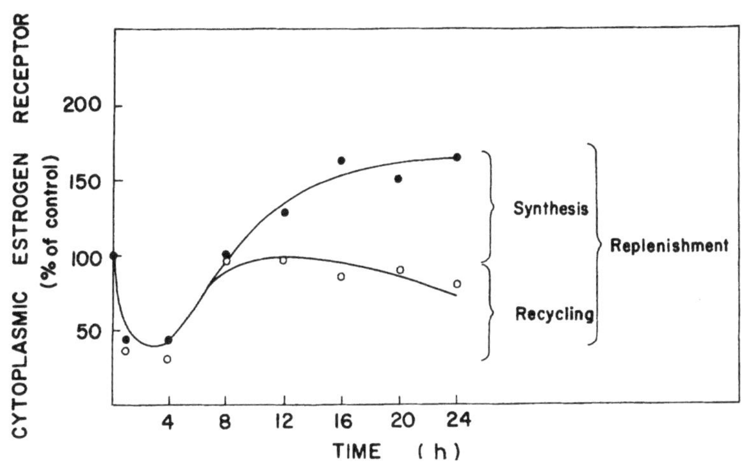

Fig. 72. The antagonism of estrogen-receptor replenishment by progesterone. Estrogen-primed immature rats received either estradiol (●) or estradiol plus progesterone (○) at time zero. The quantity of cytoplasmic estrogen receptor was measured by exchange at various times after injection (Hsueh et al., 1976)

source of its antagonist effect does not lie at these levels. As discussed in Sect. 1 this chapter above, one of the ways that progesterone could antagonize or modify estrogen effects is by interfering with estrogen receptor availability, i.e., reduced replenishment of R_c^E and thus could effectively eliminate the ability of the uterus to respond to estrogen. This has been suggested by several laboratories (Oka and Schimke, 1969; Means and O'Malley, 1971; Bronson and Hamilton, 1972; Jacobson et al., 1972; Milgrom et al., 1973a; Leung and Sasaki, 1973; Brenner et al., 1974; Mester et al., 1974; Pavlik and Coulson, 1976b). In the rat, progesterone alters the level of R_c^E and decreases the ability of estrogen to stimulate the uterine growth (Hsueh et al., 1975). In these experiments estrogen-primed immature rats were injected with progesterone or estradiol plus progesterone, and the quantity of R_c^E was measured as function of time after treatment (Fig. 72). The quantity of R_c^E is depleted rapidly in both groups, and this is accompanied by a concomitant accumulation of the receptor in the nucleus (Hsueh et al., 1976). After the initial depletion the level of R_c^E gradually increases between 4 and 8 h. However, between 8 to 24 h after injection, the R_c^E level continues to increase in the estradiol-treated group but remains constant in the estradiol-plus-progesterone treatment group. These observations, recently confirmed by Saiduddin and Zassenhaus (1977a) suggest that two separate processes are involved in the replenishment of R_c^E. In addition, progesterone inhibits the replenishment process occurring between 8 and 24 h after estradiol injection. Coulson and Pavilik (1977) have shown that the inhibitory effect of progesterone is not attributable to a general decline in protein synthesis and hence may be an effect directed specifically to R_c^E replenishment.

To study further the time sequence of progestational inhibition of R_c^E replenishment, progesterone was injected at various intervals after estradiol treatment, and the level of R_c^E was measured 24 h after estradiol injection. Progesterone injection at 4–8 h after estrogen decreased the quantity of R_c^E to that level observed when progesterone was injected simultaneously with estradiol. However, when progesterone was injected between 8 to 24 h after estradiol, the level of R_c^E 24 h after estradiol treatment was progessively increased, suggesting a time-dependent escape from the antagonistic effect of progesterone.

These data support the existence of two separate processes for receptor replenishment. The first phase, 4–8 h after injection, may represent a recycling of R_c^E which is not blocked by progesterone. The quantity of R_c^E replenished equals that which is depleted, and hence recycling through some deactivation-reactivation mechanism could account for this phase of replenishment. The second phase, 8–24 h after injection, may involve the synthesis of R_c^E blocked by progesterone treatment. The additional R_c^E appearing during phase 2 is approximately 50 percent greater than that initially present (Fig. 72). This 50 percent increase in R_c^E is suppressed even when the injection of progesterone is delayed up to 8 h, implying that progesterone may act on some late event in the sequence that leads to new R_c^E synthesis.

Thus, R_c^E replenishment in estrogen-primed animals may be accomplished by the following mechanisms: (a) a recycling of existing receptor following nuclear accumulation, which is not blocked by the presence of progesterone; and (b) a synthesis of new R_c^E, which is blocked by progesterone. The decrease in R_c^E after progesterone treatment does not appear to be a pharmacologic effect since the half-maximal dose (0.6 mg progesterone) as defined in this study is well within the generally accepted "physiologic" range for this hormone (Zarrow et al., 1964).

The ability of progesterone to cause a reduction in R_c^E appears to be specific for estrogen antagonists. A second antagonist, testosterone propionate, caused a 25% reduction in R_c^E and a slight reduction in estrogen-stimulated uterine weight. Other steroids such as hydrocortisone acetate have no effect on R_c^E replenishment (Clark et al., 1977b). Progesterone has little effect on R_c^E in the nonestrogen-primed uterus. Presumably this reflects an insensitivity of the uterus to progesterone as a result of the low quantities of progesterone receptor present in the nonprimed uterus (as discussed earlier).

b) Nuclear Accumulation and Retention of the Receptor-Estrogen Complex

As discussed above, any interaction interfering with nuclear retention of the R_nE complex will be antagonistic. Therefore, progesterone possibly reduces the length of time that R_nE complexes remain in the nucleus. Preliminary evidence indicates that this is not the case when estradiol and progesterone are injected simultaneously. However, nuclear retention time does appear to be reduced in animals pretreated with estradiol and subsequently treated with estradiol plus progesterone (Hsueh et al., 1976). Since the level of R_c^E is also reduced in pretreated animals, the observed decrease in nuclear retention of R_nE may result from a decreased quantity of R_c^E that is available for translocation.

In conclusion, the reduced ability of the uterus to respond to estradiol after progesterone exposure probably results from decreased cytoplasmic estrogen receptors. The possibility that progesterone interfers with the nuclear retention of the R_nE complex remains to be resolved.

c) Antagonism of Estrogen Receptor by Progesterone in the Adult Rat Uterus

As in the immature rat, progesterone causes a significant decrease in cytoplasmic estrogen receptor content in the adult (Fig. 73). A reduction is also observed in the amount of nuclear-bound estrogen receptor although these quantities were small in all groups and hence have little influence on the total amount of receptor. The decrease in quantity of R_c^E does not result from a general depressive action of progesterone on uterine protein synthesis; thus the depression is significant when the data are expressed as sites per mg protein (25.9 ± 1.9 pmol/100 mg protein for estradiol pretreatment group; 11.1 ± 1.7 pmol/100 mg protein for estradiol plus progesterone pretreatment group, $p < 0.01$). We have shown that testosterone has a weak effect on R_c^E replenishment and that this effect is correlated with its weak anti-estrogenicity (Hsueh et al., 1976). To further examine the specificity of the depressive effect of progesterone on uterine R_c^E content, estrogen-treated adult castrate rats were injected with hydrocortisone acetate. This treatment caused no significant decrease in either cytoplasmic or nuclear estrogen receptor content (Fig. 73). These data suggest that the depressive effect of progesterone on estrogen receptor levels in the adult rat is hormone specific.

To demonstrate that the decreased level of estrogen receptor does reflect a decrease of uterine responsiveness in progesterone-treated animals, animals from estradiol and estradiol-plus-progesterone pretreatment groups were injected with estradiol, and the uterine weights measured 24 or 48 h later. The uterine weights

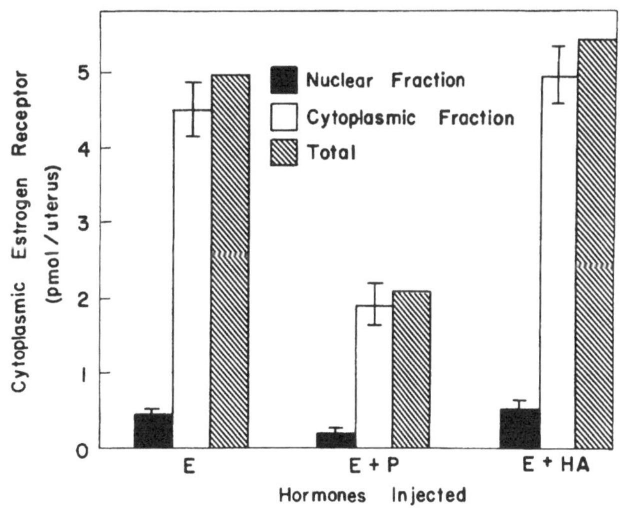

Fig. 73. The effect of progesterone and hydrocortisone acetate on the quantity of estrogen receptor in uteri of adult castrate rats. Three groups of castrate rats were given 10 μg estradiol. After 24 h, the animals were injected with 10 μg estradiol (*E*), 10 μg estradiol plus 6.5 mg progesterone (*E* + *P*), or 10 μg estradiol plus 5 mg hydrocortisone acetate (*E* + *HA*). The animals were killed 24 h later, and their uteri were analyzed for cytoplasmic, nuclear, and total estrogen receptor content by the [^{3}H]-estradiol exchange assay. The data represent the mean ± SEM of six - eight experiments with three or four animals per experiment (Clark et al., 1977b)

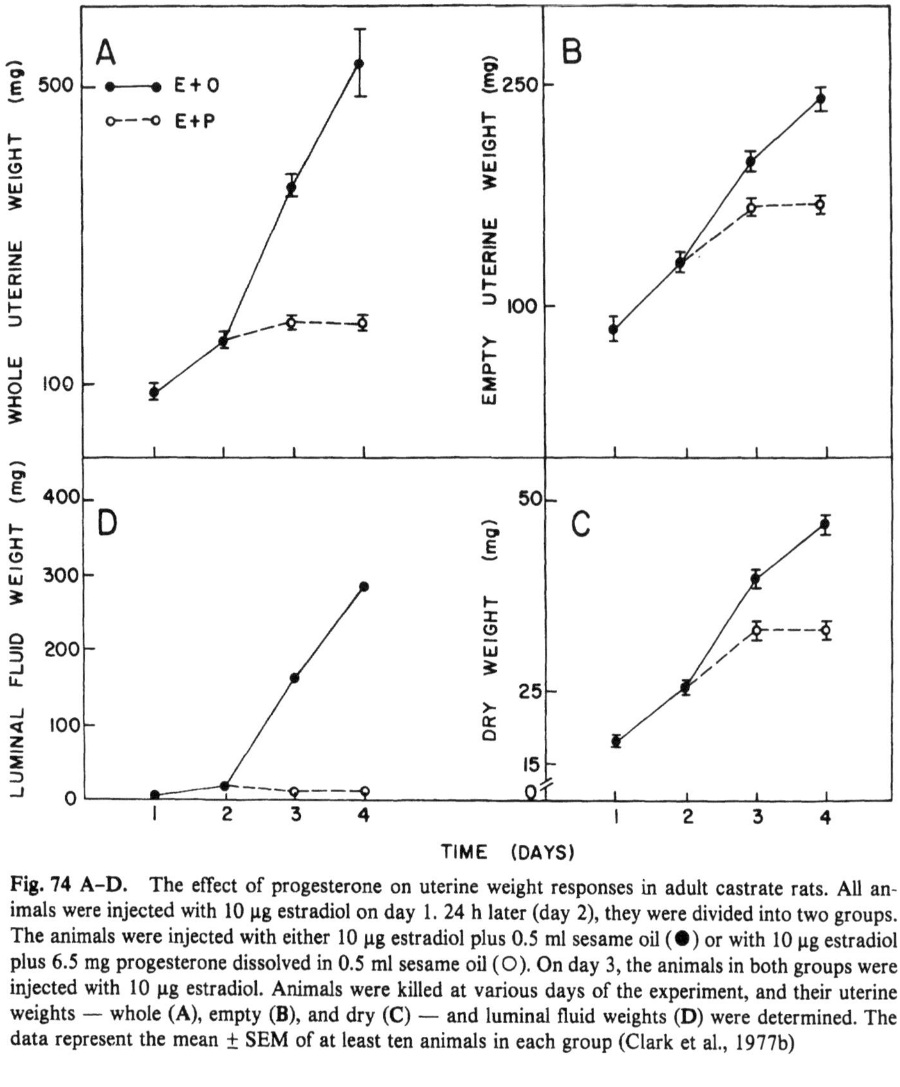

Fig. 74 A–D. The effect of progesterone on uterine weight responses in adult castrate rats. All animals were injected with 10 μg estradiol on day 1. 24 h later (day 2), they were divided into two groups. The animals were injected with either 10 μg estradiol plus 0.5 ml sesame oil (●) or with 10 μg estradiol plus 6.5 mg progesterone dissolved in 0.5 ml sesame oil (○). On day 3, the animals in both groups were injected with 10 μg estradiol. Animals were killed at various days of the experiment, and their uterine weights — whole (**A**), empty (**B**), and dry (**C**) — and luminal fluid weights (**D**) were determined. The data represent the mean ± SEM of at least ten animals in each group (Clark et al., 1977b)

(whole, empty, and dry weight) as well as the weight of luminal fluid at various stages of treatment were measured and are shown in Fig. 74. Progesterone depressed uterine weight dramatically within 24 h after treatment. A significant decrease in the weight of luminal fluid was observed. This reduction appears to be responsible for the major portion of the reduction in whole uterine weight. A slight reduction in empty uterine weight and dry weight was observed 24 h after estradiol in the progesterone-pretreatment group. A second injection of estradiol increased the difference in whole uterine weight and weight of luminal fluid between the two pretreatment groups as seen on day 4 (Fig. 74). Estradiol treatment on day 3 did not increase the uterine weight responses in the estradiol-plus-progesterone pretreatment group but increased significantly that of the estradiol pretreatment group, thus indicating a decreased responsiveness of estradiol-plus-progesterone pretreatment group on further estradiol treatment.

Table 9. Progesterone antagonism of estrogen-induced uterine growth

Pretreatment (day 2)	Treatment (day 3)	Whole uterine weight (mg)	Percent of saline control	Empty uterine weight (mg)	Percent of saline control	Dry weight (mg)	Percent of saline control
E	E (10)	528.0 ± 51.0	187	237.0 ± 6.4	157	47.1 ± 1.8	144
	S (6)	282.0 ± 51.8	100	151.0 ± 9.6	100	32.6 ± 3.2	100
E + P	E (9)	176.0 ± 3.7	114	163.0 ± 3.6	116	32.9 ± 1.1	109
	S (6)	155.5 ± 11.8	100	141.8 ± 8.8	100	30.3 ± 1.6	100

No appreciable difference among treatment groups was observed when empty uterine weights were compared in saline-injected controls (Table 9). Estradiol treatment on day 3 increased the uterine weight responses in both estradiol and estradiol-plus-progesterone pretreatment groups when compared with saline-injected controls. However, the ability of estradiol to increase uterine weight was signficantly depressed ($p < 0.01$) in the estradiol-plus-progesterone pretreatment group when compared with the estradiol pretreatment group. Estradiol treatment after estradiol pretreatment caused dramatic uterine growth over the saline-control group, whereas estradiol treatment after estradiol-plus-progesterone pretreatment causes a very slight stimulation. These data, together with those in Chap. VI. 5a, indicate that the antagonistic effect of progesterone on uterine growth correlates with its depressive effect on the replenishment of estrogen receptor.

The effect of hydrocortisone acetate on uterine responsiveness was also tested. Animals in estradiol-plus-hydrocortisone-acetate pretreatment groups were injected with estradiol, and the uterine weights were measured 24 h later. Hydrocortisone acetate did not cause any appreciable change in uterine growth parameters when compared to estradiol pretreatment (data not shown).

The possible influence of progesterone on uterine DNA content was examined since it has been shown that progesterone antagonizes estrogen-induced cell division in the chick oviduct (Oka and Shimke, 1969). Employing the paradigm described in Fig. 74, uterine DNA content was measured 24 h after pretreatment with estradiol or estradiol plus progesterone. No antagonistic effect of progesterone on uterine DNA synthesis could be demonstrated. Therefore, it is not likely that progesterone exerts its antagonism by inhibiting estrogen-induced cell division as has been reported for the chick oviduct (Oka and Shimke, 1969). Since the rat uterus is fully differentiated whereas the chick oviduct is not, these dissimilarities in progesterone action may reflect basic differences in the level of differentiation of these organs.

To establish that receptor replenishment is similar in both endometrium and myometrium, these tissues were separated from adult castrate rats as described by Clark et al. (1977b). The estrogen receptor content was measured in nuclear and cytoplasmic fractions of these tissues (Fig. 75). A significant inhibition of cytoplasmic receptor content in estradiol-plus-progesterone treated animals was observed in both myometrium and enriched endometrium. In addition, a similar decrease was also demonstrated in nuclear fractions of both tissue types. The results indicate that progesterone has no differential effect in these two tissues on the reduction of R_c^E. This does not, however, rule out the possibility that progesterone causes a reduction in R_c^E in one or more cell types within these tissues while having no influence on others.

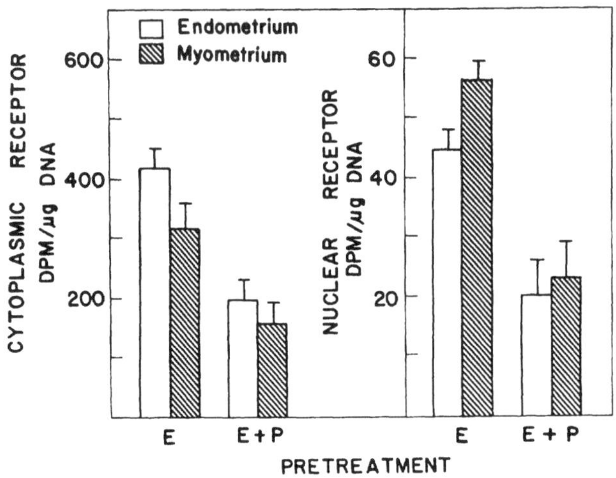

Fig. 75. Concentration of cytoplasmic and nuclear estrogen receptor in the endometrium and myometrium of the adult rat uterus. Estrogen-primed ovariectomized adult rats were treated with estrogen (*E*) or estrogen plus progesterone (*E* + *P*), and the quantity of estrogen receptor was measured by ^{3}H-estradiol exchange in the endometrium and myometrium 24 h later

Indeed, it seems quite possible that differential effects do occur, and these could account for the observations of Martin and Finn (1969), Finn and Martin (1970), Finn (1971), and Tachi et al. (1972). These authors noted that estradiol caused a surge of mitotic activity in the uterine stroma of progesterone-pretreated mice and had no effect on cell division in the epithelial cells. In nonprogesterone-treated animals estradiol had the reverse effect, i.e., mitosis was greatly stimulated in the epithelial cells, while only slightly increased in the stromal cells. Thus, one can envision that progesterone reduces the quantity of R_c^E or the amount of R_nE retained in nuclei of epithelial cells while having no effect on stromal cells. Hence, the subsequent exposure of the animal to estradiol fails to stimulate epithelial cells, but causes cell division in the stroma.

These suggestions are supported by the observation that progesterone pretreatment in the rat and mouse enhances the uptake of ^{3}H-estradiol by the stroma while reducing uptake by the luminal epithelium (Smith et al., 1970; Tachi et al., 1972, and Tchernitchin, 1976). This scheme is obviously oversimplified since pretreatment with progesterone not only reduces the ability of epithelial cells to respond to estradiol, but it enhances the ability of stromal cells to respond. Further complications are added by the failure of progesterone pretreatment to reduce ^{3}H-estradiol uptake by glandular epithelium in the rat uterus (Tachi et al., 1972). Obviously the interrelationships between estrogen, progesterone, and the control of uterine function involve many levels of interaction, of which the modulation of receptor levels constitutes only one.

Progesterone has also been shown to enhance the activity of estradiol-17β-dehydrogenase (Tseng and Gurpide, 1975; Tseng et al., 1977). This enzyme catalyzes the conversion of estradiol to estrone and hence could influence the quantities of estradiol bound to receptor. Since the affinity of the receptor for estrone is about 10 percent that for estradiol, if estradiol is converted to estrone then the amount of effective estrogen would decline. Tseng et al. (1977) and Gurpide et al. (1976) suggest that the ability of progesterone to increase the activity of estradiol-17β-dehydrogenase is a major mechanism by which progesterone acts as an estrogen antagonist. It seems likely that both increased enzymatic conversion of estradiol and decreased estrogen receptors are involved. The metabolism of estradiol may also form an important step in the mechanism by which the R_nE complex dissociates from nuclear binding sites.

Not withstanding the above-mentioned considerations, it is still helpful to picture the interaction of estrogen and progesterone as shown in Fig. 76. Estrogen binds to the cytoplasmic estrogen receptor, R_c^E, to form a complex, R_c^EE, which translocates to the nucleus, R_nE. This complex is responsible for the stimulation of events, leading to an increase in both R_c^E and the cytoplasmic progesterone receptor, R_c^P. Estrogen also stimulates uterine hypertrophy and hyperplasia. The elevated levels of R_c^P augment the ability of the uterus to respond to progesterone. Progesterone binds to its cytoplasmic receptor to form R_c^P P, which undergoes translocation to the nucleus to form R_n^P P. The R_n^P P complex elicits the characteristic progestational responses that prepare the uterus for implantation. Progesterone also reduces R_c^E and perhaps the nuclear retention time for R_n^E E, thereby decreasing the ability of the uterus to respond in a totally estrogen-directed fashion. Thus, progesterone reduces and/or redirects the ability of the uterine cells to respond to estrogen in such a way as to produce an appropriate uterine environment for implantation and pregnancy. This scheme does not rule out other possible interaction and control points, such as membrane effects, events not mediated by receptors, steroid effects on pituitary hormone secretion, etc. It is presented only as a model providing a framework for testing ideas.

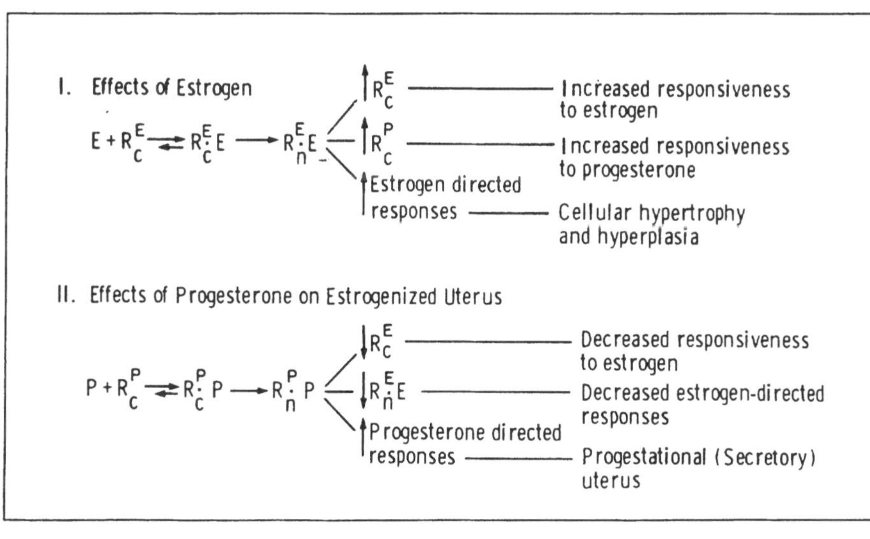

Fig. 76. Interaction of estrogen and progesterone in the control of receptor levels and uterine growth. *E*, estrogen; R_c^E, *cytoplasmic estrogen receptor;* R_c^EE, receptor-estrogen complex in the cytoplasm, R_n^EE, receptor-estrogen complex in the nucleus; *P*, progesterone; R_c^P, cytoplasmic progesterone receptor; R_c^PP, receptor-progesterone complex in the cytoplasm; R_n^PP, receptor-progesterone complex in the nucleus. See text for details

We conclude that progesterone antagonizes and redirects the ability of uterine cells to respond to estradiol by decreasing the quantity of R_c^E. This reduction of R_c^E decreases the number of receptor-estrogen complexes that are translocated and retained in uterine nuclei. Thus, the ability of estrogen to stimulate uterine growth is altered and greatly reduced. These interactions may occur during the estrous cycle of the rat and probably form a basic control mechanism by which the reproductive tract oscillates each cycle. These studies indicate that the traditional measurement of hormone levels in the blood offers an incomplete picture for understanding the action of steroid hormones under various physiologic states. Responsiveness of target organs may vary as the result of changes in receptor concentration, an important variable in the control of reproductive function, which can be evaluated only by measurement of tissue levels of hormone receptors. These interactions of two ovarian hormones at the receptor level provide a basis for the cyclic changes observed in uterine tissues during the estrous cycle of the rat. These mechanisms may also function during the menstrual cycle in the primate (Brush et al., 1967; Evans and Hähnel, 1971; Brenner et al., 1974; Gurpide and Tseng, 1974; Tseng and Gurpide, 1974) and will be discussed further in Chap. VIII.

6. Estradiol Antagonism by Estriol and Other Short-Acting Estrogens

Estriol has been considered a "weak" estrogen and an estrogen antagonist for many years (Hisaw, 1959; Huggins and Jensen, 1955; Wotiz et al., 1968). This conclusion is based on experiments in which estriol was administered by injection. In Chap. V we demonstrated that the "weak" uterotropic action of estriol when administered by injection is correlated with its inability to promote long-term retention of the estrogen receptor. Therefore it is reasonable to speculate that estriol may antagonize estradiol action by reducing the number of R_nE_2 complexes bound to nuclear acceptor sites. It is also possible that R_nE_3 complexes fail to stimulate replenishment of R_c^E and act in a fashion similar to progesterone. These two possibilities were examined in the following studies.

a) Effects of Estriol on the Replenishment of the Cytoplasmic Estrogen Receptor

An injection of estradiol and/or estriol causes a rapid accumulation of receptor in the nucleus and a concomitant depletion of receptor in the cytoplasm. Both hormones share equal potency in this regard, establishing that the antagonistic effect of estriol cannot lie at this level (see Fig. 76, Chap. V, and Anderson et al., 1972b, 1973a, 1975). However, the depletion of cytoplasmic receptor by estradiol injection is followed by the gradual replenishment of this receptor species. Estriol could act as a partial agonist-antagonist by failing to stimulate the replenishment of R_c. Such a failure to replenish R_c would decrease the receptor sites available for estrogen binding and produce a uterus relatively insensitive to subsequent estrogen treatment.

This possibility was examined by injecting immature rats with estradiol, estriol, or estradiol+estriol as described in Fig. 77. The quantity of uterine cytoplasmic receptor and uterine weight were measured 24 and 48 h after injection. Estriol clearly an-

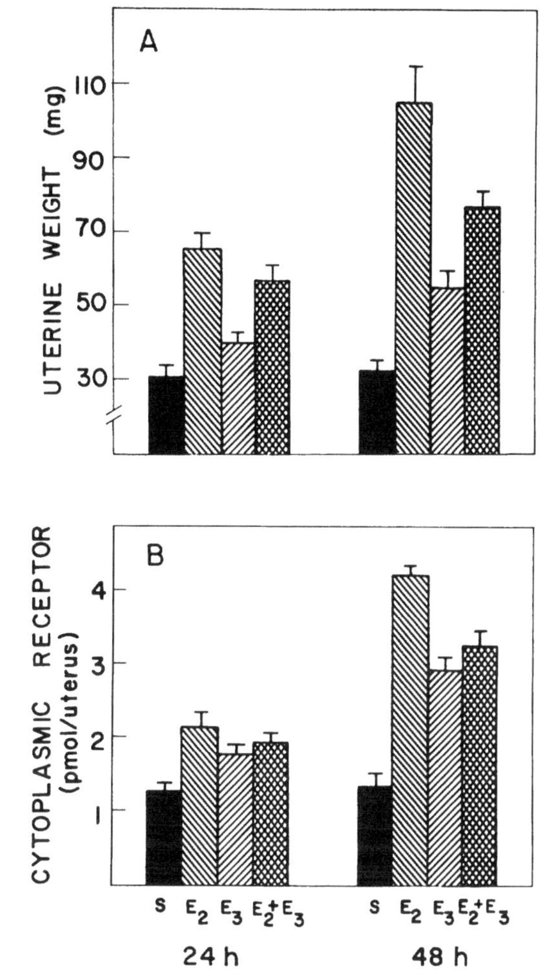

Fig. 77 A and B. Antagonism of estradiol by estriol when administered by injection. Immature rats were injected with saline (*S*), 0.1 µg estradiol (E_2), 1.0 µg estriol (E_3), or a combination of these hormones (E_2, 0.1 and E_3, 1.0 µg). Rats were either killed 24 h after treatment or were reinjected at 24 h and killed at 48 h. The uteri were weighed (**A**) and the quantity of cytoplasmic estrogen receptor (**B**) was determined by the [^{3}H]-estradiol exchange assay. The *vertical bars* represent the mean ± SEM of four separate experiments with five–seven rats per experiment (Clark et al., 1977c)

tagonized the effect of estradiol on uterine weight and cytoplasmic receptor (Fig. 77). Estriol alone was clearly a partial agonist, since it stimulated uterine growth as well as increased the cytoplasmic receptor level. Although cytoplasmic receptor levels in the estriol group are low compared to estradiol alone, the levels in the estriol or estradiol+estriol groups were above control, and thus a reduced number of receptors was not likely to be the source of antagonism. In the experiment shown in Fig. 77 rats were injected with 0.1 µg estradiol, 1.0 µg estriol, or a combination of those hormone doses. Similar experiments in which estradiol and estriol were injected at identical dose levels (0.1 µg) produced similar patterns of uterine growth and R_c levels.

b) Estriol Effects on R_nE_2 Retention and Uterine Growth

The effect of estriol on nuclear retention of R_nE_2 complexes was examined in immature rats injected as described in Fig. 78, and the quantity of receptor in the uterine nuclear fraction was assessed as a function of time. Both estrogens caused a

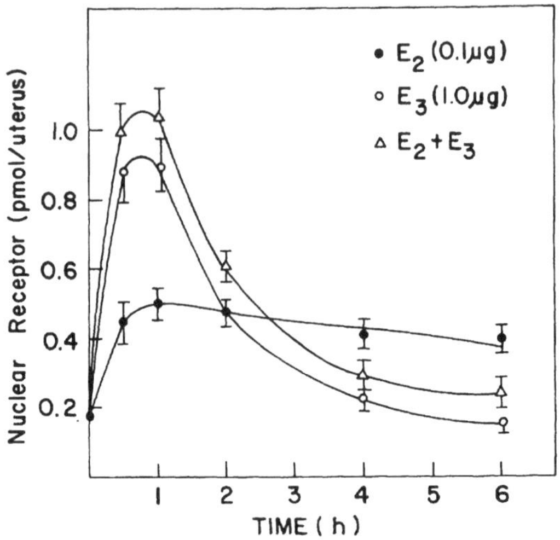

Fig. 78. Nuclear retention of the receptor-estrogen complex after treatment with estradiol, estriol, or a combination of these hormones. Immature rats were injected with estradiol, estriol, or a combination of the two steroids, and the quantity of nuclear receptor-estrogen complex was measured by ^{3}H-estradiol exchange at various times. Each point represents the mean ± SEM of three separate experiments with four rats per experiment (Clark et al., 1977c)

rapid accumulation of receptor by the nucleus. Estriol was injected as a 1 µg dose, whereas estradiol was injected at the 0.1 µg level. This difference in dose accounts for the greater initial nuclear accumulation of receptor following estriol treatment. These two hormones are not significantly different with respect to initial nuclear accumulation when administered at the same dose (Anderson et al., 1972b). Estradiol treatment resulted in long-term nuclear retention of the receptor, whereas a rapid decline in nuclear-bound receptor was observed following estriol treatment. With estriol treatment nuclear-bound receptor returned to control levels within 6 h. When estradiol and estriol were administered simultaneously, the quantity of receptor that exhibited long-term retention was reduced.

The results above suggest that estriol would not be an antagonist if it were present in a continous fashion, which would result in a constant occupancy of the receptor. In our initial experiments to test this proposal we injected rats every 3 h for 12–15 h and observed full estrogenic stimulation of the uterus (Anderson et al., 1975). In order to extend these observations immature intact or ovariectomized rats received paraffin implants of estradiol, estriol, or estradiol+estriol as described in Fig. 79. Each hormone treatment produced identical increases in uterine weight, and no antagonism by estriol was noted. Likewise no differences were observed in the effects of these treatments on the compartmentalization or quantities of the various estrogen receptor species. Each hormone or combination of hormones caused an accumulation and maintenance of nuclear R_nE at relatively high levels (0.8–1.2 pmol/uterus) (Fig. 79C). This accumulation and retention of nuclear R_nE was accompanied by a depletion of R_c, which gradually increased to very high levels by 48–72 h (Fig. 79B).

Lippman et al. (1977b) have shown that estriol is fully capable of stimulating metabolic activity of human breast cancer cells (MCF-7) in culture. Martucci and Fishman (1977) have confirmed these results and demonstrated that estrone is also capable of causing true uterine growth under these conditions. A similar effect has been observed by Lan and Katzenellenbogen (1976) with 17α-ethinylestriol and estriol cyclopentylether. These steroid derivatives extend the biologic half-life of the

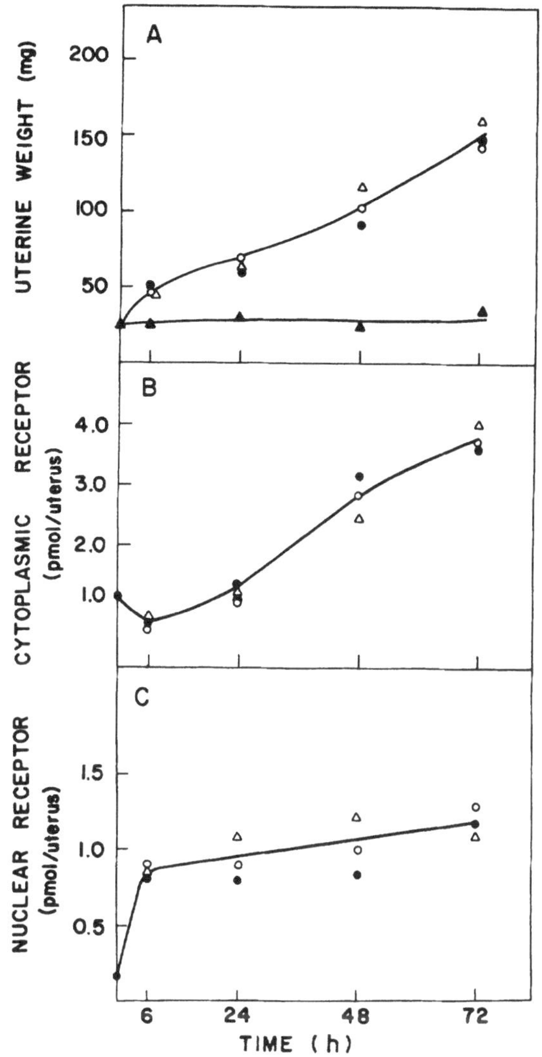

Fig. 79 A–C. Failure of estriol to antagonize the effects of estradiol on uterine growth, replenishment of the cytoplasmic receptor, and nuclear estrogen receptor. Immature rats were implanted with paraffin pellets containing estradiol (●), estriol (○), estradiol + estriol (△), or no hormone (▲). The animals were killed at various times, and the uterine weights (A), cytoplasmic receptor (B), and nuclear-bound receptor (C) were determined

estrogen and result in long-term nuclear occupancy and true uterine growth, analogous to the implant system.

These results demonstrate that estriol is neither a "weak" estrogen nor an estrogen antagonist when present in a continuous or chronic fashion. However, estriol does manifest these properties when injected. This paradox relates to the concept, previously suggested by Emmens and Miller (1969), Martin (1969), and Miller (1969), that "weak" estrogenicity correlates with short-term receptor occupancy. What is clear from the present results is that "weak" estrogenicity in the case of estriol arises from competition between R_nE_2 and R_nE_3 complexes for nuclear retention sites and from the rapid clearance of estriol from uterine tissue (Jensen et al., 1966). Thus, estradiol and estriol promote the translocation of cytoplasmic receptors to the nuclear compartment where R_nE_3 complexes compete with RE_2 complexes for those nuclear sites involved in long-term retention and promotion of uterotropic responses (Anderson et al., 1972b, 1973a, 1975). Since estriol is cleared rapidly and because R_nE_3 is in equilibrium with receptor and estriol, the competition between R_nE_2 and R_nE_3 reduces the number of receptor-estrogen complexes retained in the nuclear compartment. Because long-term retention is causally related to the stimulation of true uterine growth, this reduction in the number of effective receptor-estrogen complexes could account for the observed antagonism. However, when estriol is present in a continous fashion, as in the pellet implant experiments presented here, estriol promotes long-term nuclear retention and true uterine growth equivalent to that of estradiol. Since long-term retention of the receptor-estrogen complex by the nucleus appears to cause true growth regardless of the estrogen occupying the receptor, estriol acts as an estrogen agonist under these conditions. These considerations of estradiol antagonism will be discussed later in this chapter.

7. Triphenylethylene Derivatives and Estrogen Antagonism

In Chap. V we demonstrated that Nafoxidine causes long-term nuclear retention of the estrogen receptor and stimulates RNA polymerase activity and uterine growth. These observations are paradoxical since triphenylethylene derivatives such as Nafoxidine are generally considered to antagonize the effects of estrogen on uterine growth, vaginal cornification, and ovulation (Emmens, 1970). Triphenylethylene derivatives, nonsteroidal compounds with some structural similarities to estrogens, will bind to the cytoplasmic estrogen receptor (Fig. 80). Other investigators have proposed that the mechanism of action of estrogen antagonists resides in their ability to compete with estrogens for cytoplasmic receptors, thereby reducing the number of receptor-estrogen complexes in the cytoplasm of estrogen target tissues (Jensen et al., 1966; Terenius, 1971; Rochefort et al., 1972a, b). With subsequent translocation to nuclear sites, considered the primary event in the mechanism of estrogen action, this reduction in receptor-estradiol complexes would lead to decreased physiologic responses. Implicit in this hypothesis is that the receptor-antagonist complex should have a lower intrinsic biologic activity than that of the receptor-estradiol complex. That is, the ability of the receptor-antagonist complex to stimulate estrogenic responses should be less than that of the receptor-estradiol complex. In the following sections of this chapter we will show that this hypothesis does not explain the mechanism of action of triphenylethylene derivatives and that the estrogenic properties of these compounds has been greatly underestimated.

a) Effects of Triphenylethylene Derivatives on Uterine Growth

Based on the above stated assumptions, one would predict that estrogen antagonism should be observed following a single injection of estrogen and the antagonists; however, as is shown in Fig. 81, this is not the case. Immature rats, 21–22 days old, were injected daily with estradiol, Nafoxidine hydrochloride, or a combination of the two compounds, and uterine wet weights were determined 24 h after the last injection. The increases in uterine weight following a single injection of estradiol, Nafoxidine, or estradiol plus Nafoxidine are identical at 24 h after the injection. Therefore, following a single injection of estradiol plus Nafoxidine, there is no apparent

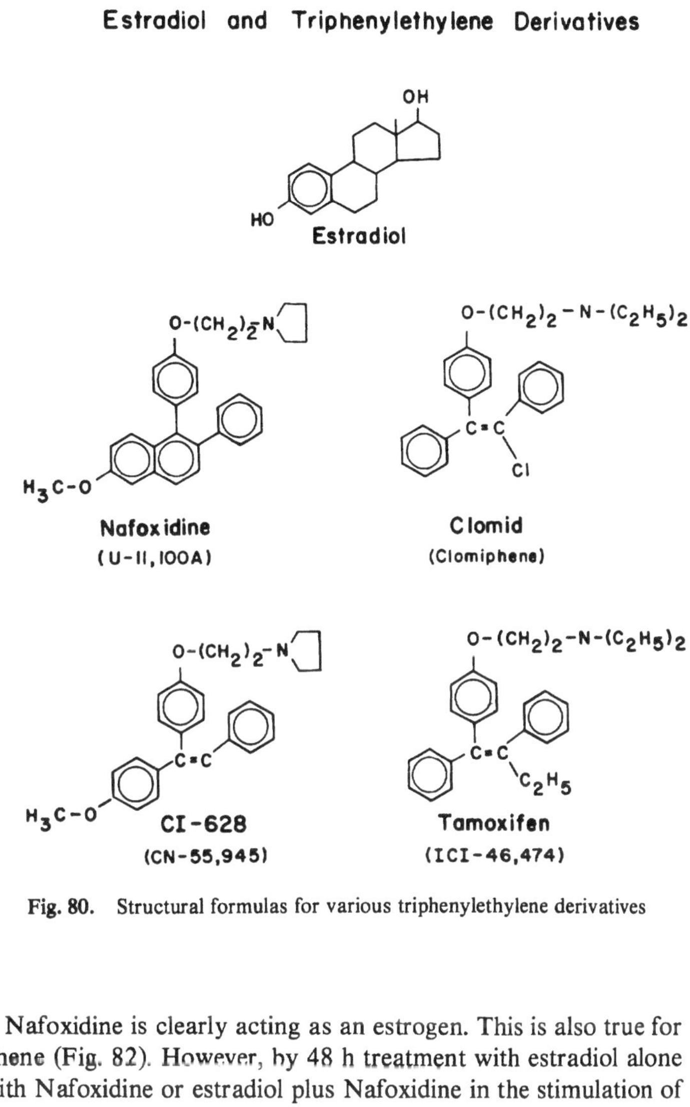

Fig. 80. Structural formulas for various triphenylethylene derivatives

antagonism; indeed, Nafoxidine is clearly acting as an estrogen. This is also true for CI-628 and Clomiphene (Fig. 82). However, by 48 h treatment with estradiol alone is superior to that with Nafoxidine or estradiol plus Nafoxidine in the stimulation of uterine growth (Fig. 81). Thus, the antagonistic properties of Nafoxidine can be observed when the compounds are administered at 24 h intervals and the uterine weights are determined after the initial 24-h period (Fig. 81). Serial injections (3–8 in number) at 24-h intervals have been routinely used by many investigators, and it is well established that estrogen antagonism can be observed under these conditions (Fig. 81, two injections; Emmens, 1970; Dorfman, 1962; Lerner, 1964; Callantine et al., 1966). The routine application of this method has resulted in the failure to recognize the significance of the agonist properties of these compounds after a single injection. It should be noted that we have used much lower doses of these compounds with similar results; that is, doses of 1–10 μg display both agonistic and antagonistic properties depending on the injection scheme.

Anti-estrogens could act by stimulating cellular hypertrophy, thereby increasing uterine weight, but fail to stimulate uterine hyperplasia. In such a case the number of

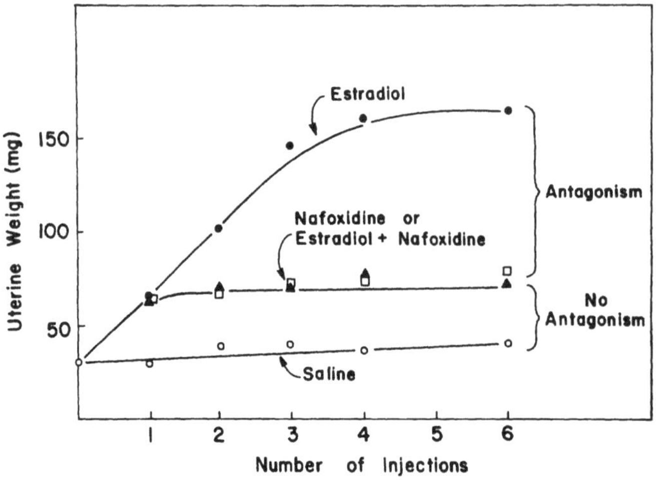

Fig. 81. Effect of daily injections of estradiol, Nafoxidine, or estradiol plus Nafoxidine on uterine weight. Immature castrate rats were injected every 24 h with 2.5 μg estradiol (●), 100 μg Nafoxidine (▲), a combination of both compounds (□), or saline (○)

cells capable of responding to subsequent estrogen treatment would be reduced, and this reduction would result in diminished uterotropic responses when compared to estrogen treatment alone. This does not appear to be the case since both estradiol and nafoxidine increase DNA content by 48 h (Clark et al., 1974a), and no differences are apparent in the capacity of these compounds to elicit uterine hyperplasia or ^{3}H-thymidine incorporation (Fig. 83).

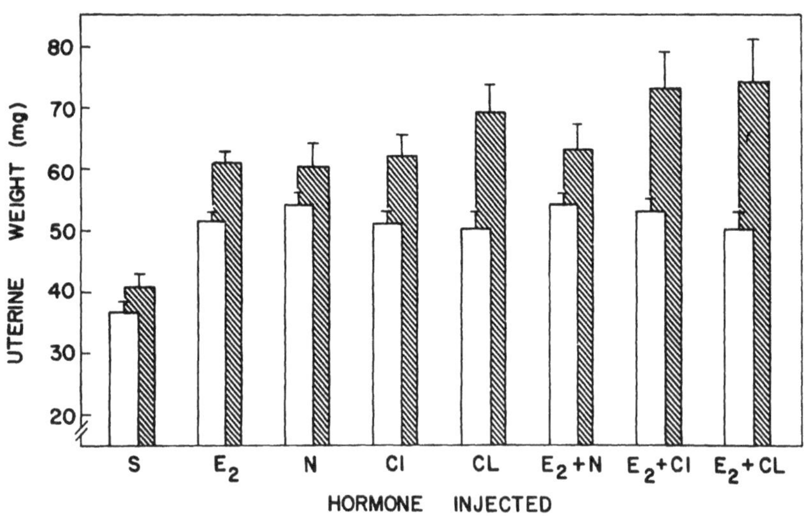

Fig. 82. Effect of estradiol and various triphenylethylene derivatives on uterine growth. Immature rats were injected with saline (*S*), 2.5 μg estradiol (*E₂*), 100 μg Nafoxidine (*N*), 500 μg CI-628 (*CI*), 10 mg Clomid (*CL*), or various combinations of these compounds. Uterine wet weight was determined 24 h (*open bars*) and 48 h (*stippled bars*) after treatment (Clark et al., 1974a)

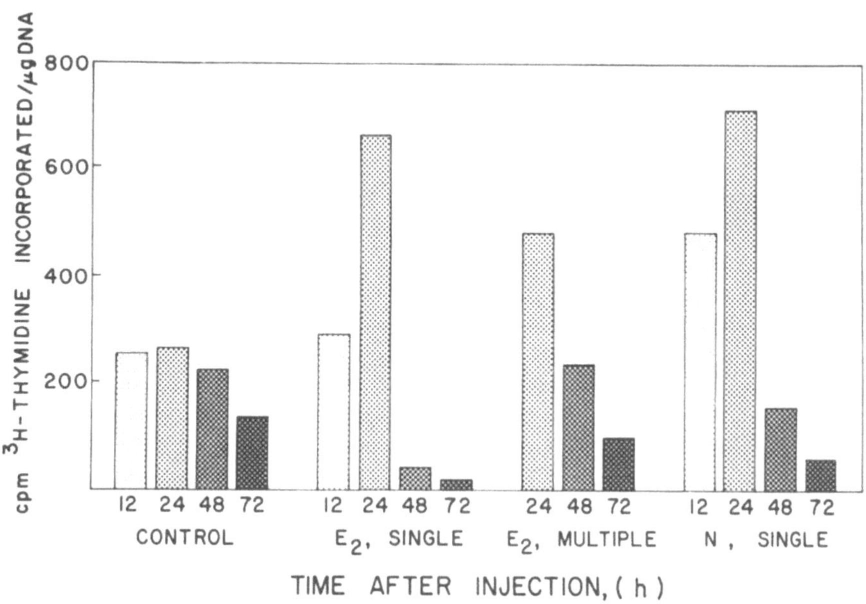

Fig. 83. Effect of estradiol and Nafoxidine on the incorporation of ^{3}H-thymidine into uterine DNA. Immature rats were injected with either saline, a single injection of 1.0 µg estradiol, daily injections of estradiol, or a single injection of 50 µg Nafoxidine. Uteri were removed at indicated times, placed in 2.1 ml Eagle's HeLa medium containing 25 µCi ^{3}H-thymidine, and incubated at 37°C for 1 h

b) Effects of Anti-Estrogens on Cytoplasmic Replenishment and Nuclear Retention of Estrogen Receptors

We have proposed that anti-estrogens antagonize estrogen-induced uterine growth as a result of their failure to stimulate the replenishment of the cytoplasmic estrogen receptor (Clark et al., 1973b, 1974). As previously described, Nafoxidine causes accumulation and long-term retention of the R_n ligand complex; however, this retention is not accompanied by the usual replenishment of the cytoplasmic receptor as is the case after estradiol treatment. This phenomenon can be observed after single or multiple daily injections of Nafoxidine (Figs. 84 and 85) or other triphenylethylene derivatives (Capony and Rochefort, 1975; Katzenellenbogen and Ferguson, 1975; Major et al., 1976; Katzenellenbogen et al., 1977; Jordan et al., 1977; Ruh and Baudendistel, 1977). In the uterus, daily treatment with estradiol doubled the quantity of cytoplasmic receptor, whereas Nafoxidine treatment depressed the level of cytoplasmic receptor during this time. The failure to stimulate the replenishment of R_c^E may render the uterus insensitive to a subsequent injection of estrogen, and therefore antagonism can be observed. Similar long-term nuclear retention and decreased replenishment of R_c^E has been observed in mammary tumors after treatment with triphenylethylene derivatives which induce partial tumor regression (Jordan and Dowse, 1976; Jordan and Jaspan, 1976; Nicholson et al., 1976; Tsai and Katzenellenbogen, 1977). These mechanisms can be visualized according to the model presented in Fig. 86. Receptor-estrogen complexes (*RE*) and receptor-antagonists complexes (*RA*) stimulate cellular hypertrophy and hyperplasia. Thus, uterine growth is

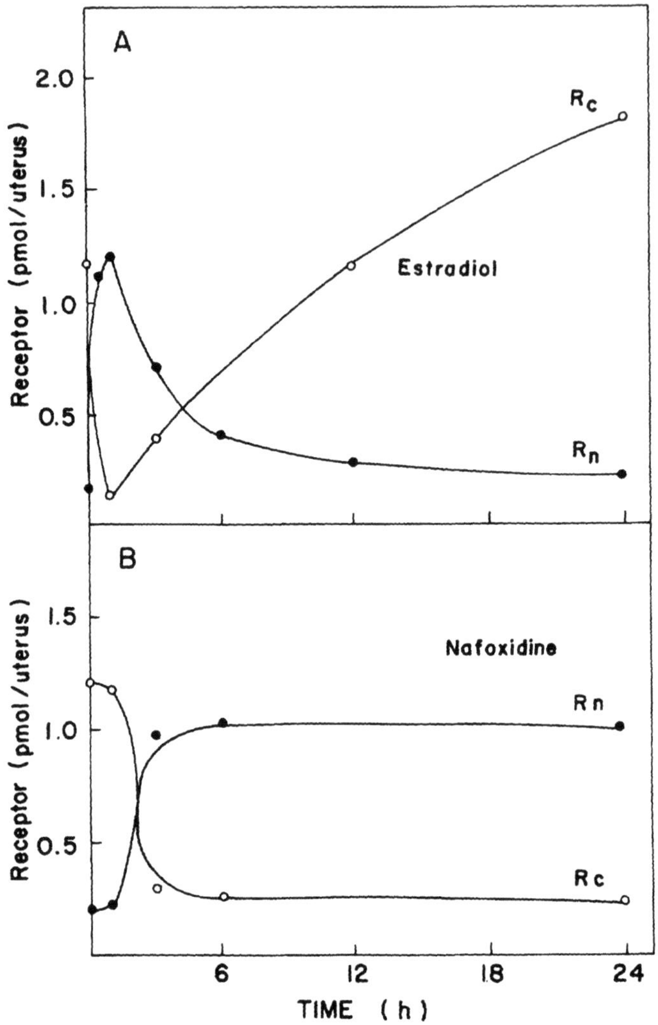

Fig. 84 A and B. Effect of estradiol and Nafoxidine on nuclear retention and cytoplasmic replenishment of the estrogen receptor. Immature rats were injected with 2.5 μg estradiol (**A**) or 100 μg Nafoxidine (**B**), and the quantities of estrogen receptor in the nuclear (R_n) and cytoplasmic (R_c) fractions were determined by exchange

observed in both cases after a single injection of the compound. However, the RA complex fails to cause the replenishment of R_c^E by 24 h. When the animal receives a second injection of estrogen at this time the uterus is nonresponsive in the animals that received the antagonists but highly responsive in estrogen-treated animals. Therefore, the uterus continues to grow in the estrogen-treated rat and remains unstimulated in the antagonist-treated animal. Although this reasoning appears logical, it fails to offer a complete explanation. One must explain how antagonism is expressed when large quantities of estrogen receptor are being retained for long periods of time in the nucleus as R_nA complexes. This should cause continued stimulation of uterine growth; instead, Nafoxidine, either as a single dose or as multiple in-

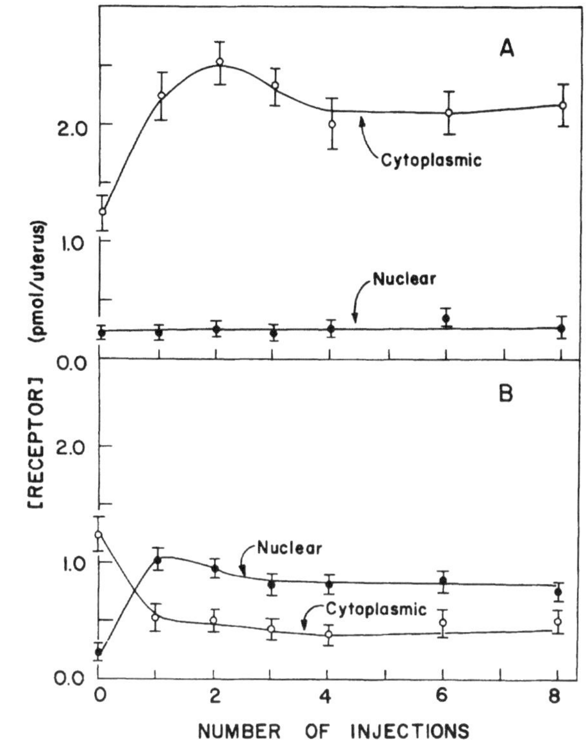

Fig. 85 A and B. Effect of daily injections of estradiol or Nafoxidine on the level of cytoplasmic and nuclear estrogen receptor. Immature castrate rats were injected daily with 2.5 μg estradiol (**A**) or 50 μg Nafoxidine (**B**), and the quantity of cytoplasmic and nuclear receptor was determined by exchange 24 h after the last injection

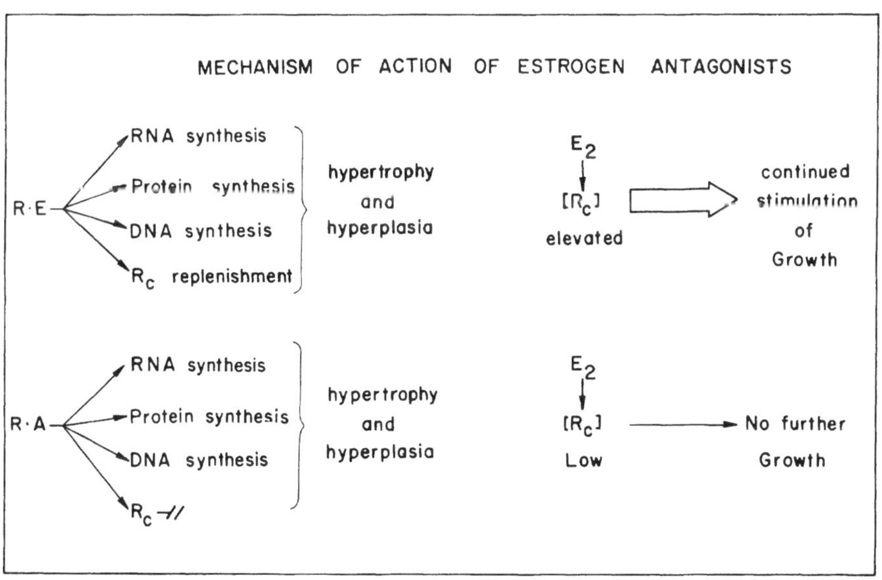

Fig. 86. One possible interpretation of the mechanism of action of triphenylethylene derivations. See text for details

jections, causes the uterus to double in size and to remain at this level for long periods of time (Fig. 81, Clark et al., 1973a). In contrast, serial injection of estradiol causes continued stimulation of uterine growth, which can produce a uterus 4–5 times that of control size.

We have already suggested that the R_nA complex may stimulate uterine genes in an anomalous manner, perhaps initiating self-limiting processes. However, the recycling of receptor to the cytoplasmic compartment may also play an important role in the overall process. To test this idea we injected rats with a large dose of estradiol 24 h after they received an injection of Nafoxidine. At this time the quantity of R_c is low, and the uterus is nonresponsive to an injection of estradiol (Fig. 84). In this manner in vivo exchange of estradiol for Nafoxidine should occur, and R_nE_2 complexes should be formed. The work of Katzenellenbogen and Ferguson (1975) indicated that this might be possible. If in vivo exchange were to occur, then the normal nuclear replenishment processes should take place and some R_c should appear in the cytoplasm. As shown in Fig. 85 the quantity of R_c was elevated by 48 h (24 h after the estradiol injection). An injection of either estradiol or Nafoxidine at this time causes significant uterine growth by 72 h (Fig. 87). Since cytoplasmic replenishment eventually occurs in rats that have received triphenylethylene derivatives, one might predict that uterine responsiveness would also return. This has been demonstrated by Katzenellenbogen and Ferguson (1975) and Katzenellenbogen et al. (1977). These investigators examined the ability of estradiol to induce IP protein synthesis and increased uterine weight after Nafoxidine and CI-628. They were able to demonstrate

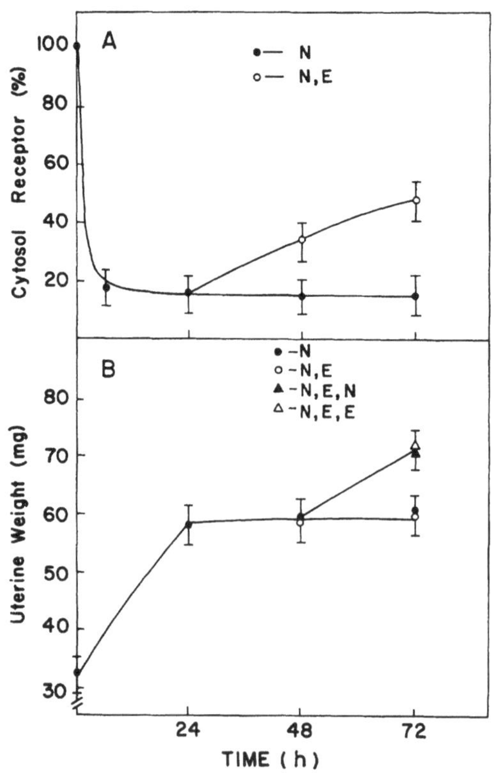

Fig. 87 A and B. Effect of estradiol on the replenishment of cytoplasmic estrogen receptors in Nafoxidine-treated rats. **A** Two groups of immature rats were injected with 50 µg Nafoxidine (●) at time zero. 24 h later one group was injected with 10 µg estradiol (○). The quantity of cytoplasmic receptor for both treatment groups was determined. **B** Rats treated with Nafoxidine and estradiol as in **A** either received no further treatment (●, ○) or were injected with Nafoxidine (▲) or estradiol (△) at 48 h. Uterine weight was measured at each time interval

a good correlation between the quantity of replenished R_c and the ability of the uterus to respond to subsequent estradiol treatment. Thus, it appears that replenishment or recycling of R_c may be an important component in the array of complex interactions that stimulate the uterus to grow. These conclusions are shown in summary form in Fig. 88. The binding of the R_nE complex to nuclear sites causes uterotropic stimulation and replenishment, possibly via recycling and resynthesis of R_c. The replenished R_c is then free to interact with estrogen to form receptor-hormone complexes that bind to nuclear sites and cause a second cycle of uterine growth stimulation. On the other hand, the R_nA complex, although capable of causing one cycle of uterotropic stimulation, is unable to effect replenishment of R_c; hence multiple cycles of nuclear binding and stimulation are not possible. One could argue that the failure to observe R_c^E replenishment in Nafoxidine-treated rats results from a contions rebinding of any R_c^E by Nafoxidine. Since Nafoxidine has a long half-life in the rat, this rebinding of R_c by Nafoxidine undoubtedly occurs. However, similar high levels of estradiol or estriol maintained for 3 days, as in the pellet implant studies discussed earlier (Fig. 79), did not block replenishment. Instead replenishment of R_c^E was greatly enhanced. Therefore, one is forced to conclude that R_nA and R_nE complexes differ in their abilities to stimulate R_c^E replenishment. This argument does not eliminate the necessity for explaining why the R_nN complex, which is bound to many nuclear sites for long periods of time, does not continue to stimulate uterine growth. A possible explanation may lie in differential cell effects as discussed in the following section.

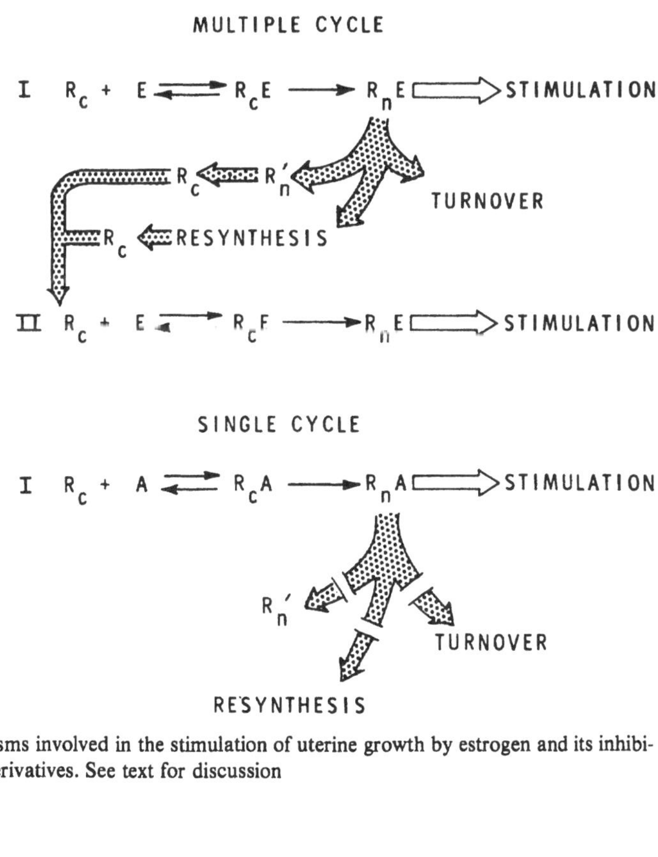

Fig. 88. Possible mechanisms involved in the stimulation of uterine growth by estrogen and its inhibition by triphenylethylene derivatives. See text for discussion

c) Triphenylethylene Derivatives and Differential Cell Stimulation

As noted above, Nafoxidine causes the uterus to double in size compared to controls, whereas estradiol causes a 2–4 fold increase in uterine weight depending upon the length of treatment (Fig. 81). Simultaneous treatment with estradiol and Nafoxidine fails to cause increases in uterine weight above that induced by Nafoxidine alone. This ability of Nafoxidine to stimulate partial uterine growth arises from its capacity to cause differential cell stimulation (Clark et al., 1978d, e). This phenomenon is characterized by extensive cellular hypertrophy of the luminal epithelium to an extent greater than that with estradiol and only partial stimulation of the stromal and myometrial tissues (Fig. 89A–D). This high level of epithelial stimulation and partial stimulation of stromal and myometrial elements accounts for the observed 2-fold increase in uterine weight that was observed in Fig. 81.

Differential cell stimulation was also demonstrated in the chick oviduct. In this system Nafoxidine does not cause any detectable increase in oviduct weight, whereas DES causes substantial increases (O'Malley et al., 1969). Simultaneous injection of both compounds results in antagonism of DES-induced oviduct growth (unpublished observations). Histologic examination of this tissue indicates that, while Nafoxidine can stimulate hypertrophy of epithelial cells, the epithelium does not progress past this point and no gland formation is noted. In contrast, in the DES-treated animals, extensive stimulation of all tissues was noted. Epithelial proliferation resulted in large quantities of glandular tissue by 48–72 h after treatment. Since the epithelial lining of the oviduct is a minor component of its total mass, the selective and limited stimulation of the epithelial cells explains why no discernable increase in wet weight was observed.

It has recently been suggested that Tamoxifen, another triphenylethylene derivative similar to Nafoxidine, is a pure estrogen antagonist in the chick oviduct (Sutherland et al., 1977). This conclusion was based on the ability of Tamoxifen to inhibit diethylstilbestrol (DES)-stimulated increases in DNA, protein, and weight of the chick oviduct. However, since no histologic examination was made, it is not possible to say with certainty that Tamoxifen is a pure estrogen antagonist. This seems unlikely to us in view of the ability of this compound to cause nuclear accumulation and long-term retention of the estrogen receptor in the chick oviduct (Sutherland et al., 1977). However, Tamoxifen does appear to be a pure estrogen antagonist in MCF-7 cells (Lippman et al., 1976). These observations may reflect the ability of triphenylethylene derivatives to act as agonists in some cell types and antagonists in others. As an example, progesterone is a pure antagonist in cultured hepatoma cells (Rousseau et al., 1972) and in the mouse mammary tumor system (Ringold et al., 1975), whereas it acts as an agonist in mouse lymphoma cells or thymocytes (Roussean et al., 1972; Munck and Wira, 1971).

The ability of Nafoxidine and other triphenylethylene derivatives to cause extensive stimulation of epithelial cells while either not stimulating or only slightly stimulating cells in other tissues is difficult to reconcile with our current concepts of steroid hormone action. Nafoxidine causes receptor binding to nuclei in the various cell types of the uterus (Clark and Hardin, unpublished observations). However, we have not ruled out the possibility that the nuclear retention of the receptor may differ in various cell types. Alternatively cellular responses subsequent to the binding of the receptor-Nafoxidine complex may be dictated by mechanisms that are specific for a given cell type.

Regardless of the mechanisms of action of triphenylethylene derivatives, histologic studies have partially clarified their paradoxical capacity to be estrogenic and anti-estrogenic simultaneously. Estrogenic stimulation of the uterus by these compounds is due to the extensive stimulation of the uterine epithelium and partial stimulation of the endometrial stroma and myometrium. Antagonism results from the failure of these compounds to stimulate growth of the stroma and myometrium continuously, whereas estradiol is capable of stimulating all three components maximally and continuously and hence produces a uterus of much greater size. The ability of triphenylethylene derivatives to stimulate growth of the uterine epithelium has been noted before (Kalra and Prasad, 1967; Terenius and Ljungkvist, 1972; Emmens and Carr, 1973); however, the relationship of this phenomenon to estrogen antagonism was not noted. Kang et al. (1975) have shown that CI-628 also causes epithelial hypertrophy of the uterus; however, they noted extensive stimulation of stromal and myometrial elements. In contrast, our observations with CI-628 indicate that it is very similar to Nafoxidine in its action and stimulates the epithelial layer at the expense of other tissue layers (unpublished observations). The reason for this discrepancy is not clear at this time.

Although a complete understanding of the mechanism of action of triphenylethylene derivatives remains to be elucidated, the demonstration of differential cell stimulation has important implications. These compounds have been used extensively to "block" estrogen action, and indeed this is accomplished in some cases. In other instances, partial estrogenic stimulation has been noted, as presented here. Triphenylethylene derivatives have long been recognized as compounds that could either block or stimulate ovulation (Holtcamp et al., 1960; Greenblatt et al., 1961; Kistner and Smith, 1961; Segal and Nelson, 1961; Tyler et al., 1962). This has been thought to be due to the ability of triphenylethylene derivatives to block the negative feedback effect of estrogens and thus result in gonadotropin secretion (Vaitukaitis et al., 1971). Another interpretation is that triphenylethylene derivatives may be acting as estrogens, which have a positive feedback effect on gonadotropin secretion. It is also possible that these compounds affect pituitary cells in a different fashion than brain cells. Ross et al. (1973) have reported negative (anti-estrogenic) effects of CI-628 on ovulation and positive (estrogenic) effects on sexual behavior. Harman et al. (1975) have noted that CI-628 will block estrogen-induced follicular growth but does not increase estrogen-dependent atresia. Nafoxidine has also been shown to inhibit the growth of mammary tumors; however, it never reduces the size of the tumor to that seen in the ovariectomized control animals (Gallez et al., 1973). All of the above observations may result from differential cell stimulation by triphenylethylene derivatives. Therefore, any interpretation of the action of these compounds based solely on their anti-estrogenic qualities may lead to invalid conclusions.

We have recently shown that Nafoxidine and Clomid cause extensive epithelial stimulation in the neonatal rat which is correlated with the production of reproductive tract abnormalities (Clark and McCormack, 1977; see Chap. VII). This observation probably reflects the ability of these drugs to act as long-acting estrogens in specific cell types and demonstrates the potential danger of considering them to be only estrogen antagonists.

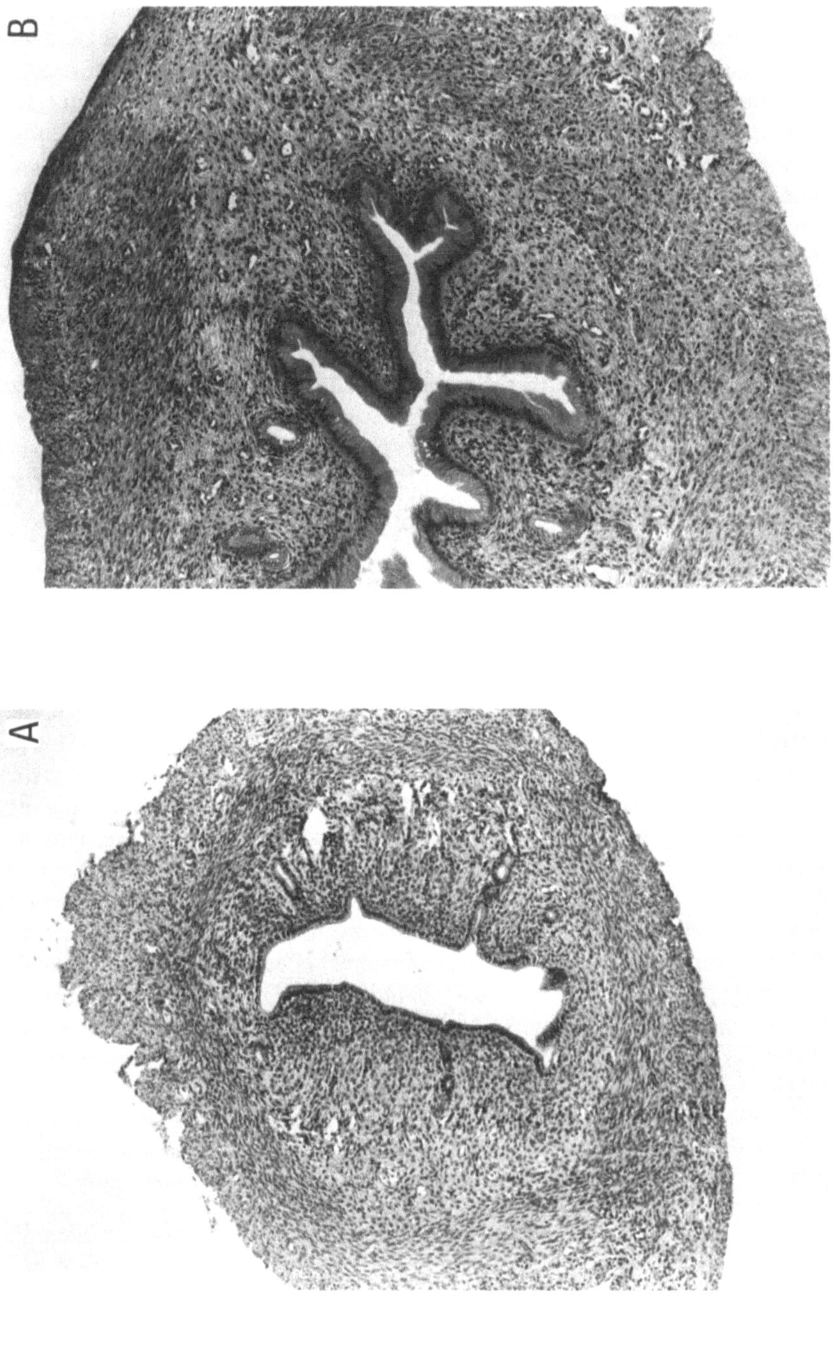
B
A

D

C

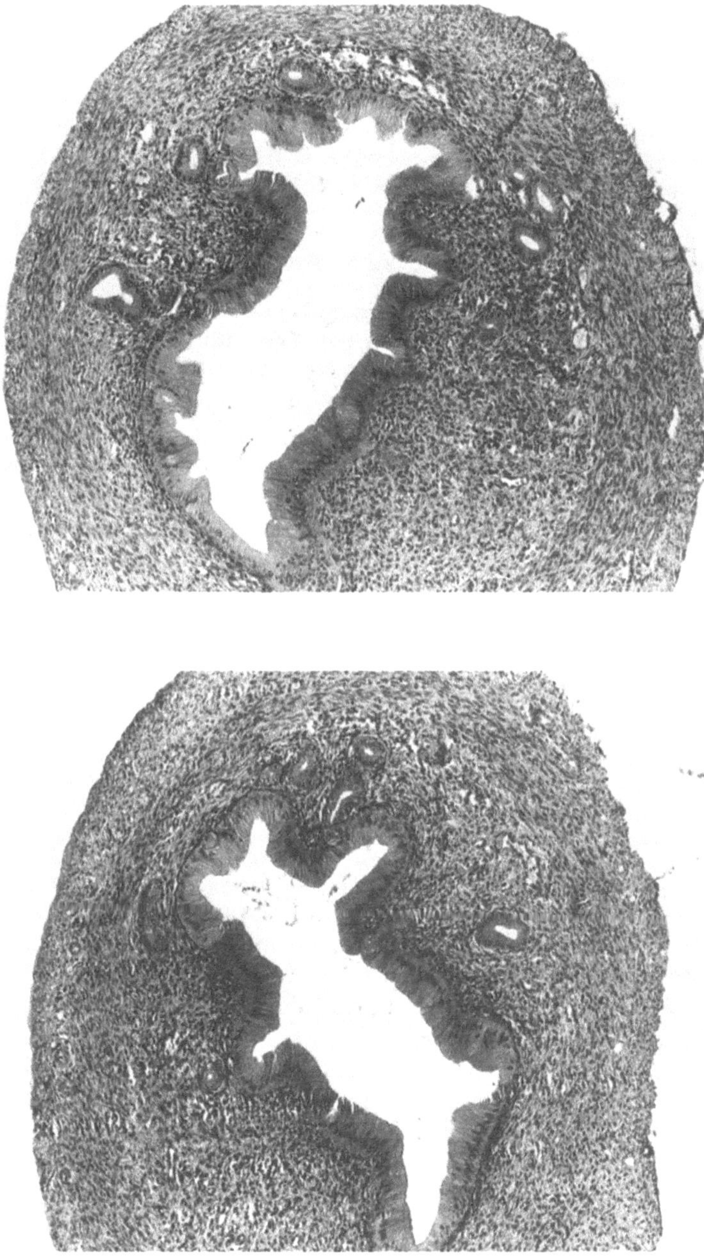

Fig. 89 A–D. Effect of estradiol and Nafoxidine on the histology of the rat uterus. Immature rats were injected daily for 3 days with saline (A), estradiol (B), estradiol plus Nafoxidine (C), or Nafoxidine (D). The animals were killed 24 h after the last injection

We conclude that Nafoxidine and probably other triphenylethylene derivatives have the ability to stimulate the growth of the epithelium of the uterus and chick oviduct, while having little to no estrogenic effect on other tissues of these organs. This differential cell stimulation may occur in all target organs for estrogen, and the end result, whether it be growth, hormone secretion, or sexual behavior, may represent the sum of the agonistic (estrogenic) and antagonistic (anti-estrogenic) properties of these drugs.

8. General Considerations of Steroid Antagonism and Hormone Potency

a) Classification of Estrogen Agonists and Antagonists

Estrogens and estrogen antagonists have traditionally been classified in the following way: estradiol and diethylstilbestrol were considered full agonists, whereas estriol and dimethylstilbestrol were classified as weak or impeded estrogens. Emmens and Miller (1969) suggested that these differences might be due to receptor occupancy and originally classified estradiol and estriol as long- and short-acting estrogens, respectively. The triphenylethylenes, such as Clomid and Nafoxidine, were considered antagonists with very weak estrogenic potency. Based on the data discussed in this chapter we have proposed a new classification of these compounds.

Terms such as "weak" or impeded imply that estriol is less effective under all circumstances. Since this is not the case, the classification of estriol as a short-acting estrogen provides a functional definition that applies in all cases. In this scheme, dimethylstilbestrol, mesobutoestrol, and 16-oxo-estradiol are similar to estriol and thus may be classified together (Martin, 1969; Miller, 1969; Emmens and Miller, 1969; Terenius, 1970; Capony and Rochefort, 1977). These compounds have in common the ability to stimulate early uterotropic reponses as shown in Table 10; however, they fail to significantly influence the long-term responses that promote uterine growth when given as a single injection (Emmens and Miller, 1969; Katzenellenbogen and Katzenellenbogen, 1973).

The long-acting estrogens are divided into two catagories, A and B, according to their retention times in the body. Subclass A includes estradiol and DES because both of these hormones are fully agonistic with no antagonistic properties and have similar nuclear retention times and biologic potencies (Ruh and Baudendistel, 1977). Subclass B includes the triphenylethylene derivatives, such as Nafoxidine, CI-628, Clomid, and Tamoxifen, because of their ability to stimulate estrogenic function over longer periods of time than estradiol or DES and yet manifest antagonistic properties when given as multiple injections. The ability of these compounds to cause differential cell stimulation adds a further complexity and sets them apart from estradiol and DES.

This classification is undoubtedly an oversimplification and will require modification as we learn more about estrogens in the future. However, this method focuses attention on the complex nature of estrogen agonists and antagonists and relates their functional activity to the retention of the receptor by the nucleus of target cells. The former classification of estriol and similar steroids as weak or impeded estrogens has led to false concepts, which have both basic and practical implications. Likewise

Table 10. Classification of estrogen agonists and antagonists[a]

Class	Examples	Nuclear retention	Cytoplasmic receptor replenishment	Pharmacologic characteristics	Uterotropic properties
1. Short-Acting	Estriol, Dimethyl-stilbestrol. (DMS)	Short (1–4 h)	Positive within 24 h	Partial agonist/ antagonist when injected[b] Agonist when implanted	Early responses[c]
2. Long-Acting	A. Estradiol, Diethyl-stilbestrol (DES), Ethinyl estriol, Ethinyl estriol cyclopentyl ether	Intermediate (6–24 h)	Positive within 24 h	Agonist	Early and late responses[d]
	B. Triphenylethylene derivatives (e.g. Nafoxidine, CI-628 Tomoxifen, Clomid)	Long (greater than 24–28 h)	Negative within 24–28 h	Agonist– one injection	Early and late responses
				Antagonist – multiple injections exhibit differential cell stimulation	

[a] This classification is based on events that occur after a single injection of the compound, except where noted otherwise.

[b] This is dependent on the MCR of the hormone. Degree of antagonism is directly related to clearance rate.

[c] Early responses include water imbibition, hyperemia, amino acid and nucleotide uptake, activation of RNA polymerase I and II, stimulation of induced protein (for references see Chap. V).

[d] Late responses include cellular hypertrophy and hyperplasia, sustained RNA polymerase I and II activity (see Chap. V).

failure to recognize the true estrogenic potential of nonsteroidal estrogen antagonists has resulted in invalid interpretations of data, and as discussed in Chap. VII, the use of these compounds for the induction of ovulation in women may have dire consequences.

b) Biologic Implications of Short- and Long-Acting Estrogen Agonists and Antagonists

The establishment of estriol as a potent estrogen when present under steady-state conditions has important implications with respect to certain hypotheses on the role of estriol in physiology and pathology. Estriol is present at high levels in the blood during pregnancy in the human (Hobkirk and Nilsen, 1962; Tulchinsky, 1973). In addition, estriol is the major estrogen present during the menstrual cycle (Goebelsmann et al., 1969). Therefore estriol is likely to be a major contributor to the total estrogen-state of the human. Indeed, Wotiz et al. (1972) and Raziano et al. (1972) have shown that antibodies to estriol administered to rats between days 11 and 14 of pregnancy cause total resorption of the fetuses. These authors have suggested that estriol may be important in maintaining the proper function of the corpus luteum during pregnancy. Its role in this capacity needs reappraisal in light of the results presented above.

Estriol has generally been considered to be an intermediate metabolite in the inactivation of estradiol. The conversion of estradiol to estriol by the liver may be a step involved in the elimination of estrogenic steroids from the body. Martucci and Fishman (1977) have proposed that enzymatic conversion to the catechol estrogen, 2-hydroxylestrone, may also serve to terminate estrogenic function in the uterus. However, the formation of catechol estrogens in the brain may serve to initiate estrogen expression in this organ (Axelrod et al., 1961; Zumoff et al., 1968; Ball et al., 1976; Naftolin et al., 1975a, b; Fishman, 1976), and the conversion of estradiol to estriol in uterine tissues certainly could allow continued estrogenic activity. Obviously the physiologic importance of various estrogens is incompletely understood. A proper assessment will come only when receptor binding and responses are analyzed under conditions that simulate the in vivo state.

A protective role has been ascribed to estriol in breast cancer. This suggestion is based on the observation that oriental women, who have a high (estriol)/(estradiol + estrone) ratio in blood, also have a low incidence of breast cancer (Wotiz et al., 1968; Goebelsman et al., 1969; Cole and McMahon, 1969; Lemon et al., 1971). This hypothesis was formulated on the assumption that estriol was a "weak" estrogen under all circumstances and that during each menstrual cycle estriol would act to reduce the "carcinogenic potential" of the more potent estradiol. Our results and those of others (Zumoff et al., 1975) indicate that this theory is suspect. In light of recent evidence showing estriol and estradiol to be of equal potency in facilitating the onset of mammary tumors in mice (Rudali et al., 1975), we suggest that the estriol theory of mammary cancer protection is untenable.

Although it has long been recognized that triphenylethylene derivatives possessed inherent estrogenicity, these compounds were generally considered to be estrogen antagonists, and most experimental results were interpreted accordingly. The demonstration that triphenylethylene should be classified as long-acting estrogens with both agonistic and antagonistic properties results in the reconsideration of previously proposed explanations of the mechanism by which they alter estrogen-stimulated responses.

The long-acting effects of triphenylethylene derivatives and their ability to cause differential cell stimulation has obvious implications for the use of these drugs in humans. Clomid has been widely used for the past 15 years to induce ovulation in anovulatory women (Greenblatt et al., 1961). Treatment consists of 50–100 mg Clomid taken orally for 5 days. If the initial trial fails and no signs of pregnancy are present, the treatment is resumed the following month, ~ 40 days later. This regimen is often continued for many months, hence the exposure of a woman to Clomid can be extensive. Under these circumstances, Clomid may stimulate some cell types while acting as an antagonist in others. The eventual effects of such stimulation may remain unknown for many years. Recently we demonstrated that Clomid administration on day 5 of pregnancy in the rat results in abnormalities in the offspring (see Chap. VIII for details). This effect probably originates from the early and intense estrogenic stimulation of developing fetal tissues and demonstrates again the extreme potency of these compounds. The possibility exists that Clomid and other triphenylethylene derivatives could cause hyperestrogenization of certain cell types and hence great caution is mandatory when these drugs are used in humans.

c) Dose Response Curves, Estrogen Potency, and Receptor Occupancy

The classic method for measuring hormonal potency has been to administer several dose levels of a hormone and examine the level of response at some optimal time after treatment. From this kind of experiment a dose-response relationship is established (Fig. 90), and other hormones can be compared to a standard hormone. Two hormones (as *A* and *B* in Fig. 90) with similar intrinsic activities, e.g., stimulate responses in similar fashions but with dissimilar potencies, will have parallel dose-response curves, and the curve for the less potent hormone (*B*) will be shifted to the right. This difference in potency is generally attributed to the different affinity of the receptor for the two hormones, i.e., the receptor in this case would have a higher affinity for hormone *A*. Hormones with differently shaped dose-response curves, as *C* in Fig. 90, are considered to act via a mechanism different from *A* and *B*. These principles are sound, and as long as care is taken to consider other important parameters that may influence their interpretation, they are very useful. However, we have introduced earlier in this book some important variables and functions that have not always been included in estimates of hormone activity. In this chapter we will elaborate on the implications of these findings.

The dose-response relationships for estradiol and estriol are similar to those shown in Fig. 90 for compounds *A* and *C*, respectively (Martin and Finn, 1971). The differences between the dose-response curves for estradiol and estriol have been assumed to reflect a difference in the mechanism of action of these hormones. From the work discussed in Chap. V, which demonstrated that the R_nE_3 complex is not retained as long in the nucleus as R_nE_2 complexes, it is possible to conclude that the different shapes of the dose-response curves are related to this difference in retention-stimulation patterns. However, because the curves in Fig. 90 were derived from responses measured after injection of hormones, another precaution must be introduced; estradiol and estriol are not present in the body as pulses but in a continuous fashion for periods of several hours to days. Under these conditions estriol is a potent estrogen and causes full uterine growth similar to that observed with estradiol (Fig. 79). Leavitt (personal communication) has recently shown that the same

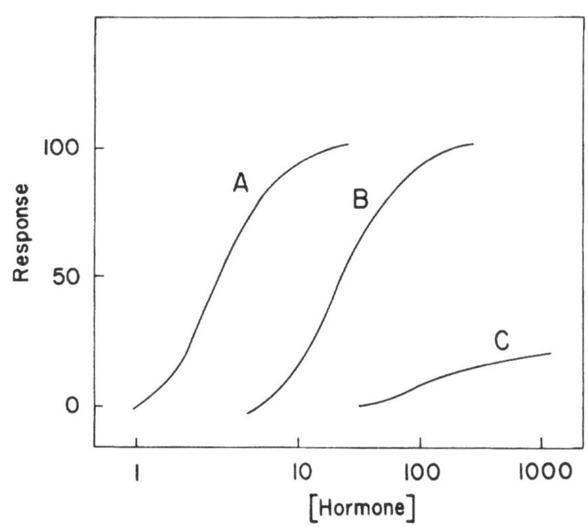

Fig. 90 Theoretical dose-response relationships. See text for discussion

maximum response and dose-response curves of a similar shape can be obtained from injection of estradiol and estriol if the level of estriol is increased sufficiently. At such high levels of injected steroid, blood levels of estriol are sufficient to maintain long-term retention of the R_nE_3 complex and thus to cause true uterine growth similar to that induced by the R_nE_2 complex.

The relationship between response as a function of time after hormone administration is recognized as an important factor in judging hormone potency. Generally a time is chosen when the biologic response is maximal, since the assay will be most sensitive. Although this is logical, important complications and considerations not usually taken into account are provided by a comparison of the uterine growth curves as a function of time after injection of estradiol, estriol, or Nafoxidine (Fig. 91). Estradiol and estriol are equal in potency at 4–6 h after injection, whereas estriol is very inferior to estradiol at 24 h. This discrepancy has been discussed in detail earlier in Chap. V as a result of different retention times of the R_nE_2 and R_nE_3 complexes. However, the important point to be derived is that a complete response vs. time curve is a necessity in the proper evaluation of hormonal activity. This is especially true (and much more complex) when one considers the uterine growth that is stimulated by Nafoxidine. In Fig. 91 animals were injected with 50 µg Nafoxidine and only 2.5 µg either estradiol or estriol; hence, the obvious conclusion that estradiol is more potent than Nafoxidine can be drawn if response is measured at 24 h. However, this conclusion is valid only if one restricts measurements of uterine growth to 24–48 h; it is not valid if measurements are made at 72 h or later. Clearly at 72 h Nafoxidine would be judged quite superior to estradiol since the stimulatory effects of estradiol have regressed by this time and the uterus has decreased in size.

These observations, coupled with the demonstration that triphenylethylene derivatives have the ability to cause differential stimulation of cell growth (see Sect. 7 this chapter and Clark et al., 1978d, e), make the assessment of estrogen potency a very complex task. Valid methods for the assay of hormone potency must include all parameters discussed above, otherwise misconceptions can arise concerning estrogenicity.

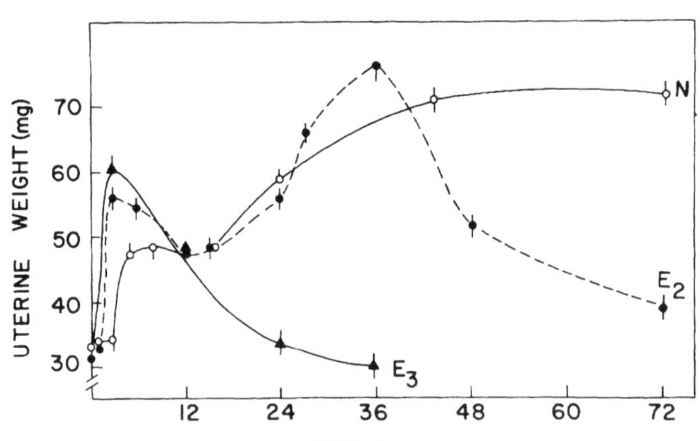

Fig. 91. Effect of various estrogens on the growth of the uterus. Immature rats were injected with estradiol (2.5 µg, ●), estriol (2.5 µg, ▲) or Nafoxidine (○), and the wet weight of the uterus was measured as a function of time

Chapter VII

Steroid Hormones and Neoplasia

1. Estrogens and Cancer

Thus far we have considered the relationship of estrogen receptors and estrogen-induced responses in normal cells. It is clear, however, that many cancerous cells are dependent on estrogen for continued growth. In addition, estrogens have been implicated as causative agents in several forms of cancer. The exact role of estrogen in carcinogenesis will probably not be defined until we understand the mechanisms by which this hormone controls growth in normal tissues (Hertz, 1974). However, because of the importance and interest in this topic, we will attempt to place estrogen receptors in proper perspective with respect to carcinogenesis and growth promotion.

Chronic exposure of mice and rats to estrogens of various types results in preneoplastic and neoplastic changes in the vagina, uterus, pituitary, and mammary gland (Cramer and Horning, 1936; Allen and Gardner, 1941; Gardner et al., 1959; Dunn and Green, 1963; Takasugi and Bern, 1964; Furth et al., 1973; Forsberg, 1972, 1975). Endometrial hyperplasia and cancer occur in women who have been exposed to either endogenous or exogenous estrogens for prolonged periods of time. These cases include women with ovarian tumors that produce estrogens (Gusberg and Kardon, 1971), women who fail to ovulate and as a result are exposed to estrogen without the normal intervention of the luteal phase of the cycle (Fechner and Kaufman, 1974; Lucas, 1974), and women who have taken estrogens for many years because they lack functional ovaries (Larson, 1954; Jackson and Dockerty, 1957; Dewhurst et al., 1975). Exposure of the human fetus to diethylstilbestrol has been associated with the development of vaginal adenosis and clear cell adenocarcinoma in the female offspring (Herbst et al., 1971; Barber and Sommers, 1974; Ulfelder, 1976).

In the studies cited above, the quantities of estrogen were very high and/or the exposure was extended for long periods. Hence, the conclusion that estrogens play some role in carcinogenesis, whether causative or permissive, must be tempered by the notion that the hormone exposure was nonphysiologic. Under *normal physiologic* circumstances estrogens may have no carcinogenic potential; however, it has been suggested that certain estrogens do function as carcinogens under physiologic conditions. Indeed diethylstilbestrol (DES) and estrone have been assigned special roles as cancer-causing agents. In contrast, estriol has been proposed to be a protective agent against the development of neoplasia (see Chap. VI). DES is considered by some investigators and many laymen to be a "monster" drug with special cancer-producing potential. This attitude results from the lay public's knowledge of the studies mentioned above, in which DES was employed in an attempt to extend pregnancy and in which vaginal and cervical abnormalities eventually developed in some of the progeny. Although DES may be linked with the production of cancer in these cases, this does not confer any special carcinogenic potential to this compound. Any estrogen not bound by α-fetoprotein probably would have produced similar results. As mentioned earlier, it has been known for many years that estradiol and DES will cause vaginal cancer in mice, and thus DES is really not different from the physiologic estrogen.

Estrone has also been indicated as the estrogen causing endometrial cancer in postmenopausal women (Siiteri et al., 1974). Estrone is formed by the aromatization of adrenal androstenedione in adipose tissue and constitutes the major estrogen in postmenopausal women. Siiteri et al. (1974) suggest that estrone may cause cancer by acting as an unopposed estrogen. However, they also suggest that estrone may differ qualitatively from estradiol in the biochemical events that result from nuclear binding of receptor-estrone complexes. The latter suggestion has led many investigators to assign special significance to estrone as an intrinsic cancer-causing hormone. Our results, discussed in Chap. V, suggest that no qualitative differences exist between the steroidal estrogens except in their abilities to promote receptor retention in the nucleus. Ruh et al. (1973) examined the ability of estrone to stimulate induced-protein synthesis in the uterus in vitro and could show no qualitative differences between estradiol and estrone. Our interpretation of the role of estrone as a carcinogenic agent in postmenopausal women is that the unopposed action of an estrogen, when present in high steady-state levels in the blood, will provide an environment wherein the predisposition to cancer can manifest itself. Hence we ascribe no special carcinogenic potential to estrone as opposed to other steroidal estrogens.

2. Triphenylethylene Derivatives, Hyperestrogenization, and Reproductive-Tract Cancer

As discussed above, hyperestrogenization can be accomplished by daily administration of estrogens, such as estradiol or DES. Triphenylethylene derivatives produce a similar effect when given as a single dose in immature rats (Clark et al., 1973b). We first showed this by administering a single dose of Nafoxidine to 21–22 day old castrate rats and observing uterine weights and receptor levels 7 and 19 days later (Fig.92). These results show that a single injection of Nafoxidine can act as a very long-lasting estrogen, presumably by the continuous occupancy and nuclear retention of the estrogen receptor. This observation led to an attempt to masculinize female neonatal rats using triphenylethylene derivatives (Clark and McCormack, 1977). The rationale for these experiments was based on the concept that androgens masculinize the hypothalamus via conversion to estrogens (Gorski, 1963; Reddy et al., 1974). Since triphenylethylene derivatives have no androgen activity and appear to act only as estrogens, administration of triphenylethylene derivatives in the neonate should cause masculinization.

Nafoxidine or Clomid were administered to neonatal rats at various dose levels on days 1, 5, 10, and 21 of life. During the 1st and 2nd week after injection, the uteri of these rats showed extensive epithelial stimulation (McCormack and Clark, unpublished observations). This extensive uterotropic stimulation was accompanied by elevated levels of nuclear estrogen receptor and depletion of cytoplasmic receptor. Control rats had a low cuboidal epithelium, and no estrogenic stimulation was observed.

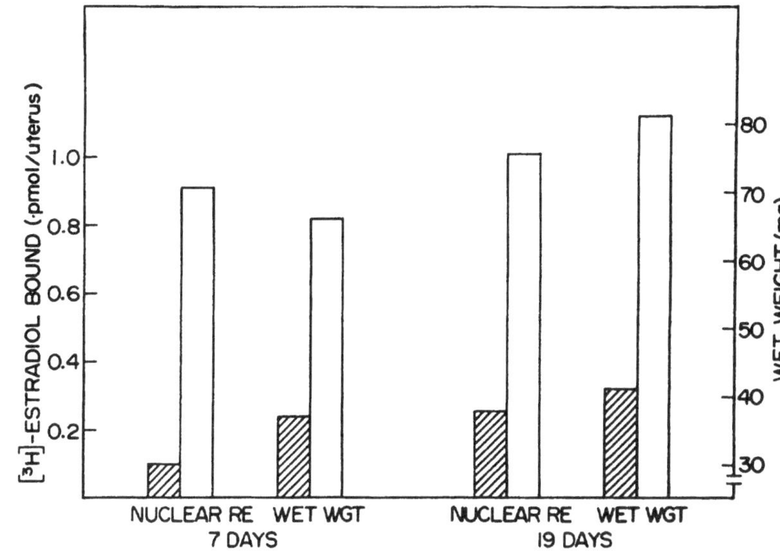

Fig. 92. Effect of a single injection of Nafoxidine on uterine weight and nuclear estrogen receptor. Ovariectomized immature rats were injected with saline (*cross-hatched bars*) or 100 μg Nafoxidine (*open bars*) and uterine wet weight and nuclear receptor were measured 7 and 19 days later

These results agree with previous studies employing 21–22 day old rats discussed in Chap. VI. Vaginal opening occurred in 86 percent of the rats injected with 100 μg Nafoxidine or 500 μg Clomid between days 21 and 34. Vaginal opening occurred in control rats between days 35 and 50. Vaginal smears of treated rats did not show normal cyclic changes; rather, a high incidence of estrus smears was noted. Therefore, these animals are probably acyclic and in a state which outwardly resembles persistent estrus. These responses are typical of masculinized female rats; however, analysis demonstrated a much more complicated syndrome. When these animals were autopsied between 60–100 days of age, they were found to have a complicated array of abnormalities of the reproductive tract. In some animals, these included atrophic ovaries with accompanying atrophic uteri while others showed cystic ovaries and enlarged uteri (Fig. 93). Hypertrophy of the oviducts was a common occurrence.

Histologic examination of the uterine tissue revealed various stages of uterine hyperplasia and squamous metaplasia. Tumors of the uterus were also observed in rats that received Clomid and Nafoxidine. Although such tumors were observed in only a few animals, it should be noted that no animals older than 100 days were used in this study. The incidence of uterine tumors would probably increase considerably with age. Other abnormalities included: hypertrophied and hyperplastic oviducts; ovarian, oviductal and uterine inflammation accompanied by pyometra; atrophic ovaries; and hylus cell tumors of the ovary. Control animals that received oil injections did not manifest any reproductive tract abnormalities and were cyclic. The types and frequency of abnormalities varied widely among the various treatment groups. The high dose of either Clomid (500 μg) or Nafoxidine (100 μg) produced some form of abnormality in 80–100 percent of the animals. Intermediate and lower doses have not been completely evaluated at this time; however, from preliminary

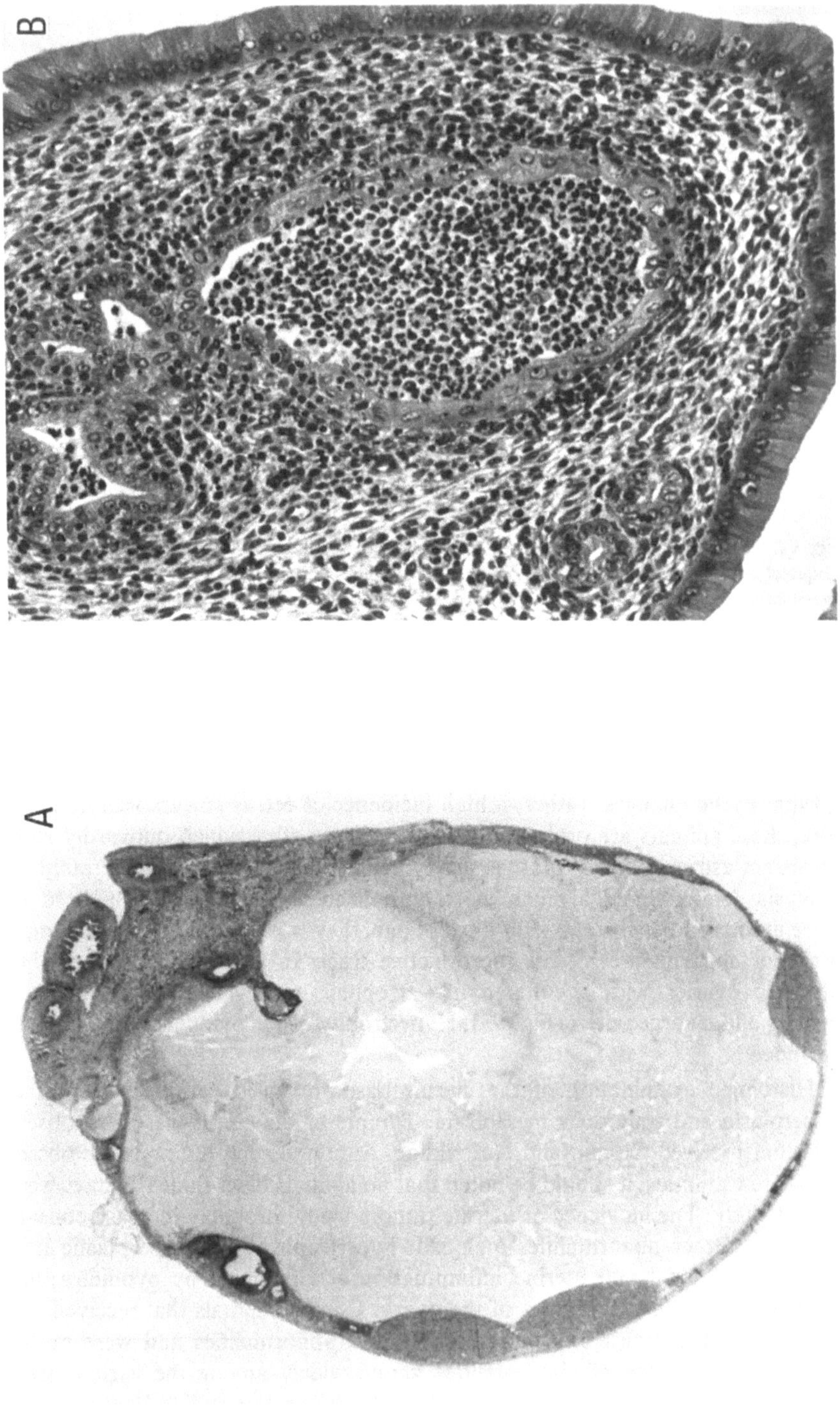
B
A

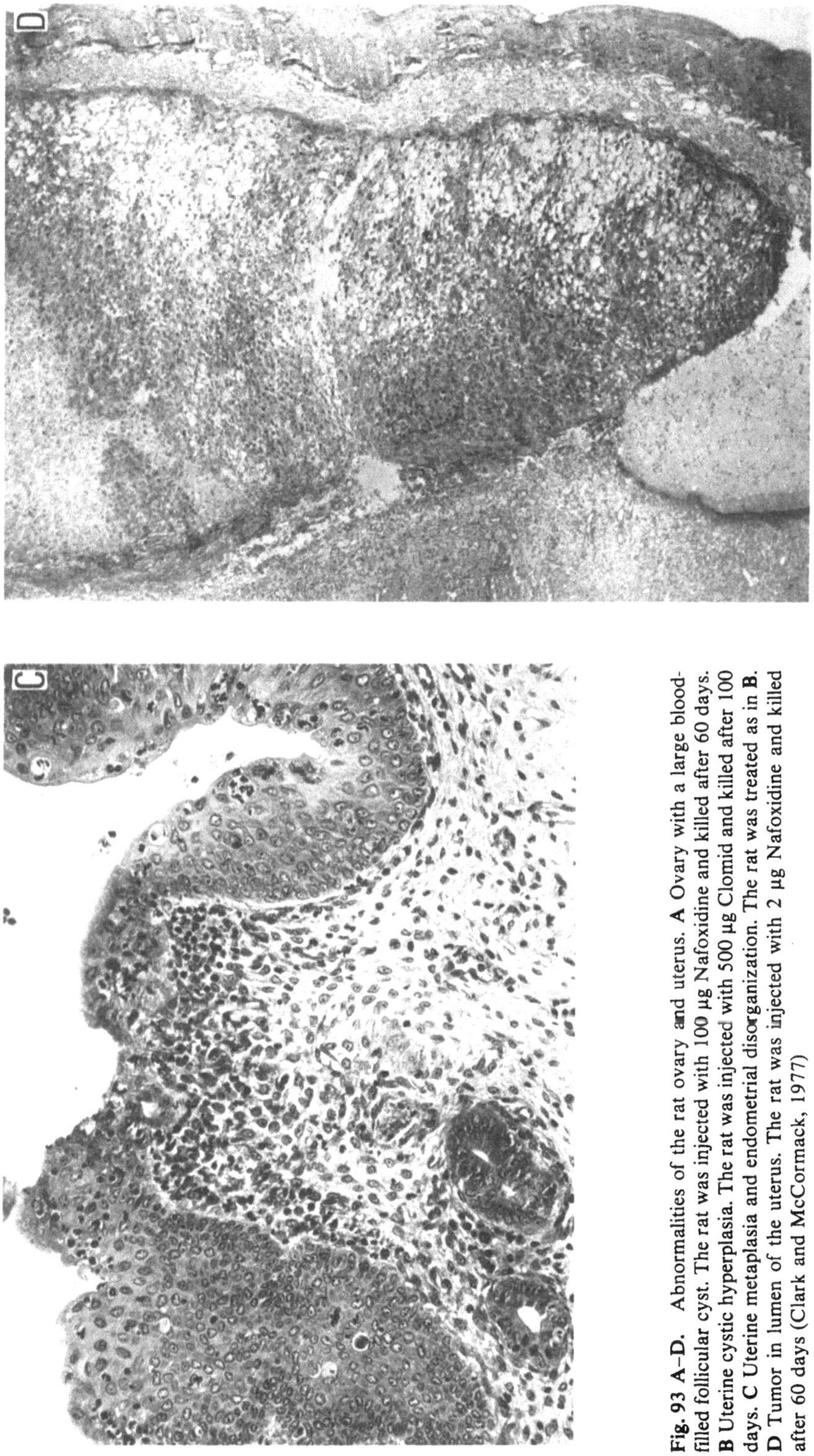

Fig. 93 A–D. Abnormalities of the rat ovary and uterus. **A** Ovary with a large blood-filled follicular cyst. The rat was injected with 100 μg Nafoxidine and killed after 60 days. **B** Uterine cystic hyperplasia. The rat was injected with 500 μg Clomid and killed after 100 days. **C** Uterine metaplasia and endometrial disorganization. The rat was treated as in **B**. **D** Tumor in lumen of the uterus. The rat was injected with 2 μg Nafoxidine and killed after 60 days (Clark and McCormack, 1977)

results it is anticipated that 10–15 percent of the animals will be adversely affected. The tumor shown in Fig. 93 was in an animal injected with only 2 μg Nafoxidine. Uterine metaplasia and infertility accompanied by polycystic degeneration of the ovary have been described by others (Fels, 1976; Gellert et al., 1977). Lobl and Maenza (1975) have observed extensive squamous metaplasia in uteri of androgenized female rats. The relationship between the latter observation and our findings is not clear at the present time.

The ability of the reproductive organs to respond to estrogenic compounds depends, in part, on the presence of estrogen receptors. Since the concentration of these receptors is known to increase with time in the neonatal rat (see Chap. VIII and Clark and Gorski, 1969), the effectiveness of the compound, as well as its mode of action, may vary with age. Thus, the variation in the kinds of abnormalities may relate to both the dose of the compound and the age at which it is administered.

Although the possibility of indirect effects of Nafoxidine and Clomid have not been eliminated, it seems likely that these drugs act directly on the various target tissues. Both drugs cause nuclear accumulation and long-term retention of nuclear receptor complex (Fig. 94) as well as excessive stimulation of the uterine epithelium in 5–10 day old rats. Since initiation of puberty via the hypothalamic-pituitary-gonadal axis is not likely at this time (Hohlweg, 1934), it seems probable that the triphenylethylene derivatives are acting in a direct fashion on the organs and tissues affected. In addition, triphenylethylene derivatives do not alter the levels of prolactin, estradiol, or progesterone in animals with regressing mammary tumors (Nicholson and Golder, 1975; Jordan and Koerner, 1976). Thus, in this case a direct action at the mammary tissue level seems likely.

Injection of Clomid on day 5 of pregnancy, the day of implantation in the rat, causes similar abnormalities of the reproductive tract of the pups when they become adults (McCormack and Clark, in press). These adverse effects of Clomid treatment are also seen in the rat mothers. Thus, exposure of either the developing fetus or the pregnant rat to Clomid results in abnormal growth of the reproductive tract, which leads to teratologic or potential preneoplastic changes.

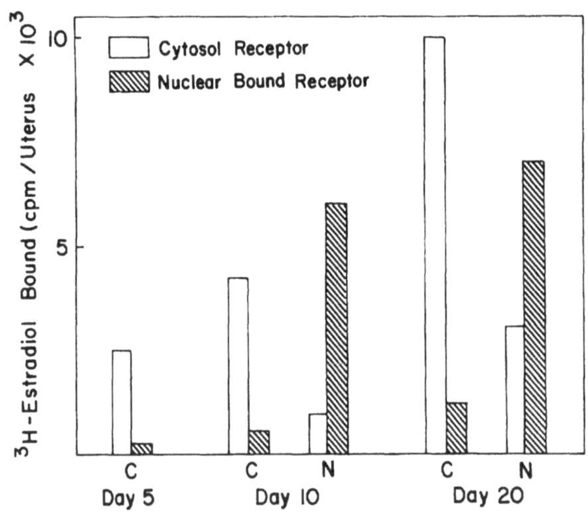

Fig. 94. Effect of neonatal exposure to Nafoxidine on estrogen-receptor compartmentalization. Neonatal rats were injected on day 1 of life with saline or 100 μg Nafoxidine, and the quantity of cytoplasmic and nuclear estrogen receptor was determined by exchange at the indicated times

The effects observed with triphenylethylene derivatives may be analogous to the hyperestrogenization observed by many investigators (see Sect. 1 above). As discussed earlier, continuous exposure of the neonatal mouse to high levels of estrogen will result in neoplastic development of the reproductive tract. Since a single injection of Nafoxidine or Clomid causes prolonged occupancy of the estrogen receptor, these drugs would mimic the effects of daily injections or implants of other estrogens. Although this is a reasonable suggestion, certain qualifications should be noted which relate to peculiar properties of these compounds. As discussed in Chap. VI, both drugs extensively stimulate the epithelial cells of the uterus, while having little or no stimulatory effect on stromal or myometrial components. This, of course, is not true for other estrogens. Therefore this selective hyperestrogenization of the epithelium places the triphenylethylene derivatives in a separate category with respect to carcinogenic potential. This point is discussed more detailed below (Clark et al., 1978d, e).

Another property that may make the triphenylethylene derivatives different from ordinary estrogens is their failure to promote cytoplasmic replenishment of the estrogen receptor. This is clearly different from the effect of continuous estradiol exposure (see Chap. VI and Clark et al., 1977c). The binding of these drugs to receptor and subsequent translocation to the nucleus does not stimulate the same processes that replenish cytoplasmic receptor after estradiol treatment. Therefore, although it is tempting to equate the hyperestrogenization following triphenylethylene derivatives with that produced by continuous exposure to estradiol or DES, sufficient differences exist to warrant caution.

Regardless of the precise mechanisms involved in hyperestrogenization of the epithelium of the reproductive tract, the fact remains that Clomid and other triphenylethylene derivatives are potential carcinogens. As discussed in Chap. VI, Clomid has been widely used for 15 years to induce ovulation in anovulatory women. Treatment can often be extensive, occurring every 40 days for 1 year or more. Thus, some women may be exposed to hyperestrogenization for long periods of time, a condition which is known to be associated with the development of neoplasia (see Sect. 1 of this chapter for references). Also, unless pregnancy is determined carefully, Clomid could be administered during early pregnancy. Such treatment could expose the fetus to hyperestrogenization, the effects of which will go unnoticed until the child reaches adulthood.

3. Hyperestrogenization and Nuclear Body Formation

As discussed in Chap. V, a single injection of either Nafoxidine or estradiol to immature rats stimulates endogenous nuclear RNA polymerase activities. The extent and the degree of stimulation correlate with the nuclear retention of the estrogen receptor. To define the relationship between nuclear retention of estrogen receptor and the stimulation of uterine RNA polymerase, the studies summarized in Fig. 95 were conducted (Clark et al., 1978d, e). Animals were either injected with estradiol or Nafoxidine or implanted with estradiol and killed at 24-h intervals. Nuclei were isolated from uteri and endogenous RNA polymerase activities as well as levels of nuclear receptor were determined. Nafoxidine-treated and estradiol-implanted animals showed elevated and sustained levels of RNA polymerase activities at all times examined. In contrast, injections of estradiol increased polymerase activities by 12 h, and these returned to control levels by 24 h. These alterations in polymerase activities closely parallel levels of nuclear estrogen receptors (Fig. 95).

To relate these biochemical observations to morphologic changes that occur in the normal immature rat uterus in response to Nafoxidine and estradiol, samples of uteri treated as above were prepared for light and electron microscopy. The ultrastructure of uterine luminal epithelial cells was examined in these samples. The height of these cells was greatly increased over that of control cells by either Nafoxidine treatment or estradiol implant (Fig. 96). The height of luminal epithelial cells 72 h after a single injection of estradiol was not significantly different from that of controls. This probably represents the regression following hypertrophy and hyperplasia 24–48 h after a single injection of estradiol. At the ultrastructural level, the hypertrophied cells resulting from Nafoxidine treatment or estradiol implant possess an abundance of free polysomes and rough cisternal endoplasmic reticulum, as well as enlarged Golgi systems, features indicative of stimulated cytoplasmic protein synthesis. Thus, a single injection of Nafoxidine or an estradiol implant causes extensive cellular hypertrophy of the immature rat uterus, which is characterized by long-term nuclear retention of the estrogen receptor and prolonged stimulation of endogenous nuclear RNA polymerase activities. This very active state of uterine cells appears to be one manifestation of the hyperestrogenized uterus.

When the ultrastructure of the nuclei of luminal epithelial cells was examined following the above treatment, several types of nuclear bodies were observed in tissues from either Nafoxidine-treated or estradiol-implanted animals (Fig. 96). Nuclear bodies were quantitated in cells that were judged to be sectioned in the longitudinal plane. Electron microscopic counts were made on ultrathin sections of nuclei referred to as nuclear "profiles." The number of nuclei that contained nuclear bodies was similar in control and estradiol-injected animals, 22 percent and 19 percent, respectively. The value observed in estradiol-injected animals probably reflects the regressed state of the epithelium 72 h after the injection. The percent of nuclei that contained nuclear bodies was significantly elevated in animals that had been implant-

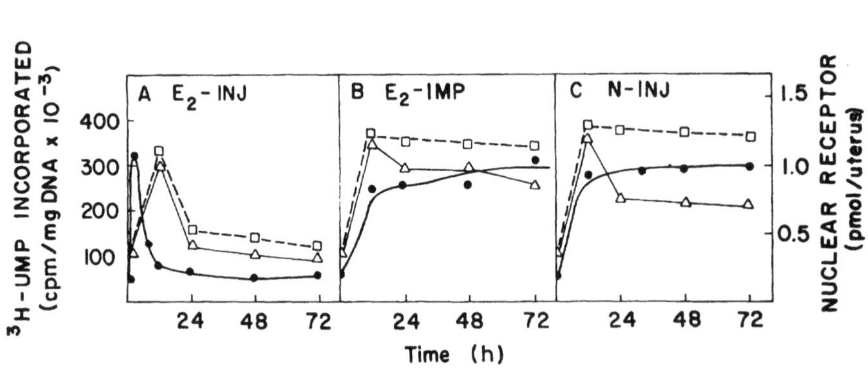

Fig. 95 A–C. Relationship of nuclear binding of receptor-estrogen complex to endogenous RNA polymerase activities. Immature rats were either **A** injected with 1 µg estradiol, **B** implanted with an estradiol-paraffin pellet, or **C** injected with 50 µg Nafoxidine. At the indicated times, animals were killed, and uterine nuclei isolated as described in Chap. II. The levels of nuclear receptor-estrogen complex (●) were determined by the ^{3}H-estradiol exchange assay. RNA polymerase I (△) and RNA polymerase II (□) activities were assayed as described in Hardin et al. (1976)

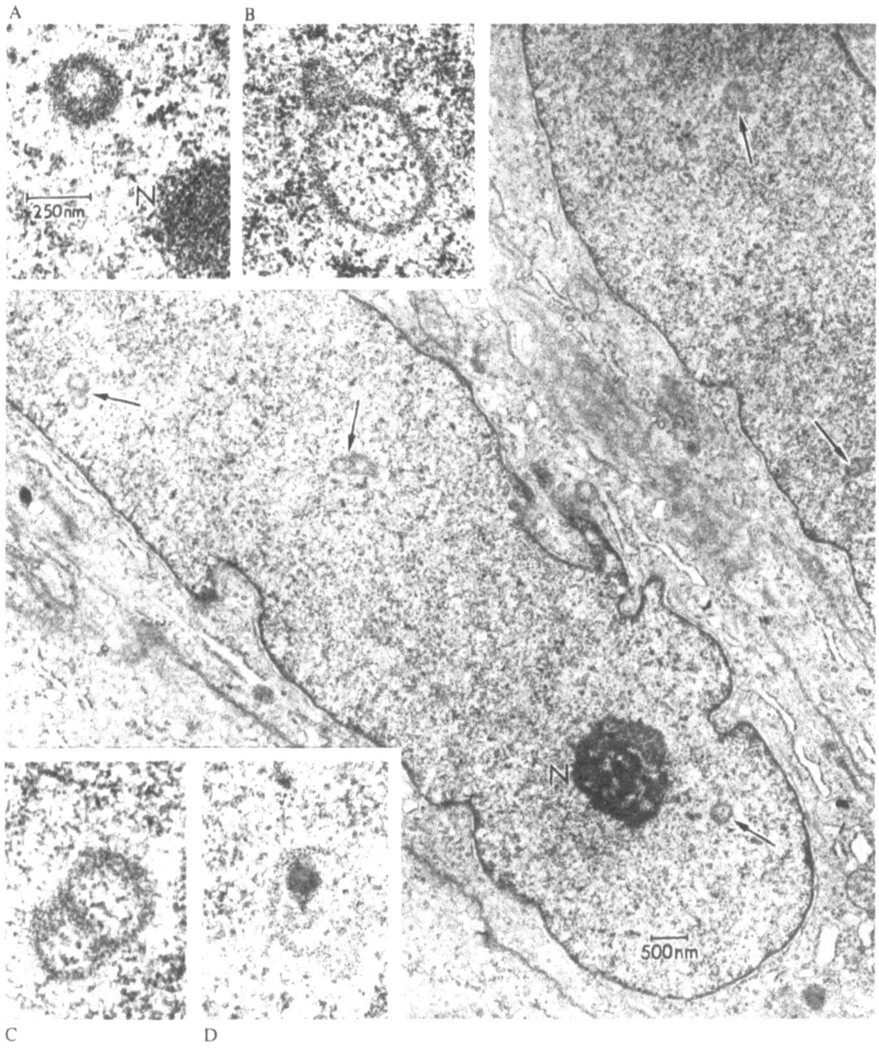

Fig. 96. Nuclear bodies in uterine luminal epithelial cells of immature rats 72 h after a single injection of Nafoxidine (50 µg). In these two profiles of luminal epithelial nuclei, three nuclear bodies occur in the left nucleus and two in the right (*arrows*). *Insets* **A–D** illustrate the principal forms of the nuclear bodies observed in these cells following either Nafoxidine treatment or estradiol implants: **A** singlet (filamentous capsule and less dense core); **B** granular nuclear body (small granules in core); **C** doublet (double core); **D** granular nuclear body with large electron opaque inclusions. N refers to nucleolus. Electron micrograph, × 14,000; *insets*, × 50,000 (Clark et al., 1978d)

ed with estradiol (37%) or injected with Nafoxidine (53%). In addition, 38 percent of the profiles from Nafoxidine-treated animals exhibited multiple nuclear bodies. No multiple nuclear bodies were seen 72 h after a single estradiol injection.

Four distinct morphologic classes of nuclear bodies were observed during the course of this study (Fig. 96). These included singlets with a filamentous capsule and less dense core (size range: 290–305 nm), nuclear bodies which contained granules in the core (size range: 700–1,000 nm), doublets with double cores, and granular bodies with large electron opaque inclusions. The latter three classes were never observed in control tissues but only in nuclei of luminal epithelium from hormone-treated animals.

These nuclear bodies closely resemble proteinaceous nuclear bodies previously described cytochemically by Bouteille et al. (1967) and Dupuy-Coin et al. (1972), who showed that the nuclear bodies in their studies consisted of a filamentous protein capsule and a core that probably contained either RNA or DNA. Such bodies have often been identified in neoplastic cells (Krishan et al., 1967).

Two interesting observations concerning nuclear bodies arise from this study. First, the number and complexity of nuclear bodies is increased in a hyperestrogenized animal; second, nuclear bodies appear predominantly in luminal epithelial cells (Padykula et al., unpublished observations). The latter point probably relates to the fact that Nafoxidine preferentially stimulates growth of luminal epithelial cells while having little effect on growth of myometrial and stromal cells (see above).

The composition and function of nuclear bodies in the hyperestrogenized uterus is not known. They are visible manifestations of hyperestrogenization. Their presence and complexity is closely related to the transcriptional activity of the uterus; i.e., nuclei of quiescent cells have fewer and morphologically simple nuclear bodies, wherease nuclei of hyperactive cells contain many bodies of varying complexity. Increased transcriptional activity and the appearance of nuclear bodies are correlated with long-term retention of the estrogen receptor in uterine nuclei. Thus, nuclear bodies may represent nuclear organelles involved in elevated transcriptional activity or may be byproducts of such activity. The relationship between these nuclear bodies, Nafoxidine, and the reproductive-tract abnormalities discussed in Sect. VII. 2 is unknown, though it is known that many neoplastic cells contain nuclear bodies (Bouteille et al., 1967; Krishan et al., 1967; Dupuy-Coin et al., 1972). Since we have shown that nuclear body formation is promoted by compounds that induce reproductive-tract abnormalities and neoplasia, the study of nuclear bodies as precursors of the neoplastic state should be pursued in the future.

4. Steroid Receptors and Neoplasia

Many cancers depend on hormones for growth, and this dependency appears to derive from the presence of some critical number of hormone receptors. This topic has received considerable attention with regard to mammary gland tumors and has been reviewed extensively by others; hence we will make no attempt to do so (Jensen et al., 1972; McGuire et al., 1975; Wittliff et al., 1976). In the preceeding sections of this chapter the possible relationships between estrogen and carcinogenesis have been discussed. In this section we will briefly examine the estrogen dependency of some cancers and relate this dependency to steroid hormone receptors.

a) Estrogen-Dependent Growth

Estrogen-dependent mammary tumors contain cytoplasmic and nuclear estrogen receptors that have the same general characteristics as those from normal mammary gland. Cytosol estrogen receptors exhibit 8*S* and 4*S* peaks on sucrose density gradients (Jensen et al., 1971; McGuire et al., 1975; Boylan and Wittliff, 1975; Wittliff et al., 1976), and nuclear extracts after estrogen administration yield 4–5*S* peaks (Wittliff et al., 1972a, b; Shyamala and Nandi, 1972). The physicochemical characteristics of receptors from cancerous tissue have not received extensive study, but generally no qualitative difference has been apparent. However, quantitative differences have been observed, and these form the basis for the method currently used for screening breast cancers (Jensen et al., 1967).

An estrogen-dependent transplantable tumor line has been studied for abnormalities in its estrogen receptors and its responses to estrogen. These tumors (MXT) are well differentiated ductal or papillary ductal carcinomas and can be transplanted subcutaneously in syngeneic female mice. The tumors (2–5 mm diameter) grow slowly, if at all, in ovariectomized mice, whereas in intact mice they grow to 18–25 mm in the same time period (8 weeks). In addition, estradiol benzoate induces significant growth of the tumors in castrates. The growth of tumors is inhibited by ovariectomy regardless of whether they are transplanted when small (< 2 mm) or large (≥ 13 mm).

This tumor line contains a relatively high level of R_c^E ($\sim$ 8–10 fmol/mg tissue), which has the general characteristics of estrogen receptors in other target tissues (Watson et al., 1977). The temperature dependence of steroid binding and exchange on cytoplasmic and nuclear receptor is similar to that for other estrogen-dependent tisssues (see Chap. II). Hormone specificity of tumor receptors is also as described for estrogen targets. Thus, displacement of estradiol can be accomplished only by competition of other estrogens. An injection of estradiol causes the translocation of R_c^E to the nucleus in a fashion that does not appear to differ from normal tissues.

The average K_d for receptor-estradiol complexes in urethan-induced tumors was 1.2 ± 0.2 n*M*, which agrees well with values reported for normal and neoplastic mammary tissues. Richards et al. (1974) reported that estrogen receptors of normal lactating mammary gland from various murine strains had K_d values ranging from 0.28 to 0.54 n*M*, while that of mammary tumors ranged from 0.35 to 0.65 n*M*. In addition, MXT tumors contain an average of 8.5 fmol/mg tissue, which is substantially more receptor than reported by Richards et al. (1974) (0.6–1.8 fmol/mg tissue).

The presence of estrogen receptor in this tumor line agrees with its apparent estrogen dependence. This is in contrast to previously described mouse mammary tumors with MTV-positive C3H female mice, which are estrogen independent. Hormone independence was correlated with the inability of the cytoplasmic receptor to undergo nuclear accumulation (Shyamala, 1972). Other estrogen-independent mouse mammary tumor lines contain little estrogen receptor in either the cytoplasmic or nuclear compartments (Richards et al., 1974; Sluyser and Van Nie, 1974; Sluyser et al., 1976; Medina et al., 1975).

This study of a new transplantable tumor demonstrates that it is estrogen dependent, contains large quantities of translocatable estrogen receptor, and that this receptor possesses characteristics similar to those described for other estrogen-dependent tissues. Thus, the binding characteristics of the estrogen receptor in the MXT system appear typical of other tumors and no different from normal tissue.

b) Interactions of Estrogen and Progesterone Receptors

In Chap. VI we discussed the ability of estrogens to elevate the level of the cytoplasmic progesterone receptor. This appears to be a general response of normal as well as some neoplastic target cells to estrogen. Horwitz and McGuire (1978) have shown that estradiol will stimulate the synthesis of cytoplasmic progesterone receptor in human breast cancer cells in culture (MCF-7). Similar observations have been made in the MXT transplantable tumor (Watson and Clark, in press). The ability of neoplastic tissues to respond to estrogen in this manner may constitute one measure of their differentiated state. The loss of the capacity of cells to synthesize progesterone receptor in response to estrogen may represent one step in the long line of changes that eventually produce estrogen-independent cancers. This rationale forms the basis for the proposal by Horwitz et al. (1975) that tumors containing estrogen receptors, but not progesterone receptors may represent a class that are nonresponsive to estrogen and hence independent of its action. Thus, mammary tumors may exist which contain estrogen receptors but which have lost their capacity to synthesize progesterone receptors in response to estrogen. The presence or absence of progesterone receptors in mammary cancers may serve as a useful marker for the prediction of clinical response to ablative therapy (Horwitz et al., 1975).

Young et al. (1976) have proposed that a loss of the ability of endometrial cells to synthesize cytoplasmic progesterone receptor also occurs during the development of cancer of the uterus, and Rao and Wiest (1974) suggested that the presence of cytoplasmic progesterone receptor be used as a guide for progesterone therapy. Pollow et al. (1977) have shown that the concentration of cytosol progesterone receptors was low in undifferentiated tumors and high in well-differentiated tumors. These authors suggest that this finding may explain why undifferentiated tumors fail to respond to progesterone therapy, whereas long remissions are noted when well-differentiated tumors are treated with progesterone. This is an appealing hypothesis in view of the observations discussed above for the mammary gland. However, other investigators have failed to find any differences in progesterone receptor in normal, hyperplastic, and neoplastic endometrium (Haukkamaa et al., 1971; Dyer et al., 1976; MacLaughlin and Richardson, 1976). Regardless of the validity of this concept for endometrial cells, it is well known that progesterone can be an effective treatment for endometrial cancer. Large doses of progestins, such as medroxyprogesterone acetate, caused remission of endometrial cancer in about 30 percent of the cases studied by Lucas (1974). In addition, progestins have been used to cause regression or remission of endometrial hyperplasia and carcinoma in situ (Kistner, 1973). As discussed above, the failure of progestin treatment to cause remission in all cases of endometrial cancer may be a function of the developmental state of the cancer.

Large doses of progesterone have also been used to treat renal carcinoma in the human (Concolino et al., 1976a). The kidney contains both estrogen and progesterone receptors, and renal adenocarcinoma can be induced by continuous exposure to estrogen (Thompson and King, 1976; Li et al., 1976). These carcinomas are estrogen dependent, and progesterone has been proposed to act as an estrogen antagonist, as discussed above, for mammary neoplasias (Concolino et al., 1976a, b). It is also possible that the progesterone used in the treatment of renal carcinoma is binding to the androgen receptor, and antagonism arises via androgen receptor-estrogen receptor interactions by unknown mechanisms (Bardin et al., 1973; Bardin et al., 1975). Although one is tempted to generalize this concept to all estrogen-dependent cancers, the work of Huggins and his co-workers has clearly shown that progesterone can stimulate mammary tumor growth (Huggins et al., 1962; Huggins and Yang, 1962; Huggins, 1965). Others have observed an accelerated rate of appearance of DMBA-induced tumors in progesterone-treated animals (Kim, 1965; McCormack and Moon, 1973). This matter is made more complex by the observation that progesterone, when combined with moderate to high doses of estrogen, will cause mammary tumor regression (Huggins et al., 1962; Muggia et al., 1968; McCormack and Moon, 1973). Obviously the effect of progesterone on mammary cancer regression is complex and unclear at the present time.

The ability of progesterone to cause remission in some endometrial cancers is presumed due to the antagonistic effect that progesterone has on estrogen-dependent growth (see Chap. VI for discussion). The mechanism of this antagonism is complex; however, one of the mechanisms undoubtedly involves the ability of progesterone to depress levels of cytoplasmic estrogen receptor and thereby decrease the capacity of the cancer to respond to estrogen. Gurpide et al. (1976) and Pollow et al. (1977) have also observed that progestin treatment increases the activity of 17β-dehydrogenase in endometrial hyperplasia and adenocarcinoma and thereby should reduce the level of effective estrogen in the tissue. The increased conversion of estradiol to less-active estrogens and the decreased level of cytoplasmic estrogen receptor probably function in concert in the remission of the cancer. Levels of cytoplasmic estrogen receptor ranging from nondetectable to very high have been described in cases of adenocarcinoma of the uterus (Muechler et al., 1975). Therefore the situation may be analogous to that discussed above for mammary tissue, i.e., the presence of adequate cytoplasmic estrogen receptor is indicative of some estrogen-dependent cancers, whereas either low levels or no cytoplasmic estrogen receptors would be found in independent tumors. Thus, the measurement of both cytoplasmic estrogen and progesterone receptors should provide a good predictive tool for clinical response to hormone therapy.

Progesterone may act under normal physiologic circumstances to control the adverse effects of estrogens on the endometrium of the uterus. In all of the cases of estrogen-induced uterine cancer discussed previously, estrogen exposure was prolonged and unopposed by the presence of progesterone. Since progesterone is known to alter or reduce the ability of the uterus to respond to estrogens, progesterone may likely play a role in protecting the uterus from excessive stimulation by estrogens. Women who have normal luteal phases and normal progesterone levels during the cycle are less likely to develop endometrial cancer than women who are anovulatory (Gusberg, 1967; Lucas, 1974). As discussed above, the ability of progesterone to antagonize the effects of estrogens probably results from the reduced levels of estrogen receptor and the elevated activity of steroid-dehydrogenase. These changes induced by progesterone during each cycle may provide a "physiologic" protection against endometrial neoplasia. Likewise these effects of progesterone may account for the protective effect of pregnancy (Sherman and Korenman, 1974).

c) Estrogen-Independent Growth

Although it is generally agreed that constant exposure to unopposed estrogen leads to the development of cancer in several tissues and organs, these neoplasias are known to eventually become independent of estrogen. That is, the cancer cell no longer requires estrogen to sustain its growth. This state of autonomy is accompanied by a loss of many estrogen-stimulated responses, and the neoplastic cell appears to undergo a process of dedifferentiation. Dedifferentiation of early endometrial carcinoma is accompanied by a loss of cilia and microvilli and the capacity for estrogen to induce their formation. In addition, changes in primary lysosomal activity are noted (Ferenczy, 1977). The inability of mammary cancer cells to synthesize progesterone receptors in response to estrogen also appears to be a manifestation of the autonomous or dedifferentiated state in some tumors (Horwitz et al., 1975). Thus, one can visualize cells that gradually change from a highly estrogen-dependent state, which is characterized by typical estrogen-induced responses, to an autonomous state, which is independent of estrogen and appears to represent a dedifferentiated state.

Zava et al. (1977) have suggested that autonomy in some mammary tumor cell lines is obtained as a result of permanent residency of unoccupied estrogen receptor in the nucleus. This suggestion arose from an evaluation of nuclear receptor content under nonexchange conditions, i.e., at 4°C. At this temperature any receptor that is unoccupied will be filled by saturating ^{3}H-estradiol, whereas those receptors that are occupied will not exchange (see Chap. II for further discussion of this point). These investigators found large numbers of unoccupied receptors in the nucleus of MCF-7 cells. Similar findings have been reported for human breast cancer by Panko and Macleod (1978).

Although this is an interesting proposal, which may well represent one mechanism by which cancer cells become estrogen independent, certain qualifications are necessary. Whether exchange of ^{3}H-estradiol occurs at 4°C depends upon the ligand occupying the receptor. If the ligand is one of low affinity, it is likely to have a rapid dissociation rate at 4°C. Under such conditions one could observe the apparent filling of unoccupied sites when in reality the assay was measuring occupied sites via exchange. This remains to be examined in the MCF-7 line.

Permanent occupancy of unfilled estrogen receptors in the nucleus of target cells is not the only mechanism by which estrogen independence could be obtained. Certainly this phenomenon should not be taken as a general one. Indeed, it is not the case for estrogen-independent hyperplastic alveolar nodules and mammary tumors derived from these nodules (Medina et al., 1975). As shown in Table 11, these

Table 11. A comparison of cytoplasmic estrogen receptors in hyperplastic alveolar nodules and mammary tumors

Tissue	Receptor (pmol/100 mg wet wt.)[a]
HAN-D1[b]	0.02 - 0.03
HAN-D2	0.01 - 0.02
Tumor-D1[c]	ND[d] - 0.01
Tumor-D2	ND - 0.01
Lactating gland (day 15)	0.53 - 0.62
Rat uterus (immature)	4.56 ± 0.4
Rat diaphragm	0.08 ± 0.01

[a] Values represent the range of determinations or the mean ± the SEM.

[b] HAN, hyperplastic alveolar nodule; D1 refers to a specific nodule line.

[c] Mammary tumor derived from a specific nodule line.

[d] ND, nondetectable.

nodules and tumors contain very low quantities of cytoplasmic estrogen receptor which correlate with their estrogen independence. We considered the possibility that cytoplasmic receptor was not detectable because it had undergone translocation to the nucleus. This was examined by exchange assays at 4°, 15°, 21°, 30°, and 47°C for varying periods of time. Such analysis provides complete information on both exchange and degradation rates as discussed in Chap. II and, of course, would detect unfilled receptor sites. The results of these experiments revealed very low quantities of receptor in the nuclear fractions of both nodules and tumor tissues (0.04–0.08 pmol/100 mg tissue weight). These values do not differ significantly from homogeneous nontarget tissues. We conclude that the hyperplastic alveolar nodule and tumors derived from them contain very little estrogen receptor in either the cytoplasmic or nuclear compartment. Therefore, the sources of estrogen independence in these tissues must lie at a different level, and is not due to the presence of unoccupied estrogen receptors in the nucleus.

Chapter VIII

Physiologic Relationships and Steroid Receptor Binding

The relationship between nuclear binding of receptor-steroid complexes and physiologic responses depends on the concentration of "free" hormone available for binding to unoccupied receptor. The factors that determine the concentration of "free" or available hormone are complex and important; hence we have chosen to discuss them as a preface to the major topic of this chapter.

1. Blood Binding and Metabolism of Steroids

A simple determination of the total quantity of a steroid in blood is not an adequate measure of its ability to stimulate a target tissue. The active fraction of steroid is that concentration of "free" steroid not bound to protein or lipid and very rarely measured. Simple measurements of total steroid concentration, which includes bound and free steroids, may lead to erroneous conclusions regarding the level of hormone stimulation in target tissue. This has been shown by several investigators for a variety of hormones (Slaunwhite et al., 1962; Sandberg et al., 1966; Oppenheimer, 1968).

Steroid-binding proteins in blood are known to have an important influence on the availability of serum steroids (for review see Westphal, 1971, Vol. 4 this series, and King and Mainwaring, 1974). The affinity of these blood binders for steroids varies from very weak ($K_d \sim 10^{-3}\ M$) to very strong ($K_d \sim 10^{-10} - 10^{-8}\ M$), and they are frequently present in high concentrations (Murphy, 1968; Soloff et al., 1971). Therefore these proteins can restrict the amount of free hormone available for receptor binding and hence can be important in the control of steroid hormone action. One such binding molecule is the sex-steroid binding protein, also called testosterone-estrogen binding globulin, which binds testosterone and estradiol with equal affinity, $K_d \sim 10^{-10}\ M$ (Murphy, 1968; Soloff et al., 1971; Burke and Anderson, 1972; Ratajczak and Hähnel, 1976; Petra and Schiller, 1977). Another important blood component binding estrogens but not androgens is α-fetoprotein (α-FP), present in both the pregnant and neonatal rat. During neonatal and prepubertal life of the rat, α-FP gradually declines from very high values in the newborn to very low values just before and after puberty (Raynaud et al. 1971; Raynaud, 1973). Therefore the quantity of free sex steroid gradually increases as the concentration of α-FP declines. In this manner increasing concentrations of steroid are available for cellular interactions. These observations have led several investigators to suggest that α-FP plays a protective role in the fetus and neonate and may be involved with the onset of puberty (Plapinger et al., 1973; McEwen et al., 1975).

Corticosteroid-binding globulin (CBG) is also of special interest and importance. CBG binds progesterone with a K_d of $10^{-7} - 10^{-6}\ M$ at 37°C and therefore may be an important factor in the control of free progesterone (for an extensive review see Westphal, 1971). This is even more important when one considers that levels of

CBG can be elevated by estrogens (Sandberg and Slaunwhite, 1959). Therefore estrogens may control the availability of free progesterone. This concept is speculative since no experimental evidence has been presented; however, this example is suggestive of the profound and complex influences that blood binders may have in the control of steroid hormone effects.

Binders of sex steroids that do not demonstrate pharmaco- or stereospecificity may also play significant roles in reproductive physiology. Serum albumin has a relatively weak affinity for estrogen ($K_d \sim 10^{-4}$ - 10^{-5} *M*) but is a significant estrogen binder. The concentration of serum albumin in blood is about 4 percent (7 × 10^{-4} *M*); at this concentration serum albumin becomes a significant competitor for estrogens. The influence of 1 percent albumin on the measurement of the estrogen receptor is shown in Fig. 97. In this experiment the presence of serum albumin results in an underestimate of both the quantity of receptor present and the affinity of the receptor for estrogen. It follows from Fig. 97 that the quantity of receptor could be measured accurately by increasing the concentration of [^{3}H]-estradiol; however, even under these conditions the calculated K_d will be inaccurate because of an overestimate of free steroid. Similar observations have been made by Wittliff et al. (1972b).

Plasma components may also affect the relative potencies of hormones. By way of example, the K_d values for complexes of estradiol and estriol with the estrogen receptor at 4°C are 10^{-10} and 10^{-9} *M*, respectively. Therefore, estradiol should be ten times more potent than estriol with respect to relative ability to cause translocation of the R_cE complex to the nucleus as well as in the induction of early uterotropic responses. However, estriol and estradiol are of equal potency in vivo with respect to both translocation and the stimulation of early uterotropic events (Anderson et al., 1973a, and Chap. V). This discrepancy between predicted and observed potencies results from differential binding of these steroids to plasma components. This is illustrated in Fig. 98 (Anderson et al., 1974a). In these experiments the effect of serum albumin on the K_d for nuclear translocation of the estrogen receptor was examined. The addition of serum albumin to the in vitro incubation dramatically altered the ability of

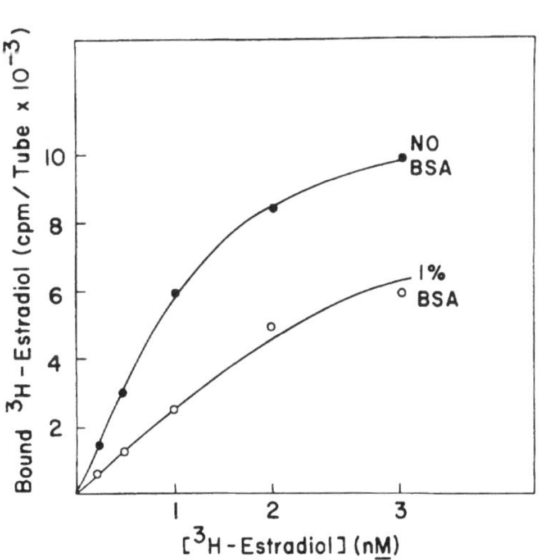

Fig. 97. The effect of bovine serum albumin on the binding of ^{3}H-estradiol to the cytoplasmic estrogen receptor. The specific binding of ^{3}H-estradiol to R_c was determined in the absence (●) or presence of (○) of 1 percent bovine serum albumin (*BSA*) at various concentrations of ^{3}H-estradiol

estriol to bind to cytoplasmic receptor and to effect translocation. This results from the relative affinities of albumins for estradiol and estriol; i.e., estradiol is more firmly bound to albumins than is estriol. Thus, the concentration of estriol required to half-saturate the translocation process in the absence of serum albumin is 3-fold greater than that of estradiol (estradiol, 2 n*M*; estriol, 5 n*M*). However, in the presence of serum albumin the concentration of estriol required to half-saturate the translocation process is one-half that of estradiol (estradiol, 18 n*M*; estriol, 9 n*M*). Therefore, while the ability of an estrogen to cause receptor translocation and uterotropic responses depends on the binding of the estrogen to its receptor, the latter is quite dependent on the amount of free estrogen available for receptor binding.

Serum steroid-binding proteins in in vitro incubations, and presumably in vivo, decrease the rate of entry of estrogen into the uterus (Peck et al., 1973a). However, they may also serve to direct hormonal signals toward organs with protein-permeable vascular beds (Keller et al., 1969), since estradiol increases the accumulation and retention of albumin by the uterus while not affecting such "nontarget" tissues as psoas muscle, liver, or brain (Peterson and Spaziani, 1971). This accumulation and retention of albumin, and presumably albumin-bound estrogen, may establish levels of total estrogen in the uterus greater than those of the systemic circulation. Thus estrogens may be maintained at higher levels for longer periods of time via the accumulation of albumin-bound steroid in target tissues.

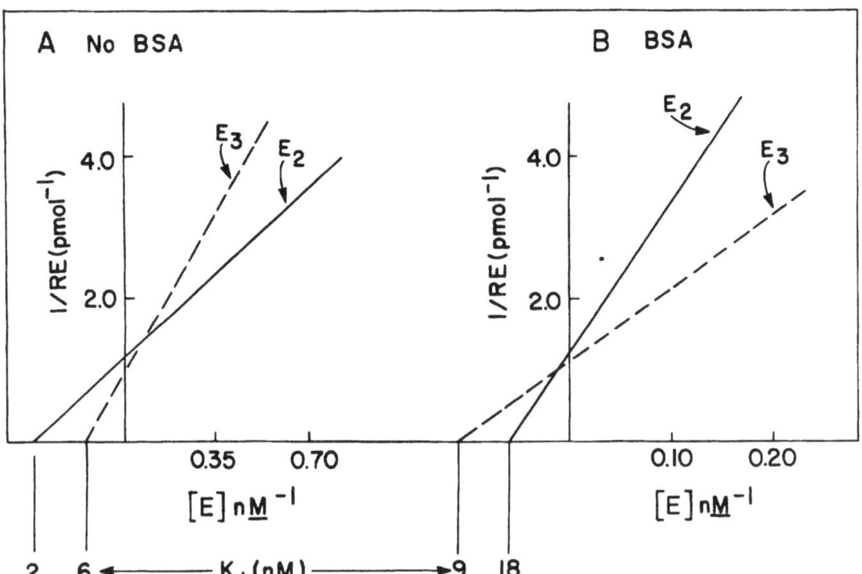

Fig. 98 A and B. Effects of serum albumin on the translocation of receptor-estrogen complexes by estradiol or estriol. Intact uteri were placed in 2 ml Eagle's medium containing various concentrations of either estradiol or estriol. The incubations were carried out in the absence **(A)** or in the presence **(B)** of 4 percent bovine serum albumin for 1 h at 37°C. The quantity of nuclear receptor-estrogen complex was determined by the [^{3}H]-estradiol exchange assay, and the data analyzed by double-reciprocal analysis. Estradiol, E_2; estriol, E_3; reciprocal of receptor-estrogen complex, $1/_{RE}$; concentration of estrogen, [*E*] (Anderson et al., 1974a)

The quantity of steroid available in vivo for receptor binding depends not only on blood-binding relationships but also on the rate of metabolism and excretion of that hormone. Therefore the metabolic clearance rate, or that volume of blood required per unit time to remove the administered hormone from the body, is very important to considerations of biologic activity (for a complete discussion of these points see Gurpide, 1975, Vol. 8 of this series). A hormone with a high affinity for its receptor and thus with a predicted high potency may also have a rapid metabolic clearance rate. Hence, the exposure time of the hormone to its receptor is short, and the prediction of high potency may be incorrect. This is exemplified by the weak estrogenic potency of estriol after an injection. The anticipated potency would be 0.1 that of estradiol if estrogenic potency were dependent solely on the affinity of the estrogen receptor for the hormone. However, as discussed in Chap. V, the uterine growth observed following these hormones is far greater for estradiol than for estriol. This is due in part to the very rapid clearance of estriol in the blood, ~ 10 min as compared to 30 min for estradiol (Jensen et al., 1966). Conversely hormones with a slow metabolic clearance rate and relatively low affinities may display an unexpectedly high biologic activity. This is the case for the long-acting estrogen antagonist, Nafoxidine. The affinity of the estrogen receptor for Nafoxidine is only 1/30 that for estradiol (Rochefort et al., 1972a); however, the effects of Nafoxidine are much more long-lasting due to the long-term retention of the receptor by the nucleus of uterine cells (Clark et al., 1973b, 1974b). Therefore, the biologic effectiveness of Nafoxidine is much greater than that of estradiol when response is considered (see Chap. VI).

Generally speaking, the correlation between the relative affinity of progestins for the progesterone receptor and their biologic activity is very high. The only exceptions are those progestins, which have a low affinity for the receptor, but have a long biologic half-life (Smith et al., 1974; Feil et al., 1976; Raynaud, 1977). This situation is analogous to the case discussed above for estrogens.

The actions of estrogen and progesterone are not generally considered to depend on the metabolic conversion of these steroids to active forms (Jensen and Jacobsen, 1962; Schrader et al., 1972; King and Mainwaring, 1974). Therefore, once receptor-steroid binding has occurred as a result of steroid entry into the cell, the receptor-steroid complex is functionally active. Caution should be exercised in assuming that metabolism is not important in some cases since, for instance, the conversion of testosterone to 5α-dihydrotestosterone is a requirement for androgen action in some male accessory sex structures (Bruchovsky and Wilson, 1968a, b; Anderson and Liao, 1968). In addition, the aromatization of testosterone is required for masculinization of the central nervous system (Naftolin et al., 1972). Gurpide and Welch (1969) have demonstrated that estradiol and estrone undergo extensive interconversion in human endometrium and that estradiol is metabolized to estrone before it is released from the tissue (see also Gurpide et al., 1971). As discussed in Chap. VI, Tseng et al. (1977) have suggested that this conversion of estradiol to estrone is a mechanism for lowering the level of estradiol in the tissue and thereby acts to control the level of estrogen-receptor binding. It is also possible that this metabolism is involved in the mechanism by which the receptor-estradiol complex dissociates from nuclear binding sites. O'Malley and Strott (1973) have shown that progesterone is rapidly converted to 5α-pregnane-3,20-dione in the chick oviduct. This steroid competes effectively for the progesterone receptor, and, in addition, Strott (1974) has shown that it is as potent as progesterone in the stimulation of avidin synthesis. 5α-

Pregnane-3,20-dione is also capable of stimulating LH release in the rat and hamster (Sanyal and Todd, 1972; Bosley and Leavitt, 1972). However, this compound is not active as a uterotropic agent (Sanyal and Villee, 1973; Blaha, 1974), a finding correlating with its lack of binding to the progesterone receptor in the uterus (Leavitt and Grossman, 1974; Leavitt et al., 1978). The metabolism of progesterone to 5α-pregnanedione and other inactive metabolites in the uterus may play an important regulatory role by reducing the effectiveness of progesterone; thus, rapid conversion to metabolites that do not bind to the progesterone receptor would provide the tissue with yet another control mechanism for hormone-induced responses (see Sect. 3.c this chapter for further discussion of this point).

2. Steroid Receptors During Development

It is clear that proper sexual development depends on the presence of steroid receptors. Bardin et al. (1973) and Gehring et al. (1971) have shown that the female phenotype will develop in genotypic male rodents with testicular feminization syndrome and that this results from a failure of the organism to produce sufficient androgen receptors. Hence, though testes develop and produce androgens, the tissues are nonresponsive to these hormones. Instead, they respond to low levels of estrogen via estrogen receptors, and the typical female phenotype develops. From these studies it is obvious that the presence or absence of receptor could have profound effects on the development and growth of target tissues.

a) Estrogen Receptors and Uterine Development

The developmental relationships between estrogen receptors and cellular growth and differentiation remain to be elucidated. However, from preceding discussions in this book (see Chap. VI) it seems possible that estrogen receptors are a constitutive function of the differentiation of estrogen target cells. As an example, the developmental program controlling the differentiation of Müllerian ducts probably controls the synthesis of the estrogen receptor.

The ontogeny of estrogen receptors in the uterus has been examined during the neonatal period in the rat (Clark and Gorski, 1970). The number of cytoplasmic receptors increases sharply between day 1 (0.14 pmol/100 μg DNA or 5,000 sites/cell) and day 10 (0.56 pmol/100 μg DNA or 20,000 sites/cell, Fig. 99). After day 10 the number of sites decreases to a value of approximately 16,000 per cell, and this quantity remains constant to day 22. This dramatic increase in cytoplasmic receptor, which occurs prior to day 10, precedes that time when uterine growth rate is maximal, i.e., between 10 and 20 days. The cytoplasmic receptor at 10 days has the same sedimentation characteristics as that of the 22-day-old rat, as judged by sucrose density gradient analysis. The 8*S* and 4*S* forms of the receptor are also observed in the Müllerian ducts of the fetal rat (Somjen et al., 1976). Dissociation constants for cytoplasmic receptors determined by the glass-binding method are similar for 5-, 10-, and 22-day-old rats (Clark and Gorski, 1969). Thus, the receptor observed during these developmental periods has the same characteristics as the uterine cytoplasmic receptor discussed previously.

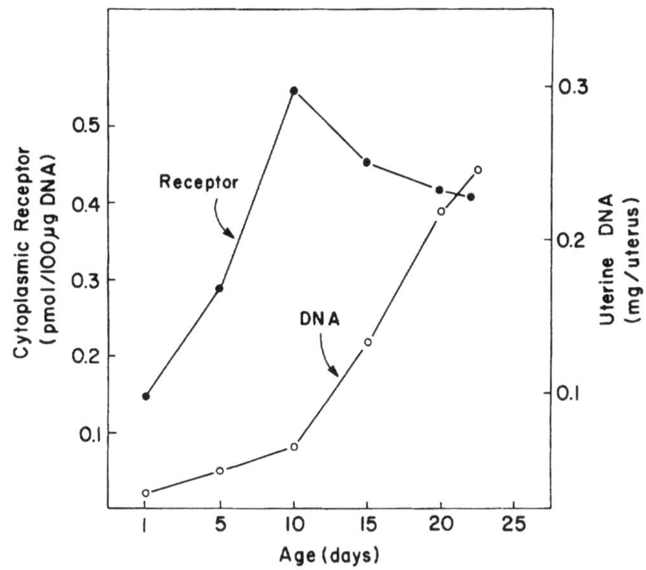

Fig. 99. Quantity of estrogen receptor in the rat uterus during early growth and development. Immature rats were killed at various ages and uterine DNA and cytoplasmic estrogen receptor were determined (Clark and Gorski, 1970)

To ascertain whether the observed increase in cytoplasmic receptor per cell was dependent on the presence of the ovary, 2-day-old rats were ovariectomized. The quantity of cytoplasmic receptor was determined when the rats were 10 days old. No significant differences existed between uteri of intact and castrate rats (Table 12). These data demonstrate that the observed increase in cytoplasmic receptor per cell during the first 10 days of life in the rat is not dependent on ovarian estrogen secretion. The decrease in cytoplasmic receptor per cell observed after 10 days probably reflects the onset of estrogen production by the rat ovaries at this time and the attendant translocation of cytoplasmic receptor to the nuclear compartment. After 8–10 days the ovary of the rat begins to secrete estrogens (Price, 1947; Falck, 1953; Presl et al., 1968), and as observed in these studies and others (Price, 1947) the uterus grows two–three times faster between days 10 and 20 than between days 1 and 10. Thus, the presence of R_c in large numbers per cell precedes this period of rapid uterine growth. Price (1947) also observed that castration of 1-day-old rats resulted in no differences in the growth of the uterus up to 8–10 days, but that after this period of ovarian independence further uterine growth was dependent on the presence of the ovaries. Also, the rat uterus is less sensitive to exogenous estrogen during the first 5–6 days of life than it is after 8–10 days (see below). This low sensitivity to estrogen at day 5 probably reflects the paucity of R_c per cell at this time. Thus, the number of sites per cell correlates with the ability of the uterus to respond to estrogen.

These observations imply that the synthesis of R_c is not dependent on estrogen but rather may be an autonomous property of uterine cells and that the receptor does not change during these developmental periods. Although other control factors from the hypothalamus, pituitary, and/or adrenals may be involved, this seems unlikely

Table 12. Comparison of estrogen receptor levels and uterine growth in intact and ovariectomized neonatal rats. Uteri were analyzed from 8–10 day old intact rats or rats which had been ovariectomized on day 2 of life. Values represent the mean ± the SEM.

Measurement	Intact	Ovariectomized
Cytosol receptor (pmol/uterus)	0.33 ± 0.03	0.32 ± 0.04 N.S.
Cytosol receptor (pmol/100 μg DNA)	0.53 ± 0.04	0.51 ± 0.05 N.S.
DNA/uterus (μg)	62.0 ± 5.6	60.3 ± 7.5 N.S.
Protein/uterus (μg)	202.0 ± 21.5	195.0 ± 19.6 N.S.
Wet wt. (mg)	10.2 ± 0.5	9.9 ± 0.6 N.S.

since Müllerian ducts will differentiate in culture and these ducts are responsive to estrogens (Price and Pannabecker, 1958).

As the number of uterine cells increases in response to ovarian estrogen after 10 days of life, uterine R_c also increases, and the ratio of R_c to cells remains relatively constant. Therefore the synthesis of receptor is coupled with an increase in cell number and is not directly related to estrogen induction. If the receptor is to interact with estrogen and produce further growth, it seems that estrogen receptors must develop prior to the time they are required and independently of estrogen induction. Subsequent to the initial developmental period, the quantity of estrogen receptor per cell probably remains constant; however, this point is not clear. Estrogen increases the quantity of R_c during cellular hypertrophy; thus, estrogens clearly influence the cellular concentration of cytoplasmic receptor. However, after cells have undergone hypertrophy, they divide and probably reestablish the original cytoplasmic receptor concentration. Whether or not this is generally the case remains unclear because these fluctuations in R_c levels have not been measured in different cell types. It is clear, however, that the uterus in the ovariectomized rat remains responsive to estrogens and contains estrogen receptors. Hence the likelihood that receptor synthesis is a constitutive property of uterine cells is high. This has also been suggested for the androgen receptor in the prostate (Sullivan and Strott, 1973).

The changing level of R_c/cell with age should provide a good model to test the relationship between receptor quantities and the ability of the uterus to respond. That is, one might anticipate that the ability of the uterus to respond to estrogen would correlate with the quantity of R_c as it increased during early development. The ability of the uterus to respond in the 2-day-old rat is reported to be minimal; i.e., stimulation of true uterine growth by estradiol was not observed at this time. As R_c numbers increase with age, the ability of the uterus to respond increases; however, the exact response profile has not yet been eludicated (see Kaye, 1978, for review). This inability to respond to estradiol might result from the low levels of R_c observed, but alternatively could result from the high level of α-fetoprotein present in the neonate (see Sect. VIII.1). The uteri from 5-day-old rats can synthesize induced protein (IP) in response to estrogen (Walker et al., 1976; Katzenellenbogen and Greger, 1974), but increases in the quantity of RNA, DNA, or protein are not observed (Sömjen et al., 1973). On the other hand, Luck et al. (1973) obtained increases in wet weight, RNA, protein, and DNA in 7-day-old-rats following estrogen. These conflicting results may stem from strain differences and have been reviewed by Kaye (1978). At present only one generalization concerning these developmental studies can be made: the ability of the uterus to respond to estrogen correlates with the quantity of cytoplasmic receptor per cell. Many other factors including the declining levels of α-fetoprotein during development (see Sect. 1 this chapter), developmental processes that are not dependent on estrogen-receptor interaction, i.e., blood – vascular changes, pituitary – gonadal maturation, and other unknown factors, complicate this question to a considerable extent.

b) Estrogen Receptors During Avian Oviduct Development

Oviduct development has been studied in the chicken and, in contrast to the situation in the mammal, the avian oviduct appears to depend on the presence of estrogen for its development (Teng and Hamilton, 1968; Teng and Teng, 1975, 1976). In most birds the right Müllerian duct regresses in the female, and the left duct develops into the oviduct and shell gland of the mature bird. The development of the left oviduct is an estrogen-dependent phenomenon. Teng and Teng (1975, 1976) have shown that the quantity of cytoplasmic receptor increases from day 8 to day 12 of incubation and that the quantity of nuclear receptor-estrogen complex increases dramatically between days 10 and 18. This nuclear translocation of receptor probably results from endogenous estrogens (Weniger, 1968). Although the development of estrogen receptors is undoubtedly required for estrogens to stimulate full development of the left Müllerian duct, the dependency of the developmental acquisition of cytoplasmic receptor on estrogens remains unresolved.

3. Steroid Receptor Binding During the Reproductive Cycle

The assumption that steroid hormone binding and response are related is difficult to prove; however, certain predictions and corollaries can be made if this assumption is correct. Fluctuation of free hormone concentration in the blood should be accompanied by concomitant receptor binding and target tissue stimulation. Since blood levels of a steroid hormone during various reproductive states are well known, it should be possible to correlate receptor occupation with these levels under various physiologic conditions. It should be noted, as pointed out earlier, that blood levels represent total, not free, hormone and hence that blood levels do not always correlate directly with available free steroid.

a) Receptor Binding in the Ovary

Many investigators have shown that the ovary is responsive to exogenous estrogens (Payne and Runser, 1958; Bradbury, 1961; Smith and Bradbury, 1961). The ability of ovarian cells to bind ^{3}H-estradiol has been demonstrated by dry-mount autoradiography (Ullberg and Bengtssom, 1963; Stumpf, 1969; Nakai et al., 1972) and by selective uptake of ^{3}H-estradiol (Saiduddin, 1971; Saiduddin and Milo, 1974; Norris and Kohler, 1974). The ovary of the immature 21–22 day old rat contains estrogen receptors similar to those of the uterus. The K_d values for both R_c and R_n are not significantly different from one another or from that of other estrogen receptors (Saiduddin and Zassenhaus, 1977a). Reiter et al. (1972) suggested that estrogen controls the rate of ovarian maturation and that ovarian estrogen receptors probably mediate this developmental response. DES treatment of the hypophysectomized female rat stimulates hyperplasia of granulosa cells, which is accompanied by an increased binding of FSH by the ovaries (Goldberg et al., 1972).

The recent work of Richards (1975), as summarized by Richards and Midgley (1976), clearly shows that the estrogen-induced maturation of follicular development involves complex mechanisms and does not represent a simple case wherein estradiol sensitizes follicular cells to gonadotropin stimulation. Estradiol treatment of the hypophysectomized rat does not increase the number of FSH receptors *per granulosa cell*; however, it does increase the number of its own receptors. Estrogen also increases the number of granulosa cells and hence elevates the quantity of FSH binding by the ovary. FSH treatment likewise increases the quantity of FSH receptors. Both hormones in concert increase LH receptors of granulosa cells and hence act to enhance the sensitivity of these cells to the ovulatory effect of LH.

Follicular atresia is associated with the loss of receptors for estradiol, FSH, and LH. The causes remain unknown; however, an intriguing suggestion by Ross and his co-workers relates follicular atresia with ovarian androgens (Louvet et al., 1975). Payne and Runser (1958) demonstrated that androgens increase follicular atresia, and Louvet et al. (1975) showed that antiandrogens or testosterone antiserum reversed hCG induced atresia in DES-treated rats. Schreiber et al. (1976) and Schreiber and Ross (1976) have recently described a testosterone receptor in the ovaries which probably mediates this atresia.

b) Steroid Receptors in the Uterus and Pituitary

1) Estrogen Receptor

The quantity of nuclear-bound receptor should reflect the fluctuation of estrogen in the blood during the estrous cycle. We have examined this proposal in the uterus of the adult rat, and, as shown in Fig. 100, the quantity of R_nE does vary in parallel with blood levels of estradiol during the estrous cycle.

These data were obtained with the ^{3}H-estradiol exchange assay in rats with 4-day cycles (Clark et al., 1972a), and the blood values for estradiol were taken from Butcher et al. (1974). The number of R_nE complexes is at a minimum during estrus and metestrus (~ 1,000 sites/cell) and increases between metestrus and diestrus (~ 3,500 sites/cell) and reaches a maximum at proestrus (~ 5,000 sites/cell). Similar observations have been made in the oviduct and uterus in the cat (West et al., 1976, 1977). The quantity of R_nE in both organs was closely correlated with ciliation and cell height in the oviduct during the estrous cycle.

Uterine weight, protein content, and the ratio of protein to DNA are all significantly higher in proestrus than in metestrus or diestrus, thus suggesting that fluctuations in protein synthetic activity of the uterus occur throughout the estrous cycle. Thus, maximal estrogenic responses are accompanied by peak concentrations of R_nE complex in the proestorus uterus.

Similar fluctuations in the quantity of R_nE were observed in the pituitary during the estrous cycle (Clark et al., 1973a). The concentrations of R_nE were not significantly different at metestrus, diestrus, or estrus but showed a highly significant increase at proestrus. Recently, Ogren and Woolley (in press) have made similar observations in the pituitary, preoptic and medial basal hypothalamus, and amygdala.

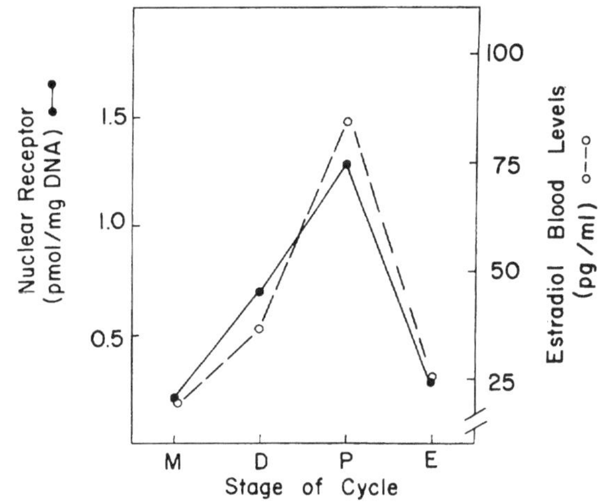

Fig. 100. The relationship between levels of nuclear receptor-estrogen complex and estradiol blood levels during the estrous cycle. Mature rats were killed at various stages of the estrous cycle (*M*, metestrus; *D*, diestrus; *P*, proestrus; *E*, estrus) and the quantity of nuclear estrogen receptor was determined by the ^{3}H-estradiol exchange assay (●). The concentrations of estradiol in the blood (O) were taken from Butcher et al. (1974) and the receptor data from Clark et al. (1972a)

The accumulation of receptor by the nucleus of pituitary cells is accompanied by a depletion of R_c^E (Greeley et al., 1975; Ginsburg et al., 1975). These investigators also observed similar depletion of R_c^E in the hypothalamus during proestrus, and Parker et al. (1976) have suggested that this depletion is associated with the preovulatory surge of gonadotropins which results in ovulation. Thus, depletion of cytoplasmic receptor and nuclear accumulation of R_nE complexes are probably important steps by which estrogens control gonadotropin secretion.

The cyclic fluctuations in R_nE during the estrous cycle probably reflect alterations in the level of estrogen in the blood. However, the concentration of R_c^E in the uterus presents a more complex picture, which probably involves an interplay with several other hormones, especially progesterone. The total quantity of cytoplasmic receptor was measured by exchange during different stages of the estrous cycle. As shown in Fig. 101, no significant differences in the content of cytoplasmic receptor was found during diestrus, proestrus, or estrus, whereas a significant decrease was observed at metestrus. In order to rule out the possibility that the observed reduction of cytoplasmic receptor was due to a reduced affinity of cytoplasmic receptor for estrogen, the dissociation constant of the receptor-estrogen complex was determined for each stage of the cycle: metestrus, 0.87 ± 0.11 n*M*; diestrus, 0.97 ± 0.11 n*M*; proestrus, 0.97 ± 0.15 n*M*, and estrus, 1.21 ± 0.20 n*M*. None of these values are significantly different. Ginsburg et al. (1975) observed a similar variation in R_c^E during the cycle. Other investigators have not observed precisely the same variations in cytoplasmic receptor levels (Shain and Barnea, 1971); however, in this study only unoccupied receptors were measured, and hence the reported values of R_c are not comparable to those of our studies.

As shown in Fig. 100, nuclear levels of estrogen receptor vary with blood levels of estrogen and may constitute a significant portion of total uterine receptor. These nuclear values have been included in Fig. 101 to show the variation in total receptor

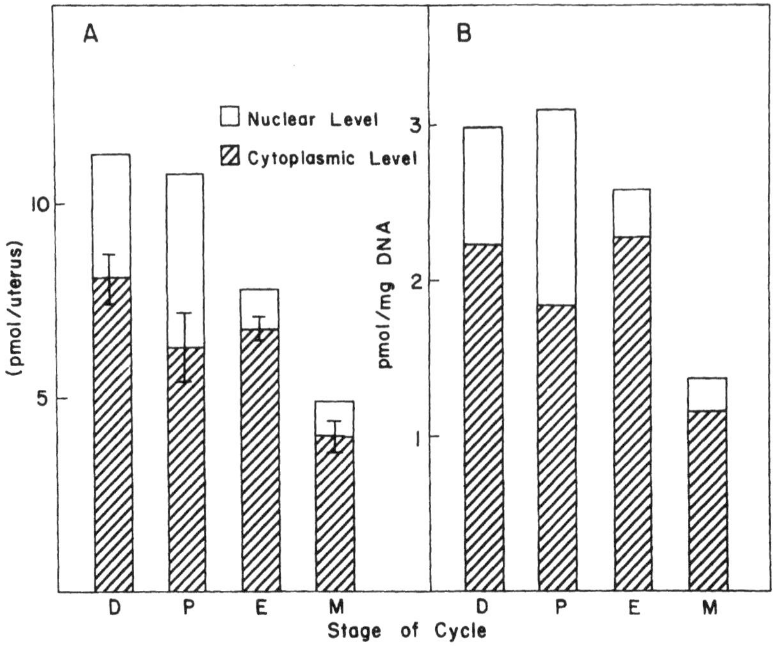

Fig. 101 A and B. Concentration of estrogen receptor in cytoplasmic and nuclear fractions of rat uterus during the estrous cycle. Uterine tissues were taken from rats on the days of proestrus (*P*), estrus (*E*), metestrus (*M*), or diestrus (*D*). Cytoplasmic and nuclear fractions were prepared, and the [^{3}H]-estradiol exchange assay was used to measure estrogen receptors in both. (Clark et al., 1977b)

content during the estrous cycle in the rat. Total uterine receptor content or receptor content per mg DNA reached high levels during diestrus and proestrus and decreased to the lowest level at metestrus.

The reason for the fluctuations in total receptor in the uterus during the cycle is not fully understood; however, as pointed out in Chap. VI progesterone is clearly involved in decreasing the level of estrogen receptors. The blood levels of progesterone increase on the eve of proestrus (Hori et al., 1968; Hashimoto et al., 1968; Shaikh, 1971; Butcher et al., 1974; Kennedy and Armstrong, 1975; Freeman et al., 1976) while estrogen decreases to low levels. The combined effect of these alterations in estrogen and progesterone may function to modulate the quantities of cytoplasmic receptors, resulting in the reduced levels observed during metestrus. As pointed out in Chap. VI other investigators have noted that progesterone lowers the levels of R^E_c. Brenner et al. (1974) showed that progesterone reduced the level of R_c dramatically during artificial menstrual cycles in the monkey. The quantity of unoccupied R^E_c was very high during the proliferative phase of the cycle and declined during the luteal phase. This decline in R_c occurs in the presence of blood levels of estradiol that do not differ from those of the proliferative phase, hence the receptor decline is not due to estradiol withdrawal. This effect of luteal phase progesterone has also been suggested to operate during the human menstrual cycle (Brush et al., 1967; Evans and Hähnel, 1971) and during the estrous cycle of the cat, hamster, and sheep (West et al., 1976, 1977, 1978; Koligian and Stormshak, 1977).

2) Progesterone Receptor

Since the biologic and physiologic interactions of the progesterone receptor in the mammalian uterus have been reviewed recently by Leavitt et al. (1978), we will not attempt a comprehensive review of this topic. The level of cytoplasmic progesterone receptor varies during the estrous cycle in all species, and estrogen appears to control its synthesis (see Chap. VI for discussion of this point). During the follicular phase of the cycle the level of cytoplasmic progesterone receptor is relatively low, and as estrogen blood levels increase the quantity of receptor is elevated (Milgrom et al., 1972b; Leavitt et al., 1978). This elevation in progesterone receptor is probably a requisite for the subsequent actions of progesterone during pregnancy or the luteal phase of the cycle.

As discussed in Chap. VI, elevated levels of serum progesterone cause the usual depletion of cytoplasmic progesterone receptor and the accumulation of nuclear receptor-progesterone complexes, but differs from estrogen in that an eventual decline in total progesterone receptor occurs after progesterone injection. This negative effect of progesterone on its own receptor was first observed by Milgrom et al. (1972b) in the guinea pig. Cytoplasmic progesterone receptor is elevated just before the preovulatory peak in plasma progesterone, probably as a result of the positive effects of estrogen on progesterone receptor synthesis. Cytoplasmic levels of progesterone receptor subsequently decline as the corpus luteum produces high levels of progesterone. The receptor also remains low during pregnancy in the guinea pig (Milgrom et al., 1972b). Similar results were observed during the estrous cycle in the hamster, rat, and mouse (Leavitt et al., 1974; Gomez et al., 1977; Philibert and Raynaud, 1977). This effect also occurs in human endometrial tissue (Pollow et al., 1977). The quantity of cytoplasmic progesterone receptor increases gradually during the menstrual cycle, reaching high levels at mid cycle. This peak is followed by a decline during the luteal phase.

The decrease in cytoplasmic progesterone receptor under the influence of elevated progesterone in the blood is not the result of nuclear accumulation (Walters and Clark, 1977a, b and see Chap. VI), but appears to be an actual decline in the total number of receptors present. This suppression of cytoplasmic progesterone receptor probably results from the serial inhibition phenomenon discussed in Chap. VI; i.e., progesterone, via its inhibition of estrogen receptor synthesis, suppresses the synthesis of its own receptor. This decline may alternatively result from a combination of low estrogen levels and a redistribution of receptors among specific cell types that constitute a small number of the total cell population. Although progesterone receptors have been described in both endometrial and myometrial tissues of guinea pig, sheep, and human (Kontula, et al., 1972, 1974a, b, 1975; Luu et al., 1975; Warembourg and Milgrom, 1977), this does not rule out the possibility of differential cell effects of progesterone.

We conclude that estrogen acts to promote uterine growth and to increase the tissue content of progesterone receptor, whereas progesterone antagonizes the action of estrogen by decreasing tissue levels of estrogen receptor. Thus, the interaction of these ovarian hormones at the receptor level provides a basis for the cyclic changes observed in uterine tissues during the estrous cycle of the rat.

c) Steroid Receptor Binding During Pregnancy and Lactation

1) Uterus

The fluctuations in number and compartmentalization of uterine steroid hormone receptors during pregnancy have received little study. The complexities of examining each of the various organs and tissues involved as pregnancy progresses obviously makes such work very difficult. One time period in which such studies are not complicated by the presence of the placenta is the first few days after conception in the rat. This period is of special interest because of the estrogen-dependent implantation of the blastocyst occurring at this time (Shelesnyak, 1959; Zeilmaker, 1963; Nutting and Meyer, 1974; Yoshinaga et al., 1969; Glasser and Clark, 1975). Blood levels of estradiol show a transient elevation in the pregnant rat between day 1 and day 4 of pregnancy. At this time the quantity of uterine R_nE increases significantly (days 2 and 3 of the implantation period) and then declines (on day 4) (Glasser and Clark, 1975). These data suggest that elevated blood levels of estradiol cause nuclear accumulation of the R_nE complex, which in turn may stimulate those events that lead to blastocyst implantation. The reduction of estrogen receptor levels may also be linked to rising progesterone levels at this time. We have previously discussed (see Chap. VI and Hsueh et al., 1976) the ability of progesterone to decrease the quantity of cytoplasmic estrogen receptor. It seems likely that both declining levels of estrogen and increasing titers of progesterone bring about this effect. Mester et al. (1974) have observed a peak in cytoplasmic estrogen receptors in the endometrium, but not the myometrium, during the time of implantation. They suggest that progesterone may differentially inhibit the level of estrogen receptor in the myometrium and consequently may decrease the capacity of estradiol to alter this tissue. Thus, elevated cytoplasmic estrogen receptor levels in the endometrium and decreasing levels in the myometrium would favor implantation.

Csapo and Wiest (1969) and Wiest (1970) have shown that pregnancy in both the rat and rabbit depends on elevated progesterone levels in the uterus. Wiest (1970) demonstrated that the uterus of the pregnant rat was capable of concentrating progesterone above peripheral blood levels throughout early and mid pregnancy. Late pregnancy (day 20–22) was associated with sharp declines in uterine progesterone levels and with the onset of parturition. Milgrom et al. (1977) have shown that the quantity of cytoplasmic progesterone receptor gradually increases during pregnancy in the rat to very high levels, while nuclear levels of receptor accumulate during pregnancy and then decline just prior to parturition. The number of cytoplasmic sites available for nuclear binding is very large when compared to the quantity of nuclear sites measured. The reason for this distribution is not clear since during pregnancy blood levels of progesterone are very high. Either all receptor sites were not measured in these studies or the metabolic conversion of progesterone to 5α-reduced steroids may result in these apparent discrepancies. Howard and Wiest (1972) showed that the ability of the rat uterus to form 5α-pregnane-3,20-dione and 3α-hydroxy-5α-pregnan-20-one increased substantially between day 11 and 21 of pregnancy. Thus, although blood levels of progesterone are high at these times, tissue levels may be much lower. This metabolic sequence could produce a cellular environment wherein the concentration of progesterone is gradually lowered during pregnancy and hence less R_nP is formed. Elevated levels of cytoplasmic progesterone receptor during the last few days of pregnancy in the rat could be due to elevated levels of estrogen that occur at this time (Yoshinaga et al., 1969). The decline in nuclear levels of progesterone receptor before birth probably constitutes an important mechanism, which decreases the control of the uterus by progesterone and increases its sensitivity to estrogen. This shift would provide a nonquiescent uterus capable of contracting and would result in parturition.

Progesterone receptors have been studied in other species during pregnancy: guinea pig (Milgrom et al., 1972b), hamster (Leavitt et al., 1974; Leavitt et al., 1978), and mouse (Philibert and Raynaud, 1977). In contrast to the studies in the rat discussed above, levels of cytoplasmic progesterone receptor either change very little, as in the mouse, or are depressed, as in the guinea pig. Obviously additional work is required to understand these species differences.

2) Ovary

The corpus luteum of the rabbit is known to require estrogen for its maintenance (Rennie, 1968a, b) and contains cytosol estrogen receptors similar to those found in other tissues (Scott and Rennie, 1971). Several observations suggest that the control of cytoplasmic estrogen receptor levels by luteolytic factors from the uterus may be important in the maintenance of corpus luteum function. Lee et al. (1971) showed a decrease in the quantity of luteal R_c in pseudopregnant rabbits prior to a decline in luteal weight. In addition, it is well known that hysterectomy can extend luteal life in the rabbit by approximately 8 days, thus implying that a uterine luteolytic factor normally terminates the life of the corpus luteum (Hilliard et al., 1974). Kimball and Hansel (1974) and Knobil (1974) have suggested that estradiol itself is luteolytic in the cow and primate. Miller and Keyes (1976) showed that estradiol treatment could extend pseudopregnancy only in the hysterectomized rabbit and suggested that a uterine factor(s), possibly prostaglandins, may control the level of cytoplasmic estrogen receptor. If so, this could provide an important mechanism for the differential control of estrogen responsiveness, i.e., the negative modulation of estrogen responsiveness in the corpus luteum, while other tissues remain fully responsive.

The corpus luteum of the pregnant rat has been studied by Richards (1974), who used the ^{3}H-estradiol exchange assay to show an elevation of R_nE between days 3 and 12 of pregnancy. Following day 12 the total amount of receptor per cell gradually declined; this was not due to a decline in endogenous estrogen (Richards, 1975). Estrone and estradiol increase toward the end of pregnancy; therefore, the decline in R_nE is not likely due to decreased levels of estradiol. However, since the ovary is the source of estradiol, blood measurements need not reflect the concentration of estrogen in the corpus luteum. In addition, Richards (1975) has noted that injections of estradiol in hypophysectomized pregnant rats cause maintenance of cytoplasmic estrogen receptors in luteal tissue. The relationship between these observations and luteotropic or luteolytic control by estradiol is not clear at present.

3) Mammary Gland

The mammary gland is a target organ for various hormones, and steroids stimulate the development and growth of this tissue. Several investigators have demonstrated that the mammary gland will accumulate estrogenic compounds in a specific manner (Pearlman et al., 1966; Sander, 1968a, b; Sander and Attramadal, 1968; Puca and Bresciani, 1969). As expected, a cytoplasmic estrogen receptor similar to that of the uterus exists in lactating and tumorous mammary tissue (King et al., 1969; Shyamala and Nandi, 1972; Jensen et al., 1972). Progesterone receptors also exist in normal mammary gland (Horwitz and McGuire, 1977; Leclercq et al., 1977; Markland and Hutchens, 1977; Rao and Meyer, 1977; Wittliff et al., 1977); however, they have not been studied as a function of the reproductive state. Shymala and Nandi (1972) have demonstrated that cytoplasmic estrogen receptors of the mammary gland of the lactating rat undergo translocation to the nucleus in a fashion similar to that observed in the uterus (Shyamala and Gorski, 1967; Jensen et al., 1968; Giannopoulos and Gorski, 1971a). Wittliff et al. (1972a) confirmed the presence of cytoplasmic estrogen receptors in the lactating mammary gland and suggested that levels of receptor vary with the differentiation of the mammary gland. These cytosol receptors have the same general physical characteristics as those of other tissues (Wittliff et al., 1972a).

The major difficulty in receptor detection in the mammary gland is the relatively low numbers of receptors per unit mass of tissue. This probably results from a small number of receptor-containing cells in the presence of a large number of cells that contain no receptor. Sar and Stumpf (1976) showed that only the epithelial cells of the mammary gland accumulated ^{3}H-estradiol while other cells do not. This problem has made the study of nonlactating mammary tissue very difficult, and hence information on this system is limited. The quantity of cytoplasmic estrogen receptor is very low in the mammary gland of the prepubertal mouse; however, it is detectable at the time of puberty as a 4*S* macromolecule (Hunt and Muldoon, 1977; Muldoon, 1978). During pregnancy the receptor levels increase, and the sedimentation pattern gradually shifts from the 4*S* form to an 8*S* form (Muldoon, 1978). These qualitative and quantitative changes appear to be controlled by prolactin (Muldoon, 1978). Several investigators have shown that prolactin does elevate the level of estrogen receptors in various tissue (Leung and Sasaki, 1973; Vignon and Rochefort, 1976; Muldoon, 1978). However, it is also known that the cellularity of the gland changes during pregnancy, and this change could simply produce more epithelial cells/unit mass of tissue.

We have partially characterized the estrogen receptor during lactation in the rat. In agreement with the studies mentioned above, cytoplasmic and nuclear receptors are similar to those of other target tissues (Hsueh et al., 1973). Cytoplasmic receptor as well as R_nE were examined on days 1, 10, and 21 of lactation. As shown in Fig. 102 the concentration of cytoplasmic receptor increases dramatically by day 10 and is even higher by day 21. Increases in cytoplasmic receptor during lactation in the rat have also been observed by Leung et al. (1976). This is in contrast to R_nE complexes, which remain low throughout lactation. This is anticipated because estrogen levels in the blood are very low during lactation (Hsueh et al., 1973). These low quantities of R_nE do not result from a failure of the translocation process since an injection of estradiol promotes the accumulation of R_nE to levels similar to those found in the cytoplasm before injection.

The elevation of cytoplasmic receptor number in lactating tissue does not depend on the presence of the ovary. This was demonstrated in animals that were castrated on day 1 of lactation and assayed for mammary receptor on day 21 (Fig. 103). The ovarian independence of this phenomenon is similar to that observed for the neonatal rat uterus (Table 12, Clark and Gorski, 1970); however, elevated levels of prolactin in the pregnant animal probably make these physiologic states very different. The

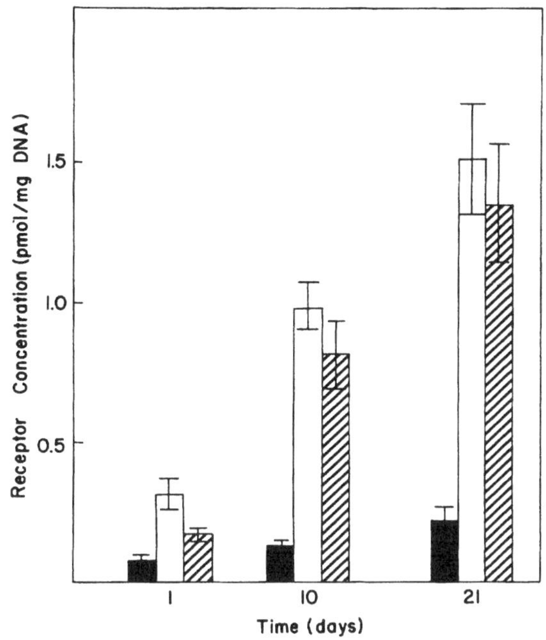

Fig. 102. The concentration of cytoplasmic and nuclear estrogen receptor in the rat mammary gland during lactation. The quantities of cytoplasmic (*open bars*) and nuclear (*solid bars*) estrogen receptor were determined in saline-injected rats at various times during lactation. The quantity of nuclear estrogen receptor 1 h after an injection of 25 µg estradiol is given by the *cross-hatched bars* (Hsueh et al., 1973)

small but significant difference in the concentration of R_nE in the intact lactating rat as compared with the ovariectomized rat (Fig. 103) indicates that ovarian estrogens are involved in the translocation process. Ovariectomy removes the primary source of estrogen in the blood and in turn decreases the quantity of nuclear receptor-estrogen complex. Thus, the small number of R_nE complexes in the ovariectomized rat indicates that estrogens of nonovarian origin, perhaps adrenal estrogens, are present. The sensitivity of the [^{3}H]-estradiol exchange assay is demonstrated by these studies. This method could be used to detect even slight alterations in the concentration of estrogen in blood.

A physiologic role for the increased concentration of cytoplasmic estrogen receptor during lactation is difficult to assess. Bruce and Ramirez (1970) have shown that exogenous estrogens inhibit the lactating mammary gland; the estrogen receptor may be involved in this response. However, our results suggest that the concentration of estrogen in the blood of the lactating rat must be very low since the concentration of cytoplasmic estrogen receptor is high. Therefore, it seems unlikely that the inhibitory effect of estrogen on mammary function has a physiologic role in the lactating rat.

4) Placenta and Fetus

Estrogen receptors are present in the maternal placenta during pregnancy in the rat (Feherty et al., 1970). McCormack and Glasser (1978) have shown that the basal zone of the rat placenta contains large numbers of cytoplasmic estrogen receptors during a limited period of mid pregnancy. The number of cytoplasmic receptors on day 9 of pregnancy is 30,000/cell while levels of R_nE are relatively low (3,400/cell). This distribution of receptors is probably due to the low concentration of estrogens in the blood at this time. However, the reason(s) for such high levels of cytoplasmic

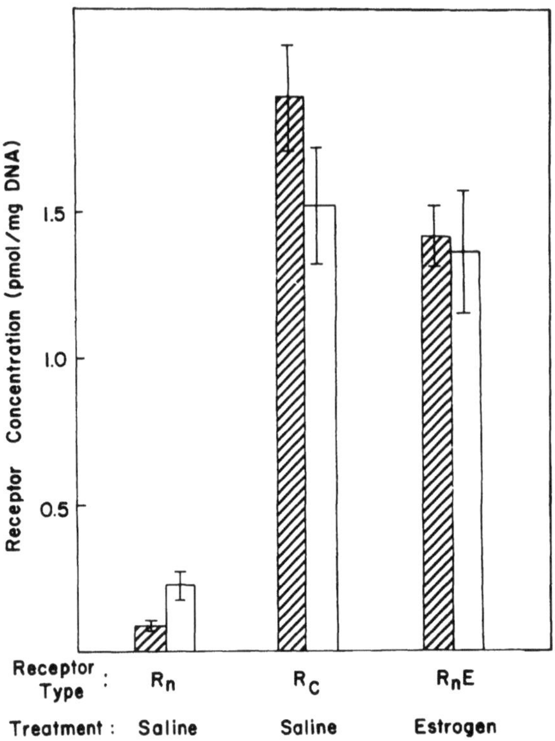

Fig. 103. Comparison of the quantity of cytoplasmic and nuclear estrogen receptor in ovariectomized and intact lactating rats. Intact (*open bars*) rats or rats ovariectomized on day 1 of lactation (*cross-hatched bars*) were injected on day 21 of lactation with saline or 25 μg estradiol. The quantity of estrogen receptor in cytoplasmic and/or nuclear fractions was determined 1 h later (Hsueh et al., 1973)

receptor during early pregnancy in this tissue are unknown. By day 15 of pregnancy the levels of receptor fall dramatically to 600 and 200 sites/cell for cytoplasmic and nuclear receptors, respectively. The reason for such a dramatic decline in estrogen receptor level in this tissue during this time period is not evident. However, the secretion of progesterone by trophoblast giant cells may bring about this decline. The effect of progesterone on the modulation of cytoplasmic estrogen receptors has been discussed in Chap. IV. It seems likely that progesterone could control the level of estrogen receptors by antagonizing the replenishment process.

Pasqualini et al. (1976) have demonstrated specific estrogen binding sites in the fetus of the guinea pig. The uterus, testes, kidney, and lung also contained significant quantities of estrogen binding sites while other organs, such as the heart, did not. The quantities of cytosol and nuclear receptor gradually increased during pregnancy to reach a maximum at term. After parturition cytoplasmic receptor remained high while nuclear complexes decreased significantly. Since these experiments did not use exchange conditions, exact receptor distributions and concentrations are unknown; however, it seems that several fetal organs contain specific receptor-like sites, the function of which is also unknown. They may represent a binding macromolecule that "protects" fetal organs from estrogens and hence may not be receptors at all.

Progesterone binding in the fetus and placenta has received little attention. Leavitt et al. (1978) have reported that no progesterone binding sites exist in these tissues in the hamster. The effective level of progesterone in the fetal environment is low, therefore the fetus may react as though progesterone had been withdrawn. The recent observations of Schwartz et al. (1976, 1977), however, offer intriguing evidence for an involvement of specific progesterone binding sites in the onset of parturition. These investigators have shown that the capacity of the amnion to bind progesterone increases dramatically near term in the human and have suggested that this results in withdrawal of progesterone from the "active" compartment. Thus, although progesterone blood levels may be high near term, the ability of the fetal tissues to respond to progesterone is diminished. Although highly speculative and offered by Schwartz et al. (1976) only as a hypothesis, the above serves as an example of steroid binding sites that are not receptor-like and may serve a protective function.

Chapter IX

Gonadal Steroids and Brain Function

Evidence suggesting that the brain is a target for gonadal steroids has existed for more than a century, since the experiments of Berthold on castration and gonadal transplantation in roosters (Berthold, 1849). Although of great importance to our understanding of sexual behavior and reproduction, progress toward establishing and understanding the relationship of steroids to the central nervous system was nonexistent until the late 1950s or early 1960s. At that time a series of implant, lesion, and stimulation experiments were conducted in a number of laboratories, all of which pointed to certain regions of the hypothalamus, limbic system, and midbrain as specific targets for gonadal steroids. Since steroid targets possess specific receptors responsible for their responses to hormonal stimuli, these can be assumed to exist in central nervous system, as well as peripheral, targets. These receptors will be the subject of this chapter. In addition we will examine some biochemical and physiologic responses that relate temporally to changes in subcellular localization of these receptors (i.e., to the translocation of R_c to the nuclear compartment to form an R_nS complex). This subject has been reviewed recently by several investigators (McEwen, 1978; Feder et al., 1978; Zigmond, 1975). Certainly we will not review the field of steroids and central nervous system function as extensively or intensively as these authors. Instead we will use their conclusions as a framework on which to base some of our own observations on alterations in brain function with steroid treatment.

1. Distribution of Targets Within the Central Nervous System

a) In Vivo Uptake and Retention of ^{3}H-Steroid

To study the regional distribution of estrogen receptors and thereby determine targets for this steroid among central nervous system structures, a number of investigators have injected ovariectomized rats with ^{3}H-estradiol with or without excess unlabeled steroid (Attramadal, 1964; Eisenfeld and Axelrod, 1965; Kato and Villee, 1967a; Eisenfeld, 1967; McGuire and List, 1969; McEwen and Pfaff, 1970; Ogren and Woolley, 1978; for reviews see McEwen et al., 1972; Kato, 1973; Zigmond, 1975). After the peripheral administration of such steroid mixtures, preferential retention of label can be observed in the hypothalamus, preoptic area, amygdala, septum, and midbrain. In such studies cerebral cortex is often used as a control for the purpose of reporting ratios of retained steroid (such as hypothalamic ^{3}H-steroid per unit mass of tissue/cerebral cortical ^{3}H-steroid per unit mass of tissue). The use of competing unlabeled steroid in studying receptors and tissue uptake is an absolute necessity. Thus, Whalen and Maurer (1969) studied the accumulation of ^{3}H-estradiol in brain and uterus as a function of the estrous cycle but did not employ competition with unlabeled estrogen for the detection of specific or saturated sites. Because of low numbers of receptors in brain and high levels of nonspecific binding, the brain was not demonstrated as a target for estrogens. The authors concluded that if estrogen receptors exist in the brain, they must respond in a manner different from uterine receptors. However, Eisenfeld (1967) had demonstrated earlier that labeled estradiol was perferentially *retained* in the hypothalamus despite the fact that initial *uptake* was similar for both hypothalamus and cortex, suggesting the existence of high-affinity binding or retention sites for estradiol in the hypothalamus. Studies of competition with unlabeled estrogens as well as other steroids (testosterone, corticosterone, 17α-estradiol) also support the suggestion that retention in these areas of the brain results from the formation of receptor-steroid complexes. Shani et al. (1971) showed that not only estrogens but also phenothiazines can reduce the accumulation and retention of labeled estradiol in median eminence and anterior pituitary, but not in a nontarget region such as cerebral cortex. The implications of these observations have yet to be pursued. Kato and Villee (1967b) demonstrated that the retention function within central nervous system tissues could be "saturated." Thus, by injecting various doses of estradiol, retention in the anterior hypothalamus was shown to reach a maximum while uptake in cerebral cortical tissue was linearly related to dose. Taken together, these studies argue strongly for the existence of saturable, specific receptor sites localized within specific regions of the central nervous system. Since these are among our criteria for receptor identification, it can be stated that by 1970 there was strong evidence to support the existence of estrogen receptors within the central nervous system.

Using similar techniques it has not been possible to demonstrate saturable or preferential retention of progesterone in the hypothalamus. For progesterone the major site of uptake is the midbrain as shown for rats, guinea gips, and hamster (Whalen and Luttage, 1971a; Wade and Feder, 1972a; Luttage and Wallis, 1973). However, uptake of progestins appears to be inversely related to their polarity; i.e., the less polar the steroid, the more it is accumulated (Wade and Feder, 1972b). Such observations, the lack of saturability, and the relationship of accumulation to polarity, could suggest that specific progesterone binding sites do not exist in the central nervous system or that they occur in numbers too limited to allow detection by these relatively crude methods.

The experimental model used to examine progesterone receptors may also present a problem. Most of the studies discussed above were in the intact rat or guinea pig. However, adrenalectomy increases the tissue uptake of ^{3}H-progesterone by brain (Whalen and Luttage, 1971b; Whalen and Gorzalka, 1974). Since the adrenals are a source of progesterone as well as of glucocorticoids, adrenalectomy could either reduce serum levels of unlabeled progesterone, thereby effectively increasing the specific activity of the label, or alternatively, adrenalectomy might unmask glucocorticoid-binding sites to which progestins may bind. Complications of this type essentially invalidate the use of tissue uptake and retention studies for the demonstration of progesterone receptors. The question of their existence will be pursued in succeeding sections.

b) Autoradiographic Localization of Estrogen and Progesterone Receptive Sites

Early experiments employing autoradiography demonstrated that ^{3}H-hexestrol or ^{3}H-diethylstilbestrol was accumulated (as evidenced by grains on radiosensitive films) in cell bodies of certain regions of the central nervous system (Glasscock and Michael, 1962; Attramadal, 1964; Michael, 1964, 1965). More recent studies have shown that neurons, not glia, concentrate or retain the greatest amount of labeled steroid. In addition these neurons, as shown earlier via dissection and radiometric analysis, occur primarily within the so-called hypophysiotropic region (anterior, preoptic, and medial-basal hypothalamus) (Stumpf, 1968, 1970; Anderson and Greenwald, 1969; Warembourg, 1970 a, b; Pfaff and Keiner, 1973). Other regions with mechanisms for retaining steroid are the amygdala (Pfaff and Keiner, 1972; Stumpf, 1972), midbrain central gray (Pfaff and Keiner, 1973), ventral hippocampus (Stumpf, 1970; Pfaff and Keiner, 1973), and the spinal cord (Keefer et al., 1973) as determined by autoradiography following systemic administration of ^{3}H-estradiol. In a study of the retention of ^{3}H-estradiol as a function of time after injection, Anderson and Greenwald (1969) showed maximal grain counts in the preoptic-ventromedial-arcuate regions at 2 h post injection; by 4 and 6 h grain counts had decreased 30–40 percent and 40–70 percent, respectively; at 8 h postinjection, grain counts were no different than control. Thus, it would appear that receptor-estrogen complexes in this region are labile or short-lived, a point to be discussed below.

Saturable nuclear retention of ^{3}H-progesterone has been demonstrated in the hypophysiotropic zone via autoradiography (Sar and Stumpf, 1973; Stumpf and Sar, 1976). This uptake and retention was much less conspicuous, however, than that previously reported for estrogens. Since this technique does not exclude the interaction of progesterone with glucocorticoid receptors, it cannot be taken as a priori evidence of the existence of progesterone receptors in this region. The conclusive demonstration of receptors requires their study in vitro to be discussed below.

2. Cytoplasmic Receptors for Estrogens and Progestins in the Central Nervous System

The existence of cytoplasmic receptors for estrogens in the brain was first established by reports of Eisenfeld (1969, 1970) and Kahwango et al. (1969, 1970). These investigators showed that ^{3}H-estradiol at low concentrations associated with hypothalamic cytosol to form macromolecular complexes that could be separated from free steroid by gel filtration or sucrose density gradient centrifugation. The proteinaceous nature of these complexes was confirmed by their inactivation with chymotrypsin and papain; ribonuclease and deoxyribonuclease were without effect. In keeping with the nature of uterine estrogen receptors, an essential sulfhydryl group in the active site was suggested by the inhibition of complex formation by p-chloromercuriphenylsulfonic acid (Kahwanago et al., 1970). In addition, stereospecificity was evident from competition studies employing unlabeled estradiol and clomiphene (which displaced labeled estradiol from cytosol complexes) as well as progesterone and testosterone (which had no effect on complex formation).

Ginsburg et al. (1972) measured the fluctuation in available cytoplasmic receptors of hypothalamus and pituitary during the estrous cycle. The number of cytoplasmic receptors available in either tissue decreased dramatically on the day of proestrus, as expected if the binding of endogenous estrogens and translocation of receptor-estrogen complexes is a physiologic process. Their estimated K_d values, 0.35 n*M* and 0.26 n*M* for hypothalamus and pituitary receptors, respectively, is quite similar to that of others and did not vary throughout the cycle.

Other investigators have compared cytoplasmic estrogen receptors from the uterus, pituitary, and brain and found them to be quite similar. Thus, on the basis of sedimentation coefficient (8*S* in low-ionic strength media), pharmacospecifity, and affinity for ^{3}H-estradiol, these receptors from the various target organs seem identical. Fox (1975) has used DNA-cellulose to assay estrogen receptors from mouse preoptic/basal hypothalamus and demonstrated the high affinity expected for cytoplasmic estrogen receptors. Shown in Fig. 104 is a saturation and Scatchard analysis of the binding of ^{3}H-estradiol to cytoplasmic receptors prepared from the preoptic/basal medial hypothalamus (Kelner and Peck, unpublished). Using protamine sulfate to precipitate R_cE complexes and mild detergent washes to reduce nonspecific binding, it is possible to examine such receptors in a reproducible fashion. The measured K_d is ~ 10^{-10} *M*, and the number of high-affinity sites per hypothalamus is ~ 25 fmol. To date no comprehensive survey exists establishing that all estrogen receptors in each target region of the brain are the same. The possibility exists that differences in the physiologic action of estrogens on brain regions may result from differences in the character of their respective estrogen receptors. This remains an interesting question yet to be answered.

Progesterone receptors from central nervous system tissues have proven more difficult to demonstrate. Seiki and Hattori (1973) and Iramain et al. (1973) reported the existence of progesterone-saturable cytoplasmic binding sites which were not altered

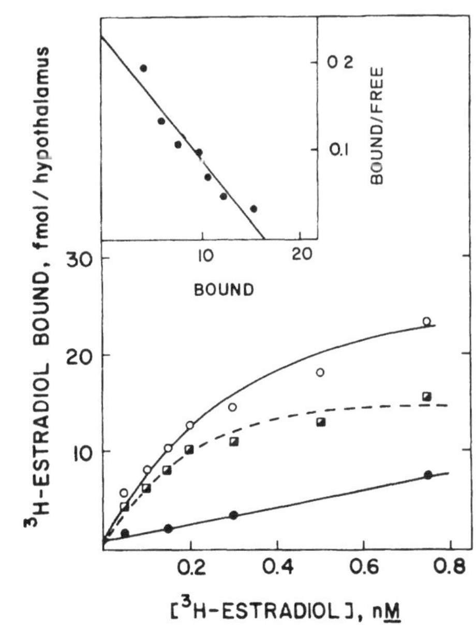

Fig. 104. Saturation and Scatchard analysis of cytoplasmic Receptors of the Hypothalamus. Hypothalamic cytosol from ovariectomized rats was incubated with various concentrations of ^{3}H-estradiol in the presence (●) or absence (○) of excess DES. Specific binding (□) was determined by difference and analyzed by the method of Scatchard (*inset*)

by glucocorticoids. On the other hand, McEwen and co-workers were unable to demonstrate a saturable progesterone receptor system using labeled progesterone (McEwen, 1978). Atger et al. (1974) also reported that saturable binding of ^{3}H-progesterone was not demonstrable in brain cytosols; however, the guinea pigs employed in the latter study were not adrenalectomized, which makes these negative results less than convincing.

These conflicting reports undoubtedly stem from very high levels of nonspecific binding, the interaction of glucocorticoids and progestins, and low levels of progesterone receptor in hypothalamic tissues. Very recently Kato and Onouchi (1977) demonstrated the existence of specific progesterone receptors in the hypothalamus and anterior pituitary of the rat by using the synthetic progestin, R5020. Using ^{3}H-R5020, these investigators demonstrated a specific, 7*S* receptor species present in estrogen-primed animals but not in unprimed ones. In addition, no specific receptor could be found in cerebral cortex, amygdala, or reticular formation. This receptor has a K_d of $\sim 10^{-9}$ *M* and is present in very low concentrations, about 10 fmol/mg protein. These results have been confirmed and extended by McEwen and co-workers (MacLuskey and McEwen, 1978) using ^{3}H-R5020. Thus, the use of a synthetic progestin has allowed the demonstration of estrogen-dependent cytoplasmic progesterone receptors, which are present in low numbers in the hypothalamus.

3. Translocation of Estrogen Receptors to Nuclear Compartments

Zigmond and McEwen (1970) injected ovariectomized rats with ^{3}H-estradiol and examined various subcellular compartments for radioactivity 2 h later. Purified nuclear fractions contained the highest concentration of label at this time, and the label largely cochromatographed with authentic estradiol. Preoptic-basal medial hypothalamus and amygdala retained the largest amounts of label in their nuclear fractions. Retention of ^{3}H-estradiol was maximal for the first 30 min – 1 h and had declined significantly by 4 h.

Chader and Villee (1970a, b) studied the subcellular localization of ^{3}H-estradiol after in vitro incubation of hypothalamus. After 1 h in the presence of 10^{-10} *M* ^{3}H-estradiol, the radioactivity extractable from various fractions was authentic estradiol localized primarily in the nuclear fraction, not the mitochrondrial, microsomal, or soluble fractions of the hypothalamus. Following dialysis of each fraction, only the nuclear fraction was capable of retaining radioactivity (Chader and Villee, 1970).

In 1972 we studied the interaction of ^{3}H-estradiol with the hypothalamus in vitro, using unlabeled diethylstilbestrol to control for nonspecific interactions (Clark et al., 1972b). Crude nuclear pellets were isolated from hypothalamus after incubation in varying concentrations of ^{3}H-estradiol with or without excess unlabeled diethylstilbestrol, and the radioactivity associated with this fraction was measured. In this way the apparent dissociation constant of estradiol and hypothalamic receptor in situ was estimated to be $\sim 10^{-9}$ *M* with 10–50 fmol of sites per hypothalamic equivalent.

Using the immature female rat and the ^{3}H-steroid exchange procedure, we subsequently studied the accumulation and retention of R_nE following an injection of estradiol (Anderson et al., 1973b). As shown in Fig. 105 physiologic doses

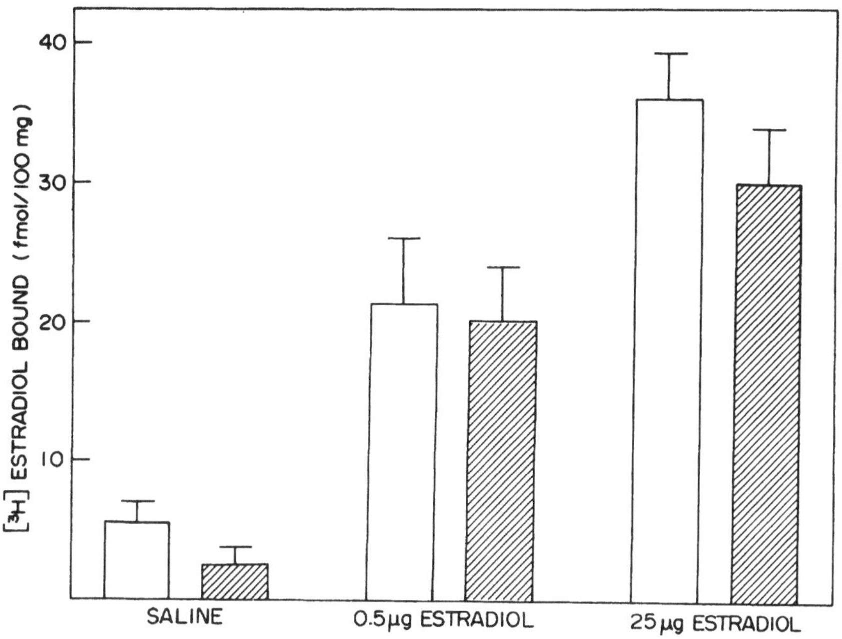

Fig. 105. Accumulation of receptor estradiol complexes by nuclear fractions of immature rat hypothalami; 21-day-old female (*open*) and male (*hatched*) rats were injected with saline or 0.5 or 25 µg estradiol and killed 1 h later. Hypothalamic nuclei were prepared and R_nE measured by ^{3}H-estradiol exchange. (Anderson et al., 1973b)

of estradiol (0.5 µg) increased the number of measurable nuclear complexes from ~ 5 fmol/100 mg tissue to about 20 fmol/100 mg tissues. Pharmacologic doses of estrogen (25 µg estradiol) resulted in a maximum of 30–40 fmol nuclear complex/100 mg tissue, similar to our previous estimates via uptake and retention of ^{3}H-estradiol in vitro (see above). We estimated a dissociation constant for R_nE complexes of about 0.4×10^{-9} *M*. The time course for nuclear accumulation and retention of receptor estrogen complexes was similar to that observed for the uterus (see Chap. V); thus, maximal levels of R_nE were observed within 1 h and remained above those of saline-injected controls for at least 6 h. With respect to region, the majority of the receptor was found in the basal hypothalamus-median eminence region, a tissue of ~ 2–3 mg wet wt. immediately adjacent to the pituitary stalk (Fig. 106). In this region concentrations of receptor ranged up to 250 fmol/100 mg tissue. It should be noted, however, that this area includes hypophyseal cells of the pars tuberalis known to contain estrogen receptors. Thus, these high levels of receptor may reflect the receptor content of cells of other than central nervous system origin.

The studies discussed above employed the immature female rat. Unfortunately, the study of R_nE complexes in the adult animal by ^{3}H-steroid exchange has proven a more difficult task. The major problem has been the number of nonspecific binding sites present in crude nuclear fractions of the adult hypothalamus, probably the result of contamination of this fraction with glial membrane. To circumvent this difficulty, we have recently adopted new procedures for exchange studies of hypothalamic nuclei. Protamine sulfate precipitation of nuclear-chromatin fractions combined

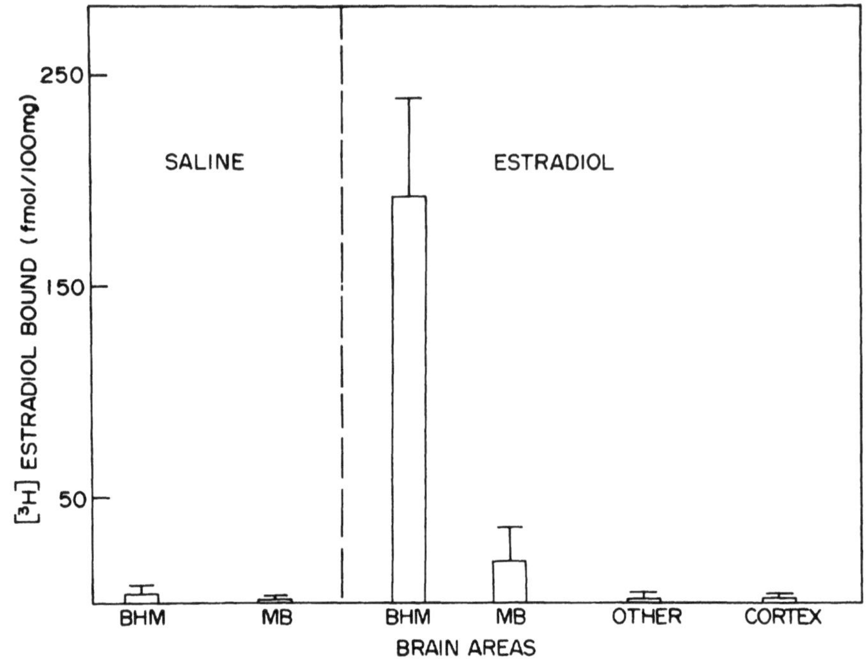

Fig. 106. Regional distribution of estrogen receptors. Rats were injected with saline or estradiol as in Fig. 105 and various regions of the hypothalamus dissected for subsequent determinations of R_nE by exchange. *BHM*, basal hypothalamus-median eminence; *MB*, mamillary bodies; *Other*, rest of hypothalamus. (Anderson et al., 1973b)

with detergent washes of the resultant pellet allows the quantitative measure of R_nE complexes by exchange (Kelner and Peck, unpublished). With this procedure, as many receptors can be measured in hypothalamic nuclei of estrogen-treated adult animals as exist in the cytoplasmic compartment of saline-injected controls (see Fig. 104 and below).

Such ^3H-steroid exchange studies suggest that the estradiol retained in the nuclear fraction after in vivo or in vitro exposure is bound to translocated cytoplasmic receptors. Further support derives from the demonstration by sucrose gradient centrifugation of a macromolecular peak labeled with ^3H-estradiol, which was extracted from hypothalamic nuclei with 0.4 *M* KCl (Vertes and King, 1971; Mowles et al., 1971). In addition, Cidlowski and Muldoon (1974) have shown that injected estradiol reduces the number of cytoplasmic estradiol binding sites. However, the ultimate demonstration of translocation requires the simultaneous study of cytoplasmic and nuclear compartments as a function of time after estrogen injection. Figure 107 is taken from unpublished work of our lab (Kelner and Peck, unpublished) and shows that with the disappearance of hypothalamic cytoplasmic receptors, there is an attendant and concomitant increase in the number of nuclear receptor-estradiol complexes as measured by ^3H-estradiol exchange. This study on the adult ovariectomized rat establishes nuclear translocation of cytoplasmic receptor as a fact.

There do appear to be differences, however, between estrogen receptors of the uterus and brain. The majority of cytoplasmic receptors of uterine cytosol bind to DNA-cellulose and are readily and rapidly converted to the so-called activated 5*S*

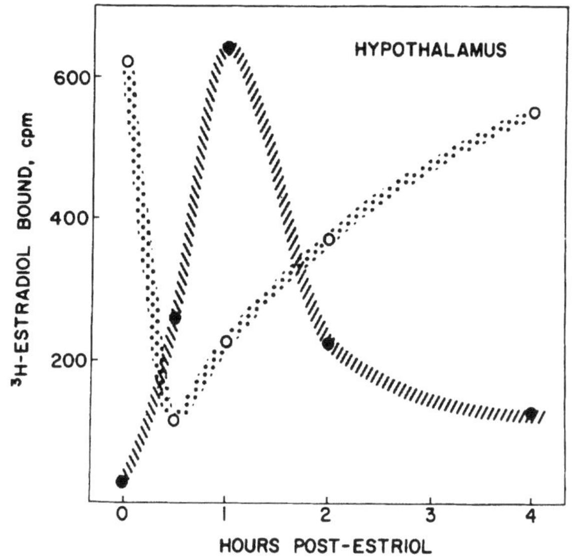

Fig. 107. Translocation of hypothalamic estrogen receptor. Castrate adult female rats were injected with saline (0 time point) or 5 μg estriol and killed at the times indicated. Cytoplasmic (○) and nuclear (●) receptors from hypothalami were determined by exchange and protamine sulfate precipitation

state in vitro (see Chap. IV and Fox and Johnston, 1974). In contrast, cytosol receptor from the hypothalamus appears to exist in two forms with only 30–50 percent of total receptor capable of binding to DNA-cellulose. There also appears to be little or no conversion of this cytoplasmic receptor to the 5*S* state in vitro. In addition, Linkie (1975, 1977) has noted that nuclear 5*S* receptors can be observed very quickly (within minutes) in rat uteri after ^{3}H-estradiol injection, whereas in the hypothalamus nuclear 5*S* complexes are only seen after a 30 min – 1 h lag. Thus, the appearance of "activated" nuclear complex in the hypothalamus is delayed relative to the uterus. This lack of in vitro conversion of cytoplasmic receptor from a 4*S* to 5*S* form and the delay in appearance of 5*S* receptor in nuclei of the hypothalamus suggest to us that receptor activation, nuclear translocation, and/or processing of receptor–estrogen complexes differs between central nervous system targets, such as the hypothalamus, and peripheral targets, such as the uterus.

When ^{3}H-progesterone is injected into rats, almost no radioactivity is found associated with the nuclear fraction of midbrain or hypothalamic cells (McEwen et al., 1976). Although Karavolas and Herf (1971) found substantial ^{3}H-progesterone retained by crude nuclear fractions of hypothalami exposed to hormone in vitro, they also observed similar retention for their control tissues, cerebral cortex, and boiled hypothalamus. Given the difficulties inherent in the exchange assay for progesterone receptors in the uterus (Hsueh et al., 1974) and the inability of many investigators to demonstrate preferential retention of injected label, it is doubtful that nuclear receptor-progesterone complexes will be demonstrated readily in central nervous system tissue. The data relating polarity of progestins to their rate of accumulation also suggests that the quantitation of this receptor will be difficult, as do the complexities introduced by the interaction of adrenal steroids with this system. Perhaps the use of adrenalectomized, ovariectomized animals that are estrogen-primed for varying periods in combination with exchange procedures employing ^{3}H-R5020 will allow the demonstration of specific progesterone receptors in nuclear fractions of the brain. It seems certain that the animal models employed to date have been inadequate for such a demonstration. In addition, as mentioned above, the use of the synthetic progestin, R-5020, may be of value in resolving these problems.

4. Age- and Sex-Related Differences in Receptor Content or Function

Most studies of the maturation or differentiation of the hypothalamus with respect to estrogen receptor content have employed the retention of injected ^{3}H-estradiol or the measurement of cytoplasmic binding sites. Presl et al. (1970) and Kato et al. (1971a; 1974) demonstrated the preferential retention of ^{3}H-estradiol in median eminence and preoptic-anterior hypothalamic regions of the hypothalamus at about 15 and 22 days, respectively. Prior to these times no preferential retention was observed (tissue/ cortex $\leq$ 1). These ages are not in agreement with studies of the appearance of cytoplasmic receptors. A progessive increase in R_c number has been observed in the rat hypothalamus from 6 days of age to about 33 days of age (Kato et al., 1971b; Plapinger and McEwen, 1973). On the other hand, retention of ^{3}H-estradiol within the nuclear fraction increased dramatically only between days 20 and 26 in the rat (McEwen et al., 1974). In contrast, Sheridan et al. (1974) used autoradiography and large doses of ^{3}H-estradiol to demonstrate nuclear accumulation and retention in neonatal rat brains similar to that of adults for specific neurons of the basal hypothalamus, preoptic region, and amygdala. The seeming disparity between appearance of cytoplasmic binding sites and that of nuclear binding capacity as well as the disparity between biochemical and autoradiographic studies probably result from the availability of free estrogen to these neural targets.

Certain steroid binding proteins exist which are characteristic of fetal serum. Originally described by Raynaud et al. (1971) as well as by Nunez et al. (1971), these fetal proteins characteristically exist at high concentration in the plasma of neonates and disappear from plasma by 22 days of age. The early studies of Kato et al. (1971b) showed rudimentary 8.5*S* binding sites at 7 days of age but much larger numbers of a slower migrating binding site. In these studies the 8–8.5*S* estrogen binding site began to predominate in number around 21 days of age. McEwen and co-workers have shown that the estrogen binding site measured in the neonatal brain differs from that of the uterus and may be the α-fetoprotein discussed above (Plapinger et al., 1973). Thus, neonatal binding macromolecules have different pharmacospecificity (do not bind diethylstilbestrol), different sedimentation coefficients (in agreement with Kato et al., 1971b), and a very high capacity — properties more indicative of α-fetoprotein than of a classic receptor. McEwen et al. (1974) suggested that these might protect the neonatal hypothalamus from the masculinizing effects of maternal estrogen, a suggestion with which we agree. Quite recently McEwen and co-workers have reviewed the development of steroid receptor systems in the rodent brain and included a discussion of sexual differentiation as well as of the maturation of these systems (MacLuskey and McEwen, 1978). All interested in this subject should refer to that work.

If age-related differences in receptor level or nature have proven difficult to detect because of complications such as those above, sex-related differences have proven almost impossible to detect. The literature abounds with reports of differences, most of which have usually resulted from poorly controlled experiments. In a recent review McEwen concludes that it may be well to emphasize "the similarities, rather than the possible sex differences, in neural estrogen receptor distribution and content" (McEwen, 1978). Among the best-controlled studies in this area of research have been those of Woolley and her associates. Maurer and Woolley (1975) referenced 19 publications on the accumulation and retention of estradiol in the brains of normal female, androgenized female, and male rats. In approximately half of these, the hypothalamus of males or androgenized females accumulated less estradiol than did those of normal females; however, the rest reported no difference in accumulation and/or retention. These conflicting reports utilized a number of different techniques for examining accumulation and a variety of animal models, making the resolution of their differences impossible. In a very recent paper Ogren and Woolley (1978) have examined nuclear retention of ^{3}H-estradiol 1 h after injection at a dose of 5 or 50 μg/kg body wt. in adult intact, gonadectomized, or gonadectomized-adrenalectomized male and female rats. Nuclear retention of ^{3}H-estradiol in preoptic-anterior hypothalamic and median eminence-basal hypothalamic regions was lower in castrate males than in castrate females; however, little or no difference was observed with injection of physiologic (0.1–0.5 μg) doses of steroid (Anderson et al., 1973b; Maurer and Woolley, 1974; Ogren et al., 1976; Lieberburg and McEwen, 1977). Since no consistent difference has been observed in total receptor content or in physical characteristics of cytoplasmic receptors (Korach and Muldoon, 1974; Talley et al., 1975), we conclude that hypothalamic tissues of male and female rats do not differ markedly in their ability to bind exogenous estrogen.

Interestingly, Ogren and Woolley (1978) have shown that the number (R_nE) and percent of total receptor ($R_nE \cdot 100$/total R) filled by *endogenous* steroids in the anterior hypothalamus and amygdala are the same for both males and females, suggesting that males and females are quite alike with respect to receptor occupancy. Since the adult rat hypothalamus and limbic system convert testosterone to estradiol (Naftolin et al., 1972; Naftolin et al., 1975b) we conclude from the work of Ogren and Woolley that estrogens derived from aromatization of testosterone occupy "masculine" estrogen receptors in the male hypothalamus and limbic system to essentially the same extent as circulating estrogens occupy "feminine" estrogen receptors in the female. This would explain the capacity of estradiol to maintain male sexual behavior (Pfaff, 1970a, b; Christensen and Clemens, 1974; Paup et al., 1975) as well as the similarities in negative feedback effects of estradiol on gonadotropin secretion in the two sexes (Mahesh et al., 1975; Paup et al., 1975).

Cidlowski and Muldoon (1976) have reported one fascinating sex difference that will require additional study — the rate of replenishment of cytoplasmic estrogen receptor differs between males and females. Additional study will be required to ensure that this observation does not simply relate to differences in rate of clearance of free steroid from the neural tissues of the sexes.

In summary, a systematic and controlled study of sexual differentiation or maturation of brain function has yet to be completed. Care must be taken in such studies to avoid interference from neonatal plasma proteins with high affinities for estrogens. Certainly the approaches outlined by the recent review of McEwen and co-workers should provide the rational approach to this problem (MacLuskey and McEwen, 1978). For the present we conclude that appreciable differences do not exist between males and females with respect to hypothalamic estrogen receptor content.

5. Gonadal Steroids and Nervous System Function

As with peripheral targets, the central nervous system responds to estrogens and progestins in a multitude of ways. Sex steroids direct the homeostatic regulation of gonadotropin secretion, modify behavioral components important to reproduction, and in addition direct the sexual differentiation and maturation of the central nervous system. A number of biochemical effects of estrogens on central nervous tissue have been described which might result in altered gonadotropin secretion and/or sexual behavior. Despite the plethora of reports on this subject, no definitive conclusions can be made as yet about cause-effect relationships between specific biochemical changes and estrogen-induced alterations in neural function. Early reviews of Gorski and McEwen (Gorski and Clemens, 1970; McEwen et al., 1974) as well as more recent reports of Feder (Feder and Marrone, 1977; Feder et al., 1978) are recommended as background on this subject. We will review only a few biochemical and physiologic effects of estrogens and progestins on the brain as studied in our laboratory.

a) Estrogens and Hypothalamic RNA Polymerases

Estrogens increase RNA synthesis in estrogen target tissues (see Chap. V). This increased synthesis appears to involve two temporally separate events: an early stimulation of DNA-dependent RNA polymerase II as first described by Glasser et al. (1972) followed by a secondary rise in both RNA polymerase I and II (Glasser and Spelsberg, 1973; Hardin et al., 1976; see Chap. V). Our own investigation of RNA polymerase I and II activities in estrogen-treated uteri suggests that the early rise in RNA polymerase II correlates with early uterine response but not with true uterine growth. Growth in peripheral targets such as the uterus requires the secondary rise in both polymerases which in turn appears dependent on the retention of nuclear receptor-estrogen complexes for prolonged periods.

Recently we have initiated investigations of hypothalamic RNA polymerase I and II designed to ask the following questions: do neurons respond to estrogens in the same manner as uterine cells or do they only exhibit early responses? Occupancy of estrogen receptors and stimulation of RNA synthesis do not, per se, cause true growth (see Chap. V); rather estrogen must be retained in nuclear fractions for a critical period, which correlates with the secondary increase and sustained activation of both polymerase I and II (see above). Earlier in this chapter (see Sect. 1) we pointed out that receptor-estrogen complexes in the preoptic-ventromedial-arcuate region appear labile or short-lived (Anderson and Greenwald, 1969). Furthermore, there are suggestions that the processing of receptor-estrogen complexes is different for hypothalamus as compared with uterus (Fox and Johnston, 1974; Linkie, 1975, 1977). These differences in the processing of receptors may result in differences in the immediate or early activation of polymerases by gonadal steroids in target tissues.

To amplify, since neurons do not "grow" after maturation (excluding the possibility of sprouting or increases in synaptic surface) and since nervous system tissue targets must respond rapidly to transient changes in gonadal steroids, it seems possible that only an initial rise in RNA polymerase II, associated with the synthesis of specific mRNAs, may occur in neural tissues after exposure to steroids. In Fig. 108 endogenous RNA polymerase I and II activities were assayed as a function of time after estrogen treatment. Note that polymerase II shows a transient activation not unlike the time course of receptor translocation, while polymerase I is nonresponsive to estrogen treatment. Analogous experiments examining longer times after estrogen showed no long-term activation of either polymerase. Control experiments employing dorsal-frontal cortex or occipital cortex, brain regions with no measurable estrogen receptor, have shown no change in either polymerase activity at any time after estrogen treatment. Thus, polymerase activation appears specific for steroid targets within the central nervous system. The early rise in RNA polymerase II suggests that new species of mRNA are synthesized in response to estrogen and/or that multiple copies of mRNAs are being produced. No dramatic or sustained alteration in RNA polymerase I is observed, consistent with our hypothesis that estrogen acts as a specific inducer, but not as a growth promoter, in brain targets.

b) Hypothalamic Control of Gonadotropin Secretion

Endocrine control of gonadotropin secretion is characterized by a series of feedback mechanisms: (1) "long" feedback loops in which peripheral targets (gonads) provide messages (estrogen or progesterone); (2) "short" loops in which the anterior pituitary may provide the signal (LH or FSH); and (3) "ultrashort" loops in which gonadotropin-releasing hormones directly alter their own rate of synthesis or release. For this

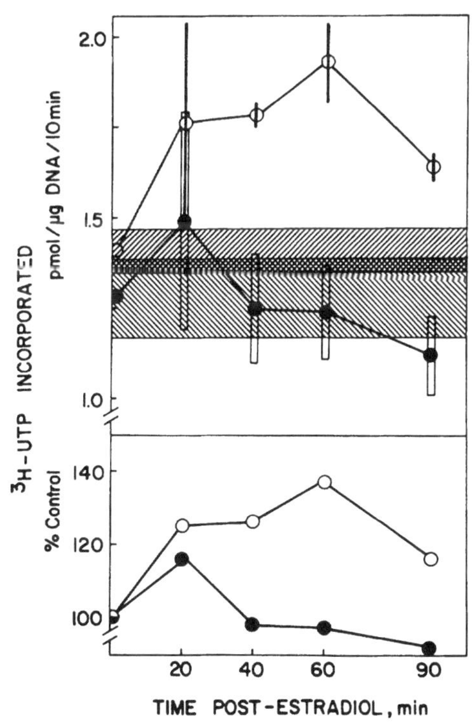

Fig. 108. Activation of hypothalamic RNA polymerase by estradiol. Castrate adult female rats were injected with saline or 5 µg estradiol and killed at the times indicated. Hypothalamic nuclei were purified and endogenous RNA polymerases I (●) and II (○) determined essentially as described by Hardin et al. (1976). Values from saline-injected controls were employed to calculate percent activation in the *lower panel.* Values are the mean ± SEM of three–six determinations

discussion we will deal only with the "long" feedback system and the effects of estrogens and progestins on the subcellular distribution and release of luteinizing hormone-releasing hormone (LHRH), also called gonadotropin-releasing hormone (GnRH).

The "long" feedback systems are of two types, negative and positive. Thus, estrogens and progestins may either inhibit or facilitate the release of gonadotropins from the pituitary (and, presumably, the release of LHRH from the hypothalamus) (Yen and Lein, 1976). The preceding parenthetic statement is not made in ignorance of reports that these loops might function primarily at the pituitary level (Hoff et al., 1977; Nakai et al., 1978; Plant et al., 1978), but rather in view of the multiple reports separating positive and negative centers within the hypothalamus itself via implant, lesion, and stimulation techniques (Schwartz and McCormack, 1972). In addition, feedback regulation of hypothalamic LHRH by gonadal steroids has been demonstrated unequivocally for the rat. Thus, castration reduces hypothalamic LHRH content (Piacsek and Meites, 1966; Araki et al., 1975a, b; Kalra, 1976; Kobayashi et al., 1978), and estrogen replacement prevents this reduction (Kobayashi et al., 1978).

Gonadal steroids could exert their influences, positive or negative, in a number of ways. Thus, they could increase (or decrease) the synthesis or degradation of LHRH, increase (or decrease) the excitatory threshold of neurosecretory elements, alter hypothalamic adrenergic mechanisms thought to be involved in control of LHRH release, or alter the mechanisms of LHRH storage and/or secretion.

We have recently studied the latter possibility using the castrate female rat and estrogen and/or progesterone replacement. Specifically, we have studied the content, subcellular distribution, and in vitro release of LHRH as a function of hormonal manipulation.

The subcellular fractionation techniques of De Robertis and co-workers (De Robertis et al., 1961) and Whittaker (Whittaker, 1959) have been applied to the hypothalamus to study the storage and secretion of peptides. Figure 109 depicts a representative fractionation scheme and defines the various subcellular fractions to be discussed. Subcellular fractionation has been used for studying the distribution of corticotropin-releasing hormone (Mulder, 1970), thyrotropin-releasing hormone (Barnea et al., 1975), somatostatin (Styne et al., 1977), and LHRH (Ishii, 1970; Fink et al., 1972; Shin et al., 1974; Taber and Karavolas, 1975; Barnea et al., 1975, 1976, 1977) in several species. Three observations are reported consistently: (1) the major portion of the releasing hormone is the equivalent of Whittaker's synaptosome-containing P_2 fraction; (2) hypotonic shock of this P_2 fraction results in the loss of releasing hormone activity from that fraction and its appearance in the P_3 or vesicular fraction; and (3) little activity was found in high-speed supernatant and microsomal fractions. Electron microscopic examination has demonstrated that fractions containing releasing hormone are enriched with synaptosomes and/or synaptic vesicles (Ishii, 1970; Taber and Karavolas, 1975; Barnea et al., 1977). Therefore the subcellular distribution of hypothalamic releasing hormones is similar to that of acetylcholine and the monoamine neurotransmitters.

Given that LHRH is contained within vesicles in synaptosome-like structures, we considered these structures to be potential sites for the expression of the regulatory actions of estrogens and progestins. The hormonal state of a rat has significant effects on hypothalamic content and secretion of LHRH. In the normal female, hypo-

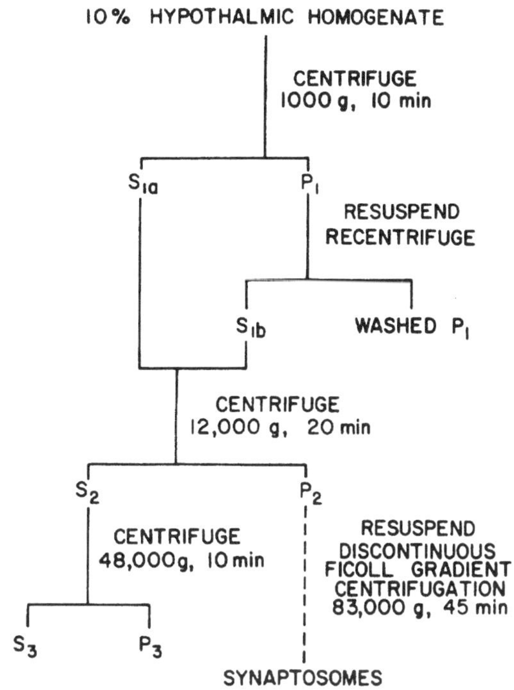

Fig. 109. Typical subcellular fractionation scheme for the study of LHRH distribution. Hypothalami are homogenized in isotonic sucrose and subjected to differential centrifugation to prepare P_1 (somal), P_2 (mitochondrial) and P_3 (microsomal) fractions. Synaptosomes are prepared from P_2 by ultracentrifugation on continuous or discontinuous gradients. S_{1a}, S_{1b}, S_2 and S_3 refer to the various supernatant fractions that are obtained during this procedure

thalamic content of LHRH changes at puberty (Araki et al., 1975a, b; Ojeda et al., 1976; Wheaton and McCann, 1976) and during the course of the estrous cycle (Araki et al., 1975). Following ovariectomy, LHRH secretion increases (Ben-Jonathan et al., 1973; Wheaton and McCann, 1976), and as expected the hypothalamic content of the peptide declines (Piacsek and Meites, 1966; Araki et al., 1975; Wheaton and McCann, 1976).

To examine the effects of estrogens on hypothalamic LHRH, we initiated studies of subcellular distribution. Hypothalami from control ovariectomized rats or ovariectomized rats injected 48 h earlier with 5 µg estradiol benzoate were homogenized and fractionated by a combination of differential and Ficoll density gradient centrifugation. The LHRH content in fractions of these gradients was determined by radioimmunoassay, and results from a typical experiment are shown in Fig. 110. Of the LHRH in the homogenate, 30 percent was in the nuclear (P_1) fraction, 50% in the mitrochondrial (P_2) fraction, 5–6 percent in the vesicular (P_3) fraction, and 3–5 percent in the high-speed supernatant, S_3. Recovery of LHRH was similar for the two groups of animals: 96.1 $\pm$ 9.7 percent and 88.7 $\pm$ 5.5 percent for control and EB-treated groups, respectively. Synaptosomes contained about 65–70 percent of the LHRH present in the P_2 fraction. The subcellular fractions were not dramatically enriched in LHRH; however, the P_2 and synaptosomal fractions from animals given estrogen had significantly more LHRH per µg protein than the homogenate (142% $\pm$ 17% and 141% $\pm$ 12%, respectively). In addition the LHRH concentration in P_2 and synaptosomal fractions from estrogen-treated animals was higher than that of corresponding fractions from castrate controls. Thus, LHRH concentrations were elevated by 52 $\pm$ 20 percent and 46 $\pm$ 26 percent ($P < 0.05$, $n = 5$) in P_2 and synap-

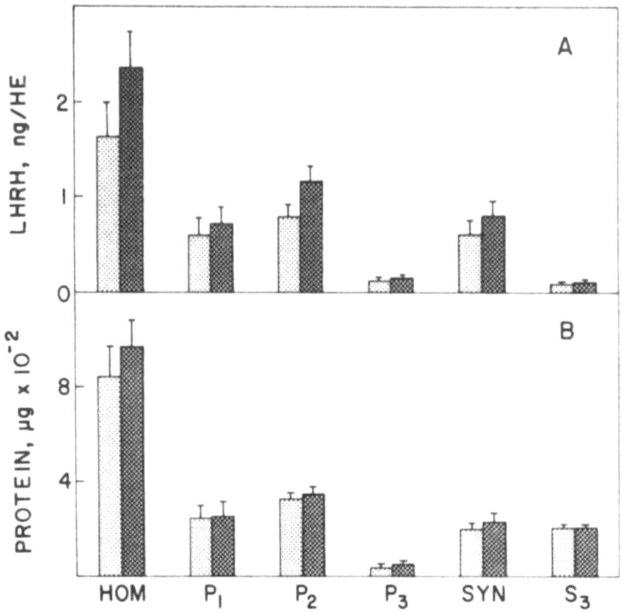

Fig. 110 A and B. Distribution of LHRH among various subcellular fractions. LHRH was assayed by radioimmunoassay for fractions of hypothalami as described in Fig. 109. Hypothalami were obtained from castrate controls (*light bars*) and castrates given 5 µg estradiol (*dark bars*). Values are the mean ± SEM of three–four determinations. Homogenate, HOM; various pellet fractions as defined in Fig. 109, P_1, P_2 and P_3; synaptosome fraction, *SYN*; supernatant fraction, S_3

tosomal fractions, respectively, when estrogen-treated animals were compared with controls.

The small changes in LHRH concentration observed 48 h after estrogen treatment and the early onset of serum LH depression (within 6 h) suggested that the effects of estrogens on subcellular distribution of LHRH might occur at an earlier time. Therefore, serum levels of LH and LHRH concentration of the synaptosome-containing P_2 fraction were examined as a function of time after estrogen injection as shown in Fig. 111. The maximal effect of estradiol benzoate on subcellular distribution of LHRH occurred between 6 and 27 h after injection with LHRH in the P_2 fraction, which was increased by about 100 percent. During or preceding the same time period, serum LH was markedly depressed, indicating the operation of the negative feedback system. As serum LH levels rose between 48 and 72 h, the LHRH in the P_2 fraction declined, remaining 50 percent above control at 57 h and returing to preinjection values after 72 h.

Following demonstration that estrogen treatment results in significant increases in LHRH in subcellular fractions of the hypothalamus known to contain synaptosomes and synaptic vesicles, estrogen-induced alterations in LHRH storage were further examined using nonequilibrium or rate sedimentation centrifugation as employed by Barnea et al. (1976) for the study of TRH- and LHRH-containing particles from the hypothalami of male rats. They have demonstrated two populations of peptide-containing particles that can be separated via their different rates of sedimentation, with the evidence suggesting that rapidly sedimenting particles are synaptosomes

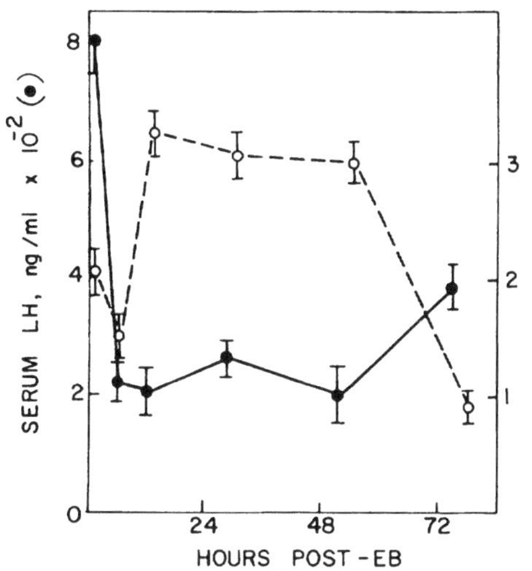

Fig. 111. Effect of estrogen on serum LH and P_2 content of LHRH in ovariectomized female rats. Adult ovariectomized female rats were injected with saline (zero time point) or 5 µg estradiol benzoate (EB) and killed at various times thereafter. Serum LH and the content of LHRH in the P_2 fraction were assayed by radioimmunoassay. All values are the mean ± SEM of three-four determinations

and more slowly moving particles are vesicles or granules (Barnea et al., 1976, 1977). With this tentative identification of the particles in mind we examined the distribution of LHRH in hypothalamic subcellular fractions of male rats as well as ovariectomized female rats with or without estrogen pretreatment.

A similar distribution of LHRH activity was observed for hypothalamic homogenates from ovariectomized female and male rats (see Fig. 112). Thus, homogenization results in the liberation of two types of LHRH-containing particles, which can be separated by nonequilibrium or rate sedimentation centrifugation but not by equilibrium centrifugation, suggesting that the particles differ in size but not in density.

Electron microscopy has been employed together with the assay of subcellular LHRH to show that the fractions containing the hypothalamic peptide consist of synaptosomes, synaptic vesicles, or both (Clementi et al., 1970; Mulder, 1970; Ishii, 1970; Fink et al., 1972; Shin et al., 1974). Barnea et al. (1977) have used electron

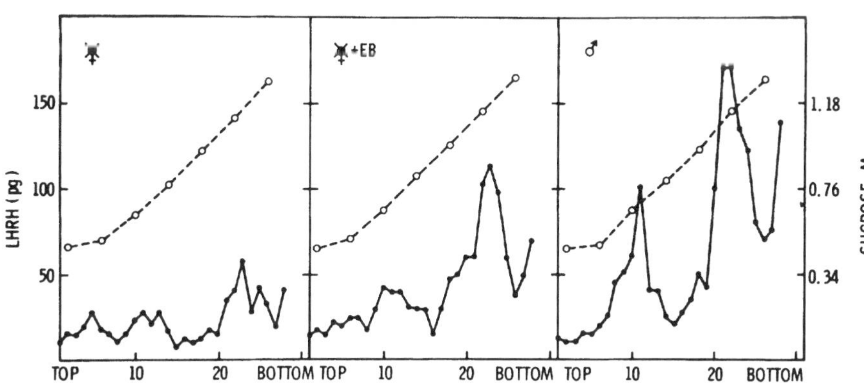

Fig. 112. Rate sedimentation analysis of the subcellular distribution of LHRH. Intact male rats, castrate female rats, and castrate female rats treated with estradiol benzoate for 24 h were killed and S_1 fractions centrifuged for a short time at high speed on linear sucrose gradients. Gradients were fractionated and analyzed for LHRH via radioimmunoassay

microscopy to show that the peak 2 region contains synaptosomelike particles while the peak 1 region contains free vesicles. Furthermore, results from our laboratory and that of Barnea et al. (1977) demonstrate that only the LHRH-containing particles that migrate with the second peak can be disrupted by hypo-osmotic conditions. Lability to osmotic shock is characteristic of synaptosomes but not of synaptic vesicles (Whittaker et al., 1964). Thus, the particles in peak 1 containing LHRH may be secretory vesicles or granules, while those in peak 2 are synaptosomes or pinched-off secretory terminals.

The similarity in the overall pattern of distribution of LHRH-containing particles for all three groups of rats (Fig. 112) suggests a common mechanism for the storage of LHRH, regardless of sex or endocrine status of the animal. The significance of these populations of LHRH-containing particles is suggested by the differences between experimental animals with respect to their content of LHRH (Table 13). In estrogen-treated ovariectomized females a significantly greater proportion of LHRH was contained in peak 2 when compared to controls. Since estrogen treatment decreases LHRH secretion, as suggested both by the increase in hypothalamic LHRH content (Kalra, 1976; Tytell et al., 1978) and by the rapid decline in serum luteinizing hormone levels (Ajika, et al., 1972; Legan et al., 1973; Ramirez and Sawyer, 1974), the increase in LHRH in peak 2 could reflect an estrogen-dependent shift in the storage of the peptide to a less easily released form. Note that in the male hypothalamus the portion of peak 2 LHRH is similar to that of estrogen-treated females. Thus, the LHRH-containing particles of peak 2 may be the site at which the effects of gonadal steroids on LHRH secretion are expressed.

Exposure to hypotonic conditions resulted in the disappearance of peak 2 but not of peak 1, in agreement with the studies of male rats by Barnea et al. (1977). We also observed free LHRH at the top of these gradients. To observe this free LHRH, bacitracin was included in the hypotonic medium to prevent degradation of the peptide (McKelvy et al., 1976). Preservation of free LHRH is important because it reveals a second difference between control and estrogen-treated ovariectomized rats: the amount of free LHRH released under hypotonic conditions is reduced by steroid treatment. Free LHRH probably arises from peak 2 since the peptide remains in peak 1 after hypotonic treatment. Thus, within the peak 2 region of the gradient there exists an osmotically sensitive, LHRH-containing compartment, the size of which is influenced by estrogen. Using repetitive freezing and thawing as a disruptive stimulus, we have confirmed that the quantity of LHRH released from a synap-

Table 13. Distribution of LHRH on rate sedimentation centrifugation[a]

Group	Total LHRH ng/hypothalamus	% Total LHRH in: Peak 1	Peak 2
Male	8.9 ± 1.2	17 ± 2	57 ± 1
Ovx female	3.8 ± 0.5	20 ± 1	48 ± 2[b]
Ovx female + estradiol benzoate	4.9 ± 0.4	18 ± 1	56 ± 2

[a] All values are mean ± SEM of four–seven experiments.

[b] $p < 0.05$ between groups.

Table 14. Freeze-thaw release of LHRH

Freeze-thaw cycles	% Total LHRH Released[a]	
	Control	Estrogen treated
0	none	none
1	18 ± 4	10 ± 2
2	21 ± 5	17 ± 3
3–4	28 ± 5	16 ± 2

[a] LHRH released ·100/LHRH content of the original synaptosomal fraction. Each value is the mean ± SEM of three or four determinations.

tosomal fraction prepared from hypothalami of ovariectomized rats is reduced when rats are pretreated with estrogen (Table 14).

To test the hypothesis that these changes indicate an estrogen-dependent shift in LHRH storage to a less easily releasable form under conditions more closely approximating the secretion of LHRH as it might occur in vivo, K^+-stimulated release of LHRH from a crude mitochondrial-synaptosomal fraction was examined after in vivo or in vitro administration of estrogen and/or progesterone.

We found that the LHRH-containing particles of the P_2 fraction were quite stable in isotonic sucrose, with basal release of LHRH rarely more than 5 percent of total P_2 content. Upon the addition of 60 m*M* K^+, the quantity of LHRH released increased by 3.5– 5.0 (Table 15). This K^+-stimulated release was significantly inhibited by the Ca^{2+} antagonists, Mn^{2+} and La^{3+}, which are used extensively to investigate Ca^{2+}-dependent release (Miledi, 1971; Levy et al., 1973; Russell and Thorn 1973; Rubin, 1974; Weiss, 1974; Normann, 1976). These ions compete with Ca^{2+} for external binding sites and prevent the movement of Ca^{2+} across the membrane (Miledi, 1971; Baker, 1972; Meiri and Rahaminoff, 1972; Rubin, 1974; Weiss, 1974; Boucek and Synderman, 1976). Warberg et al. (1977) have also demonstrated the Ca^{2+} dependence of LHRH release from a synaptosomal preparation by varying the concentration of the ion in the incubation medium. A similar Ca^{2+} requirement exists for synaptosomal release of other hypothalamic peptide hormones such as corticotropin-releasing hormone, prolactin release-inhibiting factor, and vasopressin (Bennett and Edwardson, 1975).

Table 15. Ca^{2+}-dependence of K^+-stimulated LHRH release

Ionic milieu	LHRH released, % P_2 content		
	Basal[a]	K^+-stimulated[a]	Net[b]
Ca^{2+}	2.7 ± 0.4	13.1 ± 0.3	10.4
Ca^{2+} + Mn^{2+}	6.2 ± 0.6	10.1 ± 1.0	3.9
Ca^{2+} + La^{2+}	3.6 ± 0.2	4.5 ± 0.3	0.9

[a] LHRH released (basel or K^+-stimulated) ·100/P_2 content of LHRH. All values are mean ± SEM of three or four determinations.

[b] K^+-stimulated – basal.

Table 16. Ca^{2+}-dependence of soluble and particulate LHRH release

Forms	Ionic milieu	LHRH released, % P_2 content		
		Basal[a]	K^+-stimulated[a]	Net[b]
Soluble	Ca^{2+}	1.8 ± 0.2	3.9 ± 0.1	2.1
	Ca^{2+} + Mn^{2+}	4.8 ± 0.3	5.7 ± 0.2	0.9
	Ca^{2+} + La^{3+}	2.5 ± 0.3	4.3 + 0.3	1.8
Particulate	Ca^{2+}	0.9 ± 0.2	9.6 ± 0.2	8.5
	Ca^{2+} + Mn^{2+}	1.4 ± 0.3	4.3 ± 0.5	2.9
	Ca^{2+} + La^{3+}	0.9 ± 0.3	0.9 ± 0.2	none

[a] LHRH released (basal or K^+-stimulated · 100 / P_2 content of LHRH. All values are mean ± SEM of three or four determinations.
[b] K^+-stimulated-basal.

Centrifugation of incubation filtrates revealed that a portion of the LHRH released under basal conditions or by K^+-depolarization sedimented at 48,000 *g* while a second, smaller fraction was soluble or free (Table 16). There were two major distinctions between the two forms of LHRH released under basal conditions: (1) basal relase of soluble LHRH was always significantly higher than basal release of particulate LHRH; (2) Mn^{2+} had little effect on the basal release of particulate LHRH. Following K^+-stimulation, the pattern of release of the two forms of LHRH was again quite different. Potassium depolarization caused only a 2-fold increase in the release of soluble LHRH whereas the release of particulate LHRH was stimulated more than 10-fold. Furthermore, release of soluble LHRH in response to K^+ was unaffected by La^{3+} or Mn^{2+} while K^+-stimulated release of particulate LHRH was reduced more than 50 percent by Mn^{2+} and was abolished by La^{3+}.

The two forms of LHRH released from the P_2 suspension, soluble and particulate, support our contention that the peptide resides in two pools. Soluble LHRH constituted about 75 percent of the basal LHRH release, which is LHRH that "leaked" out during an incubation under nonstimulatory conditions. With the addition of 10 m*M* Mn^{2+}, 60 m*M* K^+, or 120 m*M* sucrose, efflux of soluble LHRH increased about 2-fold, and this efflux was unaffected by the Ca^{2+} antagonist, La^{3+}. Thus, soluble LHRH appears to exist in a pool, which responds in a nonspecific fashion to changes in the ionic or osmotic environment. The pattern of particulate LHRH release, on the other hand, suggests a second compartment of tightly bound LHRH. Under nonstimulatory conditions, particulate LHRH constituted about 25 percent of that released, and this fraction was unaffected by 10 m*M* Mn^{2+}, 0.5 m*M* La^{3+}, or 120 m*M* sucrose. However, the addition of 60 m*M* K^+ caused a 10-fold increase in the release of particulate LHRH, and this stimulation of release was inhibited by both Mn^{2+} and La^{3+}.

The release of a particulate or insoluble product has little precedent in the literature on secretory systems. However, Warberg et al. (1977) also noted that a portion of the LHRH released in their experiments sedimented upon centrifugation. In the adrenal medulla secretion of catecholamines is accompanied by release of the entire contents of the secretory vesicle (Rubin, 1974). If the entire matrix of material within a vesicle is expelled as a unit, it is conceivable that the components might remain insoluble for some time. Such a phenomenon has been reported for the secretion of enzymes by polymorphonuclear leukocytes (Whittaker et al., 1964).

Table 17. Effects of estrogen on serum LH, P_2 content of LHRH, and in vitro LHRH release[a]

Time post-estrogen (h)	Serum LH (ng/ml)	P_2 LHRH (pg/μg protein)	LHRH release (% P_2)	
			Basal	K^+-stimulated
0	572 ± 82	1.3 ± 0.1	3.5 ± 0.2	15.3 ± 1.2
6	279 ± 45	2.5 ± 0.1	2.6 ± 0.3	11.7 ± 0.8
12	338 ± 49	3.3 ± 0.2	2.2 ± 0.1	12.0 ± 0.9
18	310 ± 43	2.4 ± 0.2	2.4 ± 0.3	11.1 ± 0.1
43	163 ± 28	2.4 ± 0.1	1.9 ± 0.2	11.9 ± 0.8
67	272 ± 30	1.4 ± 0.1	3.1 ± 0.2	15.3 ± 0.7

[a] Adult castrate female rats were injected with estradiol benzoate and killed at the times indicated. Serum LH and P_2 content of LHRH were determined by radioimmunoassay. All values are the mean ± SEM of three - six determinations.

The temporal relationship between gonadal steroid treatment and LHRH release was studied in ovariectomized rats given a single injection of 50 μg estradiol benzoate and killed at intervals from 0 to 75 h after injection. At each time, the following were measured: serum LH, P_2 LHRH, and basal as well as K^+-stimulated LHRH release. The data are given in Table 17. Within 6 h a dramatic decline was seen in serum LH. The decrease was most pronounced between 6 and 48 h, after 72 h serum LH had begun to rise (also see Fig. 111). The postestrogen variation in P_2-LHRH was also quite marked. In as little as 6–12 h, the concentration of LHRH was increased by 2–2.5 fold. This elevation persisted for ~ 48 h.

In vitro LHRH release from P_2 suspensions prepared at intervals during the 75-h period after estrogen is also given in Table 17. The overall effect of estrogen was to suppress both basal and K^+-stimulated release. Of physiologic significance, the overall pattern of change in LHRH release with time after estrogen is similar to that observed for serum LH and is the inverse of that observed for P_2-LHRH.

Progesterone is known to inhibit the daily LH surge in ovariectomized rats treated with estradiol benzoate (Freeman et al., 1976). To examine the interaction of progesterone with estrogen, ovariectomized rats were injected with 5 mg progesterone at 1600 hours one day after receiving 50 μg estradiol benzoate. Animals were then killed 1 and 2 days after progesterone. Basal and K^+-stimulated LHRH release and serum LH concentrations in estrogen-plus-progesterone animals, as well as those which received only estrogen, are listed in Table 18. One day after progesterone administration, estrogen-plus-progesterone treated rats showed a small elevation in basal LHRH release compared to estrogen-treated rats. Two days after progesterone administration, the K^+-stimulated LHRH release of the estrogen-plus-progesterone-treated animals showed a significant reduction ($p < 0.05$) to a level similar to that demonstrated by rats between 6 and 51 h postestrogen (see above). This reduced LHRH release coincided with a significant depression in the serum LH of those animals.

From these studies we conclude that suppression of LHRH release from the P_2 suspension of rats treated with estrogen reflects a functional alteration of the hypothalamic neurosecretory system. Thus, this suppression follows the same time course as the decline in serum LH. Furthermore, at 72 h postestrogen when in vitro release

Table 18. Effects of estrogen and progesterone on LHRH release[a]

Time after estrogen (h)	Time after progesterone (h)	Serum LH (ng/ml)	LHRH release (% P_2)	
			Basal	K^+-stimulated
51	-	201 ± 53	2.9 ± 0.2	16.8 ± 0.4
51	23	143 ± 31	4.2 ± 0.2	18.3 ± 0.7
75	-	382 ± 41	3.4 ± 0.2	21.6 ± 1.0
75	47	106 ± 20[b]	2.6 ± 0.1[b]	15.3 ± 0.6[b]

[a] All values are the mean ± SEM. For LHRH release, n = 3, while for serum LH n is 6.
[b] $p < 0.05$ compared to the 75-h estrogen-treated group without progesterone.

of LHRH returns to preinjection levels, serum LH begins to rise toward control values and the P_2 content of LHRH begins to fall.

Treatment of ovariectomized rats with single or multiple injections of estrogen results in the appearance of a daily surge in serum LH concentration, which occurs at the same time of day as the proestrous LH surge in intact rats (Legan et al., 1975; Legan and Karsch, 1975; Freeman et al., 1976). Although progesterone may either facilitate or depress this surge, our data (Table 18) and that of Freeman et al. (1976) show that it can be blocked by repeated administration of progesterone.

A daily neural signal for the LH surge may exist in the rat, which requires elevated serum estrogen (Legan and Karsch, 1975), and progesterone may limit the expression of this signal to the day of proestrus (Freeman et al., 1976). In ovariectomized rats treated with estradiol benzoate alone a significant rise in both basal and K^+-stimulated LHRH release from the P_2 suspension occurred at 1500 h for 3 consecutive days in animals which received only estrogen (data not shown). However, when the rats also received progesterone this elevation was suppressed. Although Freeman et al. (1976) observed the inhibitory effect of progesterone on the LH surge within 1 day, the suppression of LHRH release in this study did not occur until the 2nd day. The reason for this discrepancy is not known. Nonetheless, the timing of the increased LHRH release, about 2 h prior to the LH surge, and its inhibition by progesterone strongly suggests that it is the neural aspect of the signal proposed by Freeman et al. (1976).

To summarize, estrogens alter the hypothalamic content and subcellular distribution of LHRH in a manner temporally related to the suppression of serum LH. Furthermore, treatment of ovariectomized rats with 5 or 50 µg estradiol benzoate results in a significant reduction in total LHRH released under either basal or K^+-stimulated conditions in vitro. Finally, rats given 50 µg estradiol benzoate show cyclic variations throughout the day in the degree of suppression of LHRH release. There appears to be a small but persistent facilitation of LHRH release each afternoon. This facilitation is associated with an increase in a soluble pool of LHRH which is readily released (data not shown). Both facilitation of release and this soluble pool are abolished by progesterone treatment. Thus, the regulatory effects of estrogen and progesterone given in vivo on LHRH secretion can be observed in a subcellular fraction of the hypothalamus containing neurosecretory cell terminals. The predominant effect of the steroids is a suppression of LHRH release, which probably results from an alteration in the manner in which LHRH is stored within neurosecretory cells. An additional effect of estrogen on the system is a cyclic facilitation of LHRH release, which may reflect the daily neural signal for luteinizing hormone secretion, which occurs in the estrogen-treated ovariectomized rat.

Chapter X

Speculations and Conclusions

1. Interplay of Low- and High-Affinity Steroid Receptive Sites

In 1969 it was suggested that serum binding proteins might direct hormonal signals toward target organs with protein-permeable vascular beds (Keller et al., 1969). Subsequently estrogens were shown to promote the accumulation and retention of serum albumins within the uterus without effecting similar increases in nontarget tissues (Peterson and Spaziani, 1971). In 1973 we demonstrated that cytoplasmic levels of ^{3}H-estradiol at equilibrium after in vitro incubation in the presence of serum albumins are 50 percent greater for the uterus than for the diaphragm, a difference which did not result on incubation of these tissues with ^{3}H-estradiol in the absence of serum albumins (Peck et al., 1973a). Thus, serum-binding proteins, an example of low affinity estrogen receptive or binding sites, may function to increase hormone concentrations preferentially in target organs with protein-permeable vascular beds and in this manner maintain the concentration of total hormone at a level required for hormonal stimulation.

The Type II estrogen-binding sites discussed in Chap. IV may serve a similar role as an "accumulator" or "sink" to sustain relatively high levels of estrogen within specific cellular locales. Thus, complexes of estrogen with Type II might serve as an estrogen buffer within the cell just as complexes of blood binders with estrogen could serve a similar function in the extracellular milieu. Since low affinity sites of Type II class are observed in both cytoplasmic and nuclear compartments, they might serve this buffer function in both cellular locales. Since their number increases with sustained estrogen treatment, these low-affinity sites might maintain a hyperestrogenized state in target tissues during fluctuations in serum levels of estrogen.

This suggested function for serum-binding proteins and/or Type II binding sites is strengthened by our observation that only 50 percent of the ^{3}H-estradiol in the uterus after in vitro incubation is associated with receptors of the high-affinity type (Peck et al., 1973a). At low concentrations of free estradiol, such as those expected for rat serum containing multiple steroid-binding proteins and only $\sim 10^{-10} M$ estradiol, one could expect little or no occupation of the high-affinity estrogen receptor if the K_d of its complex with estradiol is nM as observed for intact uteri in vitro (see Fig. 113 and discussion below). Thus, the levels of available estrogen in mammals may be so low as to require not only soluble cytoplasmic receptors (see Chap. III, Sect. 3) but also additional binding sites such as albumins and Type II estrogen-specific sites to retain steroids for the production of hormone-induced responses.

Finally, one must consider that blood binders and/or Type II sites might function to promote the entry of steroids into target cells via a mechanism such as that demonstrated for cholesterol and low-density lipoprotein (Goldstein and Brown, 1977; see Chap. III for brief discussion). In such a model these low-affinity binding sites, or their complexes with steroid, would serve as ligands for specific receptors on

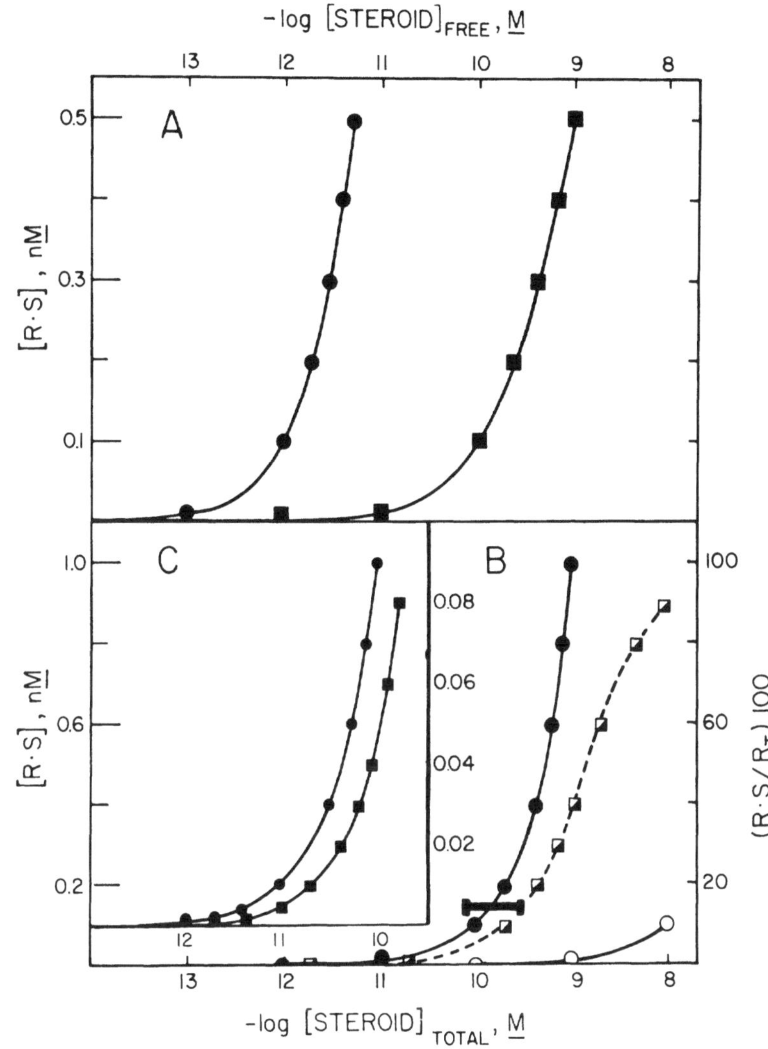

Fig. 113 A–C. Receptor steroid complexes as a function of free and total steroid. Saturation plots have been modeled for the following circumstances: $K_d = 10^{-9}M$ and $[R] = 10^{-9}M$ (■, □) or $10^{-7}M$ (●, ○). **A** Saturation is shown as a function of *free* steroid. **B** Both the level of *RS* (*closed symbols*) and the percent of receptor occupied by ligand (*open symbols*) are given as a function of *total* steroid for a simple single-binding site system. The *bar* represents the range of total estradiol observed during the estrous cycle of the rat (Butcher et al., 1974). **C** Note that the inset is merely an expansion of scale to examine extremely low concentrations of total steroid and *RS*. The sale to the right of inset **C** gives n*M* concentrations of [*RS*] observed

target cell surfaces. Subsequent to complex formation with these surface elements, the complexes would be internalized via endocytosis. At present, no data exists to establish such a route; however, its existence could explain the complexities above. As noted in Chap. III, the development of estrogen target cells in culture will likely yield the answer to this intriguing question.

A final answer to such questions, however, will require detailed knowledge of the subcellular compartmentalization or specialization of receptors and receptor-steroid complexes. Do regions exist with abnormally high densities of receptors and/or steroid? In such a case, the presence of spare receptors may serve the role suggested by Levitzki (1976) and discussed below.

2. Spare Receptors and Hormone-Induced Responses

An alternative explanation of hormone-induced responses when concentrations of steroid are far below the dissociation constant of their complex with receptor has been proposed by Levitzki (1976). In this model, high concentrations of receptor are employed to drive a linearly responsive system. This situation is entirely analogous to that we have discussed in Chap. II where $[R] >>> K_d$ in some saturation analyses of receptors. If one assumes that 20,000 receptor molecules exist in a uterine cell 1,000 μm^3 in volume, the receptor concentration is $3 \times 10^{-8} M$, assuming that receptors are free to move within the entire space. Since unoccupied receptors exist initially in the cytoplasmic space and since nuclear volume may account for 50 percent of the cellular volume in quiescent cells, the concentration of R_c may approach $10^{-7} M$ or 100 times the effective K_d of R_cS. Under these conditions, as discussed in Chap. II, saturation of R with S is a linear function of total steroid, and essentially all available S is bound to R. Figure 113B and C contrasts the saturation of R by S for the above condition, i. e., $[R] = 10^{-7} M$, with that where $[R] = 10^{-9} M$ or the effective K_d for the RS complex. Note that high concentrations of R (o) relative to K_d allows the saturation of at least twice as many receptors for a given concentration of total steroid as observed for low concentrations of R. Thus, to localize or concentrate R may allow a more responsive system despite low levels of circulating steroid.

This is an attractive hypothesis since it could explain the existence of the excess or "spare" receptors observed in target tissues such as the uterus. However, unless one invokes the compartmentalization or clustering of high-affinity receptors within restricted locales, it appears to be incomplete. Certainly under normal circumstances with soluble and mobile cytoplasmic receptors, the sizeable concentration of receptors (100 K_d) discussed above does not alter greatly the occupation of receptors by ligand when the concentration of total ligand is below or near the K_d of its complex with receptor (see Fig. 113B and C). In this figure, the number of occupied receptor sites is plotted as a function of total steroid (i. e., bound plus free or total available) for two conditions — when [R] is at its presumed K_d, nM, and when [R] is 100-fold in excess of that concentration. Note that the number of occupied receptors is only 2-fold different for a given level of total available steroid until steroid levels approach the K_d of the RS complex. Furthermore, while the change in percent of receptor occupancy is rather large for the case of low receptor concentrations (10% → 50% occupancy for 1 log unit of steroid), under conditions of high receptor concentration the change in percent of occupancy is 0 01% → 0.1% for the same change in total steroid. With such minute changes in percent of occupancy, despite the large numbers involved, it seems doubtful that such a mechanism alone can serve the case of steroid hormone action. Other concentrating mechanisms, as the restriction of receptor locale, seem necessary to explain the receptor occupancy observed during the cycle.

An additional observation suggesting the presence of multiple or cooperative concentrating mechanisms derives from a comparison of the dynamics of receptor occupancy and serum steroid levels. The blood level of estrogens varies about 4-fold during the estrous cycle of the rat, and the number of nuclear R_nE complexes varies to about the same extent, i.e., 4–6 fold. Thus, occupancy follows a linear relationship with total steroid as expected if $[R] >>> K_d$. However, 60 percent of the available receptors are occupied at proestrus (see Chap. VIII for references) while total estradiol concentrations are 85 pg/ml (approximately 3×10^{-10} M) (Butcher et al., 1974). Such low levels of total steroid could not result in the occupancy of 60 percent of the total available receptor (Fig. 113B), even in a linear system without additional mechanisms to ensure high local concentration of free hormone. Blood and/or tissue estrogen-buffering components may function to elevate or maintain estrogen at locally high levels. Alternatively, small numbers of receptors may be restricted to minute locales. Undoubtedly these disparities relate to our inadequate knowledge of levels of estrogen available from the circulation and at the tissue level.

3. Nuclear Acceptors and Salt-Insoluble Receptors

The growth of the rat uterus observed 12–24 h after estrogen stimulation has been classified as true growth. True growth is characterized by increased synthesis of DNA, RNA, and protein, resulting in cellular hypertrophy and hyperplasia. These responses are maximized by small quantities of estradiol (0.2 μg/100 g body wt.) and require the occupancy of approximately 10 percent of the total available estrogen receptors for 6 h or longer (see Chap. V for complete discussion of these points). These observations suggest to us that the retention of receptor-estrogen complexes for relatively long periods of time might result from the specific association of receptor-estrogen complexes with nuclear acceptor sites and that subsequent nuclear processing of the receptor-estrogen complex may occur. We have suggested that this acceptor site binding is of a greater affinity than the binding of receptor-estrogen complexes to nonacceptor sites and that this difference can be detected by differential salt extraction (Clark and Peck, 1976a). Receptor-estrogen complexes that resist extraction by 0.4 M KCl appear to be associated with a limited number of nuclear sites, which correlate with the number required to maximize uterine growth (Fig. 53).

Some investigators have questioned the concept of salt insoluble acceptor site complexes as discussed above and in Chap. V (Juliano and Stancel, 1976; Traish et al., 1977; Muller et al., 1977). Their experiments, with one exception, involved the injection of ^{3}H-estradiol, subsequent isolation of target tissue nuclei, and salt extraction of nuclear pellets as a function of time. In those studies 90–95 percent of the ^{3}H-estradiol was extracted from the nuclear fraction. Thus, they suggested that the remaining 5–10 percent resulted from trapping of receptor-hormone complexes by the gelatinous nuclear pellet. Although salt-resistant sites may be fortuitous, these experiments are not adequate to test the hypothesis. Our experiments involved the injection of nonlabeled estrogen with subsequent measurement of nuclear-bound receptor by the ^{3}H-estradiol exchange assay. In this manner the presence of high-affinity, specific estrogen-binding sites was measured — not simply radioactivity due to ^{3}H-steroid. These high-affinity, estrogen-specific nuclear bound sites more accurately reflect the estrogen receptor than either extracted or residual label, which may or not be due to receptor.

An additional point concerning this disparity should be mentioned. In our experiments a single KCl extraction was performed with 0.4 *M* KCl, not several extractions with 0.6 *M* KCl. Continued extraction with elevated salt is likely to remove all hormone and receptor, but this does not discount the fact that following one extraction 10–15 percent of the receptor sites remain in the nuclear fraction. The observations of Barrack et al. (1977) as well as experiments from our own laboratory have examined this point. Virtually all of the ^{3}H-estradiol present in uterine nuclei following an injection can be extracted; however, 2,000–3,000 receptor sites per nucleus can be demonstrated in these extracted nuclei by ^{3}H-estradiol exchange. A simple explanation for these results is that hormone is more readily extracted than tightly associated receptors. Alternatively, these observations may reflect the dimer state of nuclear-bound receptor (see Chap. IV for discussion of possible subunit nature of the estrogen receptor). Thus, salt extraction may remove one subunit and/or the ^{3}H-estradiol bound to that subunit but leave the other subunit (with or without its associated steroid) bound to nuclear sites. This remaining subunit could then be measured by ^{3}H-estradiol exchange. It is also possible that salt-insoluble forms are present only in certain cell types of the uterus. Perhaps salt-resistant sites only occur in cells of the uterus which constitute less than 5–10 percent of the total. If this were the case, interesting physiologic relationships might exist between those cells containing salt-resistant sites and those with readily extractable nuclear complexes (see below).

Barrick et al. (1977) have suggested that salt-insoluble sites are bound to the nuclear matrix. The nuclear matrix is made up of proteins that form the residual elements of the nuclear membrane, nucleolus, and interchromatic structures (Berezney and Coffey, 1977). The relationship between the association of occupied receptor with matrix and the control of cell function is not understood at present. The binding of receptor-estrogen complexes to chromatin in vivo may involve the occupation of acceptor sites within the nuclear matrix. This interaction could be a key function by which estrogens stimulate cell division. This hypothesis will be very attractive if salt-insoluble sites are found associated with those target cells that undergo mitosis following estrogen treatment (as in the uterus) but not to those not undergoing division (as in the brain).

4. Nuclear Processing of the Receptor-Estrogen Complex

Long-term retention or acceptorlike binding of the receptor-estrogen complex to nuclear sites may reflect some processing step(s) as discussed earlier (in Chap. IV). This concept is reinforced by in vitro studies with intact uteri and the actions of triphenylethylene derivatives in vivo. As shown in Fig. 114 the rate of R_nE dissociation is very slow and appears to be unaffected by a drastic shift in the cytoplasmic concentration of ^{3}H-estradiol. These experiments involved the incubation of intact uteri for 60 min at 37°C with 10^{-8} *M* ^{3}H-estradiol followed by exposure of the uteri to fresh medium that contained no ^{3}H-estradiol. Such an exposure should, of course, shift the equilibrium toward dissociation of R_nE complexes, a process which should occur very rapidly at this temperature. However, as shown in Fig. 114, the rate of dissociation of R_nE is very slow under these circumstances, implying that estradiol is "locked" within the nucleus in some manner, presumably at the acceptor site. This phenomenon has also been observed by Russell and Thomas (1974), who demon-

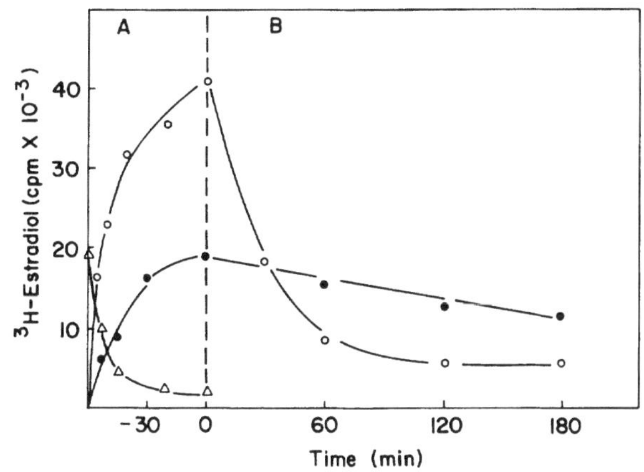

Fig. 114 A and B. Dissociation of receptor steroid complexes in intact uteri in vitro. Intact uteri from 21-day-old rats were incubated at 37°C in Eagle's medium containing $10^{-8}M$ ^{3}H-estradiol for varying times. After 60 min exposure, exogenous ^{3}H-estradiol was removed by washing with fresh medium, and the incubation was continued. Cytoplasmic receptor (Δ), R_nE (●), and total ^{3}H-estradiol (O) were assayed by scintillation spectrometry as a function of time

strated retention of the R_nE complex for 24 h in cultured rabbit uterus. Using a different approach, Gannon and Gorski (1976) were able to "trap" unfilled receptor in the nuclear fraction after exposure of uteri in vitro to hypertonic sucrose, ethanol, or mercurials. This nuclear "trapping" or binding of receptor was not followed by a redistribution of receptor to the cytoplasm when uteri were placed in fresh medium. It is possible that processing of R_nE complexes is not initiated in vitro in the uterus, or, perhaps, that processing is blocked at some step which permits the detection of an intermediate state. Thus, because of this "blockade," we observe in vitro a function that occurs in vivo and may reflect nuclear processing of the R_nE complex.

Triphenylethylene derivatives such as Nafoxidine cause the retention of receptor for very long periods of time within uterine nuclei. In addition, R_c replenishment does not occur as rapidly in animals so treated as it does in E_2-treated animals (Fig. 115). This long-term retention of receptor within the nucleus is accompanied by a sustained stimulation of uterine size, which is mainly attributable to hypertrophy of the uterine epithelium (see Chap. VI). This sustained stimulation suggests that long term nuclear retention of R_nN complexes results from rebinding of Nafoxidine to newly replenished R_c because Nafoxidine is retained in the body for long periods of time. As discussed in Chap. VI, the slow turnover of these compounds is undoubtedly involved and could account for our failure to measure replenished cytoplasmic receptor; however, slow turnover cannot account for the following observations:

1. As little as 5–10 µg Nafoxidine will promote nuclear retention without significant R_c replenishment for 72 h. It is very unlikely that such small quantities of Nafoxidine could maintain tissue levels high enough to cause constant binding to replenished R_c.
2. When estradiol was implanted at a level that caused nuclear retention for 3 days, similar to that observed with a single injection of Nafoxidine, R_c replenishment took place as usual and was elevated to very high levels (Fig. 115). Thus, R_c replenish-

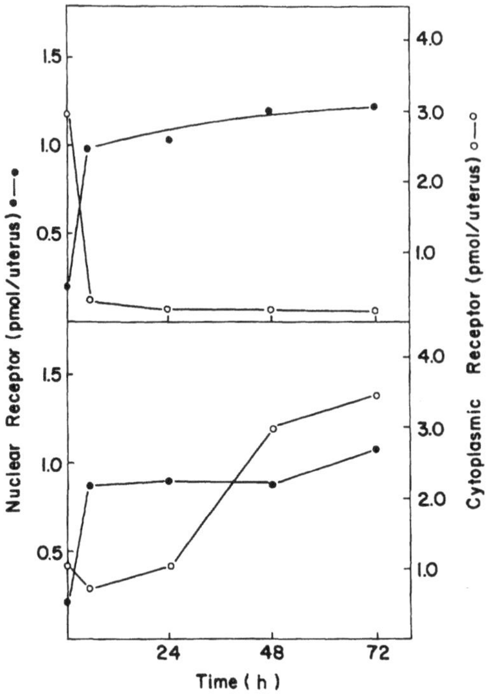

Fig. 115. Nuclear retention and cytoplasmic replenishment of R_n and R_c: triphenlethylenes vs. estradiol. Rats were injected with 100 μg Nafoxidine (*upper panel*) or implanted with estradiol (*lower panel*), and both cytoplasmic (○) and nuclear (●) receptor were measured by exchange as a function of time after treatment

ment does take place in the presence of high concentrations of estrogen and high concentrations of nuclear R_nE, but not in the presence of R_nN. Therefore some basic difference must exist in the way in which R_nN complexes interact with nuclear binding sites.

The binding of R_nE complexes to specific nuclear-acceptor sites may require processing factors that are actively involved in the mechanisms by which R_nE complexes stimulate transcriptional events and/or eventually leave the nucleus. If such processing factors exist, they might not recognize R_nN complexes. This reduced regognition by nuclear-processing mechanisms could result from differences in the receptor-ligand complex or in the ligand alone. The chemical structures of the triphenylethylene derivatives were compared with estradiol in Chap. VI, Fig. 82. These compounds have a ring structure that appears to bind to the active site of the estrogen receptor (Capony and Rochefort, 1975). However, the triphenylethylene derivatives differ considerably from estrogens in having bulky groups at the sixth position of the B-ring. The presence of this bulky side chain may account for the reduced affinity of the receptor for these compounds and may also account for the failure of nuclear processing factors to recognize these ligands when complexed with the receptor.

In conclusion, we suggest that growth-promoting steroids like estrogen bind to a limited number of receptors to form complexes that associate with acceptor sites in the nucleus. The binding of the R_nS complex to acceptor sites and the subsequent reactivation or replenishment of R_c involves a processing of R_nE in the nucleus. Thus, processing is involved in the control of replenishment of R_c as well as in the control of normal cell growth.

5. Homeostatic vs. Growth-Regulating Steroid Hormones

Estrogen and progesterone have been considered throughout, along with other steroids, as a homogeneous set of hormones. That is, their mechanism of action has been discussed without regard for possible, basic differences. Since all steroid hormones likely share the fundamental receptor-binding interactions discussed throughout this monograph, this presumption is in part justifiable. However, in this section, we will present possible differences among the binding-response relationships of various steroid hormones and elaborate on possible mechanisms for these differences.

In general, steroid hormones can be divided into two types: (a) homeostatic regulators, such as the mineralocorticoids and glucocorticoids and (b) growth stimulators, such as estrogens, androgens, and progesterones. These categories are not absolute since it is clear that glucocorticoids may act as growth promoters during certain developmental periods (Greengard, 1970, 1975) and that the sex hormones may function to produce homeostatic effects in the brain (see Chap. IX). However, in the adult animal and with the exception of the nervous system, these two categories generally apply. It seems likely that hormones that cause a tissue to grow must stimulate a greater number of processes than those that regulate a few specific cellular functions as do the homeostatic steroids. Therefore, growth-promoting hormones may display more complex mechanisms by which they stimulate cellular biosynthetic activity. Potential schemes by which these classes of hormone might act are shown in Fig. 116.

The simplest involves a reversible chemical equilibrium whereby the concentration of free hormone determines the quantity of R_nS bound to chromatin via mass action (Fig. 116A). Because the reaction is reversible, fluctuations in free steroid regulate receptor binding and cell function. This scheme represents one which should be very

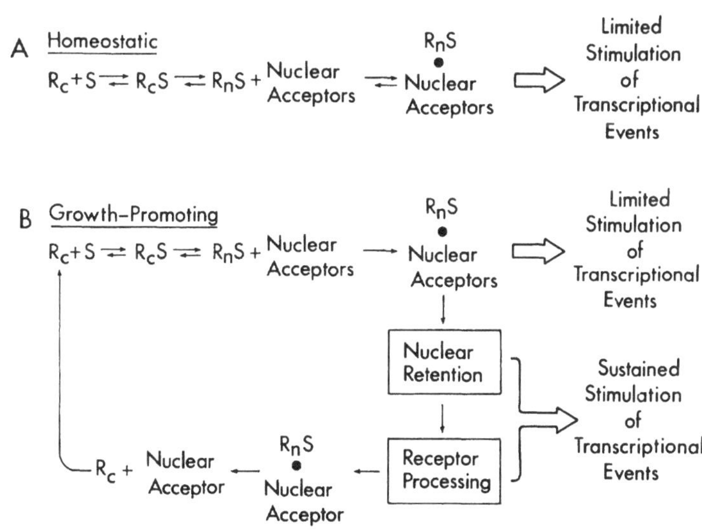

Fig. 116 A and B. Homeostatic vs. growth-promoting hormones: models of cellular interaction. This figure contrasts the nuclear interactions proposed in the test for homeostatic and growth-promoting steroids

sensitive to rapid changes in hormone levels and is characterized by brief periods of occupancy of nuclear acceptor sites by R_nS complexes. This mechanism is probably involved in the action of glucocorticoids and mineralocorticoids as well as sex steroids when they act as homeostatic regulators, as in the adult central nervous system. In fact, the response to homeostatic hormones is proportional to the amount of R_nS within the nuclear fraction, and thus a maximal response is reached when 100 percent of the available receptor sites are in the nucleus. This is the case of glucocorticoids as regulators of tyrosine aminotransferase and tryptophan oxygenase activity (Baxter et al., 1973; Beato et al., 1972) and is probably true for regulation of sodium resorption by aldosterone (Anderson and Fanestil, 1978). Thus, homeostatic regulators can be easily visualized as binding according to a reversible chemical equilibrium as shown in Fig. 116A and the level of free S in the blood is the primary effector of cell function. This, of course, assumes that receptor levels remain constant.

The second scheme (Fig. 116B) is more complex in that in addition to the mechanisms employed by homeostatic hormones R_nS complexes also bind to nuclear sites and remain for an extended period in the nuclear compartment. This is one of the characteristics of growth-promoting hormones such as estrogen. Whether this mechanism is characteristic of all growth-promoting steroids is not known.

In Chap. V we divided uterotropic responses into early and late categories relative to their time of expression after hormone administration. Early responses include stimulation of "induced protein", water imbibition, histamine mobilization, hyperemia, lysosome labilization, and transient increases in RNA, lipid, and protein synthesis. Several of these early responses are directly proportional to the quantity of nuclear-bound receptor-estrogen complex and manifest characteristics similar to the homeostatic steroids (Fig. 116A). This is also shown in Fig. 51 for the water imbibition response and in studies of the in vitro induction of IP (Katzenellenbogen and Gorski, 1972).

Late uterotropic responses, which include sustained RNA and protein synthesis and cell growth, are not proportional to the quantity of occupied receptor. Instead, these responses are maximized when 10–20 percent of the total number of available receptors are occupied and retained in the nucleus. These nuclear-bound receptor-estrogen complexes must be retained for 6 h or longer to promote true growth of the uterus. This extended period of occupancy may allow a search for nuclear-acceptor sites and/or a processing of receptor-hormone complexes, which results not only in the stimulation of RNA and DNA synthesis, but inactivation of the receptor-hormone complex. Palmiter et al. (1976) and Buller and O'Malley (1976) have suggested that receptor-steroid complexes bind to nuclear sites and move along chromatin in search of specific acceptor sites. This nuclear search function may well be a component of the total picture of nuclear retention; however, the interaction of the R_nE complex and acceptor sites could also be a pseudo-irreversible phenomenon. R_nE complexes may undergo conformational changes when they bind to acceptor sites that do not permit the ready dissociation of the complex or the hormone from the complex. For dissociation of the R_nE complex to take place, processing factors may be required which recognize the complex and bring about changes in the receptor and/or hormone, resulting in dissociation of the complex. This processing may be an important step in the replenishment of cytoplasmic receptors. Occupancy of nuclear-acceptor sites by such a pseudo-irreversible mechanism may be responsible for the sustained stimulation of RNA polymerase I and II activities and the elevation of RNA polymerase initiation sites, which appear to be required for true uterine growth.

In reality these schemes proposed for homeostatic and growth-promoting hormones may differ only in a quantitative way and the discrimination of such quantitative differences may be beyond our present limits of resolution. That is, the scheme in Fig. 116A may be a special class of that in Fig. 116B, wherein the nuclear retention of homeostatic receptor-steroid complexes within target tissues may simply be very short. The work of Munck and Foley (1976) with glucocorticoid receptors of thymus cells suggests that this is the case. They observed that the pathway for replenishment of receptor may be different from that of depletion and suggested that rapid nuclear processing occurs. Thus, one can envision a continuum of nuclear retention times varying from very short to very long depending on the hormone and the specific response to be produced.

6. Conclusions

a) Hormone Delivery to Target Cells

The interaction of cellular components with steroid hormones depends on the amount of free hormone available to the cell. This amount is determined by a complex interplay of mechanisms that involve binding components of the blood and intercellular spaces. Neither the interplay nor the mechanisms themselves are completely understood at this time. However, the following are the minimal factors to be considered in deriving a model of these systems:

1. Steroids are bound with different affinities by specific and nonspecific blood-binding components. These interactions determine the quantity of free steroid available for entry into cells. A steroid hormone that is not bound tightly to blood-binding sites is physiologically more potent than one which is bound tightly, provided that all other variables remain constant (see Chaps. III and VIII for discussion).

2. Steroids have different metabolic clearance rates, and relative physiologic effectiveness depends on this variable. Thus, steroid receptors may have a low affinity for a given hormone and yet, if this hormone has a long half-life in the body, its potency may be greater than anticipated (see Chap. VIII).

3. Steroids that are bound to blood components may enter intercellular spaces of some organs via protein-permeable vascular beds, thereby creating a local elevation of total steroid. Subsequent dissociation of these complexes will allow the maintenance of free steroid levels within the intercellular space (see Chap. III and this chapter).

4. Steroids can also enter the intercellular space in the free form. There they may bind to intercellular proteins. One may be that which we have called Type II (see Chap. IV and previous discussion in this chapter). These intercellular sites may augment the ability of blood-binding components to maintain elevated tissue levels of hormone. Alternatively, steroids bound to intercellular sites may be accumulated actively by target cells in a manner analogous to the uptake of lipoprotein-cholesterol complexes (see Chap. III). This could constitute a major concentrating mechanism in some tissues.

b) Hormone Binding to Receptors in the Cytoplasm

Steroid hormones enter cells by diffusion and, in addition, perhaps by active processes discussed above and in Chap. III. Within target cells the steroid binds to soluble macromolecules called receptors. In this section we will consider the initial binding interaction to take place in the cytoplasm; however, it should be noted that this may be an oversimplification (see Chap. III for discussion).

1. Estrogen and progesterone receptors are multimeric proteins specific for their respective hormones. These receptors may consist of two different 4*S* subunits, both of which bind steroid (see Chap. IV).

2. When cytoplasmic receptors bind steroid hormones, they form a complex subsequently found in the nucleus. The nuclear accumulation of receptor-steroid complexes is called translocation. For the estrogen receptor this process may involve a dimerization of subunits whereas the progesterone receptor may exist as a dimer, which subsequently dissociates upon translocation (see Chap. IV).

3. Other secondary estrogen-specific binding sites may exist in the cytoplasm of target cells (see Chap. IV). At present we only know that they exist in the cytosol fraction and cannot be certain of their exact compartmentalization. These Type II sites have a sedimentation coefficient of 4*S* and are protein in nature. These sites do not undergo nuclear translocation; however, their number is increased by estrogen treatment. Thus, they may be of physiologic importance. For instance, Type II sites may concentrate estrogens in a manner similar to that discussed for intercellular mechanisms. Therefore free steroid could be available for binding to receptor by dissociation from these sites, as well as by direct diffusion into the cytoplasm.

c) Nuclear Binding of Receptor-Hormone Complexes and Hormone-Stimulated Events

1. The translocation of receptor-estrogen complexes to the nuclear compartment initiates the stimulation of early or homeostatic estrogen-induced responses. The extent of these responses, such as water imbibition, transient activation of RNA polymerase and glucose utilization, appears directly related to the number of R_nE complexes. A second class of responses culminating in growth do not correlate directly with this number. As discussed in Chap. IX, some steroid targets, such as the adult brain, may only demonstrate early or transient responses. Their inability to manifest late or growth responses may relate to their high degree of differentiation and/or their lack of the appropriate processing machinery.

2. The initial binding of receptor-estrogen complexes is not sufficient to cause true uterine growth, even though a full stimulation of early uterotropic events is accomplished (see Chap. V and above). Thus, a cascade mechanism eventually resulting in cell growth is ruled out in the uterus. True uterine growth requires nuclear retention or sustained occupancy of nuclear-acceptor sites by receptor-estrogen complexes.

3. Receptor-steroid complexes bind to specific and nonspecific sites within the nucleus. Receptor-estradiol complexes bind to specific acceptor sites on chromatin with greater affinity than they bind to nonacceptor sites (see Chaps. IV and V). Acceptor-site binding of receptor-estradiol complexes is limited to approximately 1,000–2,000 sites/cell in the uterus. Receptor-estradiol complexes bound to acceptor sites are not readily solubilized by high-salt extraction, whereas those complexes not bound to acceptor sites are easily extracted with high salt concentration.

4. Nuclear retention of receptor-estrogen complexes at acceptor sites may involve a pseudo-irreversible mechanism responsible for the stimulation of transcription and biosynthetic events, which result in cell growth and receptor replenishment (see Chap. V and this chapter). This mechanism may be unavailable to steroid targets, such as the adult central nervous system, which have differentiated such that they can no longer undergo mitosis. This mechanism may also be unavailable to receptors of the "homeostatic" steroids.

5. "Receptor-processing" may result in the elevation of secondary estrogen binding sites (Type II) in the nucleus (see Chap. IV). The function of these sites is not known; however, they may be components of the nuclear processing machinery which interact with the receptor-estradiol complex and bring about recycling or replenishment of the receptor; integral components of the mechanisms which control RNA and/or DNA synthesis; and components of the "off-reaction" involved either in turning off receptor-stimulated events or in removing hormone and/or receptor from the nucleus.

6. Replenishment of the cytoplasmic estrogen receptor probably occurs both by de novo synthesis and by recycling from the nucleus. Replenishment is stimulated by nuclear binding of the receptor-estrogen complex and correlates with the extent of cellular hypertrophy and hyperplasia (see Chap. VI).

7. The level of cytoplasmic progesterone receptor is a function of the estrogenic state of the target tissue. Thus, estrogen appears to control the synthesis of progesterone receptors. Nuclear binding of receptor-progesterone complex is followed by a loss in total receptor unless sufficient estrogen is provided.

d) Antagonism of Hormone Action

1. Progesterone appears to decrease the ability of the uterus to respond to estradiol by reducing nuclear retention and cytoplasmic replenishment of the estrogen receptor. The reduced replenishment may be due to an inhibition of receptor synthesis. This ability of progesterone to antagonize estrogen action is an important component in the mechanism controlling reproductive cyclicity (see Chap. VI).

2. Since the level of cytoplasmic progesterone receptor is dependent on the estrogenic state of the target and since progesterone reduces the responsiveness of targets to estrogens, progesterone reduces the cytoplasmic levels of its own receptor via serial antagonism. This serial antagonism may be involved in resetting the biologic clock controlling reproductive cyclicity.

3. Estrogenic compounds can be broadly classified as short- and long-acting. This classification is based on their ability to maintain occupancy of the estrogen receptor in the nucleus of uterine cells (see Chap. VI). Short-acting estrogens, such as estriol, when administered by injection reduce the quantity of receptor-estradiol complexes bound to nuclear sites and thereby act as antagonists. These short-acting estrogens, if present in a continuous fashion, are not antagonists since sustained receptor occupancy results in sustained estrogenic responses. Thus, the temporal aspects of steroid clearance and receptor occupancy are important determinants in their classification as agonists or antagonists.

4. Triphenylethylene derivatives, such as Nafoxidine or Clomid, bind to estrogen receptors and cause translocation of the receptor to the nucleus. The nuclear binding is abnormal in that it is of very long duration (see Chaps. VI and VII). In addition no replenishment of the cytoplasmic estrogen receptor takes place during this period. Reduced replenishment of the estrogen receptor and/or inappropriate nuclear binding of the receptor drug complex reduces the effectiveness of estradiol and results in antagonism. These compounds have the dual capacity to act as estrogens in epithelial cells of the uterus while acting as anti-estrogens in the stroma and myometrium. This differential cell effect accounts for the ability of these drugs to partially stimulate uterine growth.

e) Neoplasia and Estrogens

1. Continuous exposure to estrogen during fetal and neonatal periods can result in cancer (see Chap. VII). Clomid and Nafoxidine given during these periods also produce abnormalities and tumors. These may result from the ability of these drugs to cause long-term nuclear retention of the receptor-estrogen complex which mimics continuous exposure to estrogens. This continuous estrogenic stimulation is observed primarily in the epithelial structures of the reproductive tract and results in hyperestrogenization of this tissue.

2. Hyperestrogenization by triphenylethylene derivatives or continuous exposure to estradiol causes extensive cellular hypertrophy of the epithelial cells of the uterus. This is accompanied by continuous residency of the receptor-ligand complex in the nucleus, elevated RNA polymerase activity, and the appearance of several types of nuclear bodies. The function of these nuclear bodies is unknown. They may be associated with an augmentation of RNA synthesis.

7. Epilogue

We realize that this chapter contains as much speculation as fact. However, throughout this volume we have attempted to assimilate our current knowledge of gonadal steroid receptors and their relationship to steroid-induced responses. In addition, where appropriate, we have interjected our own reflections concerning these relationships. The interface between hard fact and conjecture may at times seem gray indeed. However, we consider speculation to be an integral part of the model-building process. Since models are important to the design of future experiments, our efforts at integrating fact and fantasy will have been a success if it assists ourselves and others in the design of fruitful experiments.

References

Aizawa, Y., Mueller, G. C.: The effect in vivo of estrogens on lipid synthesis in the rat uterus. J. Biol. Chem. *236*, 281–286 (1961)

Ajika, K., Krulich, L., Fawcett, C. P., McCann, S. M.: Effects of estrogen on plasma and pituitary gonadotropins and prolactin, and on hypothalamic releasing and inhibition factors. Neuroendocrinology *9*, 304–315 (1972)

Alberga, A., Massol, N., Raynaud, J.-P., Baulieu, E.-E.: Estradiol binding of exceptionally high affinity by a nonhistone chromatin fraction. Biochemistry *10*, 3835–3843 (1971)

Allen, E., Gardner, W. V.: Cancer of the cervix and of the uterus in hybrid mice following long continuous administration of estrogen. Cancer Res. *1*, 359–366 (1941)

Allen, E., Smith, G. M., Gardner, W. V.: Accentuation of the growth effect of theelin on genital tissues of the ovariectomized mouse by arrest of mitosis with colchicine. Am. J. Anat. *61*, 321–341 (1937)

Anderson, N. S., Fanestil, D. D.: Biology of mineralocorticoid receptors. In: Receptors and Hormone Action. O'Malley, B. W., Birnbaumer, L. (eds.), pp. 323–351. New York: Academic Press 1978

Anderson, C. H., Greenwald, G. S.: Autoradiographic analysis of estradiol uptake in the brain and pituitary of the female rat. Endocrinology *85*, 1160–1165 (1969)

Anderson, K. M., Liao, S.: Selective retention of dihydrotestosterone by prostatic nuclei. Nature *219*, 277–279 (1968)

Anderson, J. N., Clark, J. H., Peck, E. J., Jr.: Oestrogen and nuclear binding sites: determination of specific sites by ^{3}-H-oestradiol exchange. Biochem. J. *126*, 561–567 (1972a)

Anderson, J. N., Clark, J. H., Peck, E. J., Jr.: The relationship between nuclear receptor estrogen binding and uterotrophic responses. Biochem. Biophys. Res. Commun. *48*, 1460–1468 (1972b)

Anderson, J. N., Peck, E. J., Jr., Clark, J. H.: Nuclear receptor estrogen complex: relationship between concentration and early uterotrophic responses. Endocrinology *92*, 1488–1495 (1973a)

Anderson, J. N., Peck, E. J., Jr., Clark, J. H.: Nuclear receptor estrogen complex: accumulation, retention and localization in the hypothalamus and pituitary. Endocrinology *93*, 711–717 (1973b)

Anderson, J. N., Peck, E. J., Jr., Clark, J. H.: Nuclear receptor estrogen complex: *in vivo* and *in vitro* binding of estradiol and estriol as influenced by serum albumin. J. Steroid Biochem. *5*, 103–107 (1974a)

Anderson, J. N., Peck, E. J., Jr., Clark, J. H.: Nuclear receptor estradiol complex: a requirement for uterotropic responses. Endocrinology *95*, 174–178 (1974b)

Anderson, J. N., Peck, E. J., Jr., Clark, J. H.: Estrogen-induced uterine responses and growth: relationship to receptor estrogen binding by uterine nuclei. Endocrinology *96*, 160–167 (1975)

Andre, J., Rochefort, H.: In vitro binding of the estrogen receptor to DNA: absence of saturation at equilibrium. FEBS Letters *50*, 319–323 (1975)

Araki, S., Ferin, M., Zimmermann, A., Vande Wiele, R. L.: Ovarian modulation of immunoreactive gonadotropin-releasing hormone (Gn-RH) in the rat brain: Evidence for a differential effect on the anterior and mid-hypothalamus. Endocrinology *96*, 644–650 (1975a)

Araki, S., Dominique Toran-Allerand, C., Ferin, M., Vande Wiele, R. L.: Immunoreactive gonadotropin-releasing hormone (Gn-RH) during maturation in the rat: Ontogeny of regional hypothalamic differences. Endocrinology *97*, 693–697 (1975b)

Ariëns, E. J.: Affinity and intrinsic activity in the theory of competitive inhibition. I. Problems and theory. Arch. Int. Pharmacodyn. Ther. *99*, 32–49 (1954)

Astwood, E. B.: A six-hour assay for the quantitative determination of estrogen. Endocrinology *23*, 25–31 (1938)

Astwood, E. B.: Changes in the weight and water content of the uterus of the normal adult rat. Am. J. Physiol. *126*, 162–170 (1939)

Atger, M., Milgrom, E.: Chromatographic separation on phosphocellulose of activated and nonactivated forms of steroid receptor complex. Purification of the activated complex. Biochemistry *15*, 4298–4304 (1976)

Atger, M., Baulieu, E.-E., Milgrom, E.: An investigation of progesterone receptors in the guinea pig vagina, uterine cervix, mammary glands, piuitary and hypothalamus. Endocrinology *94*, 161–167 (1974)

Attardi, B., Ruoslaht, E.: Foetoneonatal oestradiol-binding protein in mouse brain cytosol is α foetoprotein. Nature *236*, 685–687 (1976)

Attramadal, A.: Distribution and site of action of oestradiol in the brain and pituitary gland of rat following intramuscular administration. In: Proc. 2nd Int. Congr. of Endocrinology. Amsterdam: Excerpta Medica International Congress Series, No. 83, pp. 612–616, 1964

Axelrod, L. R., Pao, P. N., Goldzieher, J. W.: The conversion of 2-hydroxyestradiol-17B to 2-hydroxy and 2-methoxy metabolites in human urine. Arch. Biochem. Biophys. *94*, 265–268 (1961)

Bahr, J. M., Nalbandov, A. V.: Reproduction in poultry. In: Reproduction in domestic animals, 3rd. ed. Cole, H. H., Cupps, P. T. (eds.), pp. 529–552. New York: Academic Press 1977

Baker, L. T.: Transport and metabolism of calcium ions in nerves. Prog. Biophys. Mol. Biol. *24*, 179–223 (1972)

Ball, P., Farthmann, E., Knupper, R.: Comparative studies on the metabolism of estradiol-17β and 2-hydroxyestradiol-17β in man in vitro and in vivo. J. Steroid Biochem. *7*, 139–143 (1976)

Barber, H. R. K., Sommers, S. C.: Vaginal adenosis, dysplasia, and clear cell adenocarcinoma after diethylstilbestrol treatment in pregnancy. Obstet. Gynecol. *43*, 645–652 (1974)

Bardin, C. W., Bullock, L. P., Sherins, L. J., Mowszowicz, I., Blackburn, W. R.: Part II. Androgen metabolism and mechanism of action in male pseudohermaphroditism: a study of testicular feminization. Rec. Prog. Horm. Res. *29*, 65–109 (1973)

Bardin, C. W., Bullock, L. P., Janne, O., Jacob, S. T.: Genetic regulation of the androgen receptor – A study of testicular feminization in the mouse. J. Steroid Biochem. *6*, 515–520 (1975)

Barker, K. L.: Estrogen-induced synthesis of histones and a specific non-histone protein in the uterus. Biochemistry *10*, 284–291 (1971)

Barker, K. L., Warren, J. C.: Template capacity of uterine chromatin: control by estradiol. Proc. Natl. Acad. Sci. USA *56*, 1298–1302 (1966)

Barnea, A., Gorski, J.: Estrogen-induced protein: time course of synthesis. Biochemistry *9*, 1899–1904 (1970)

Barnea, A., Ben-Jonathan, N., Colston, C., Johnston, J. M., Porter, J. C.: Differential subcellular compartmentalization of thyrotropin releasing hormone (TRH) and gonadotropin releasing hormone (LRH) in hypothalamic tissue. Proc. Natl. Acad. Sci. USA *72*, 3153–3157 (1975)

Barnea, A., Ben-Jonathan, N., Porter, J. C.: Characterization of hypothalamic subcellular particles containing luteinizing hormone releasing hormone and thyrotropin releasing hormone. J. Neurochem. *27*, 477–484 (1976)

Barnea, A., Neaves, W. B., Porter, J. C.: Ontogeny of the subcellular compartmentalization of thyrotropin releasing hormone and luteinizing hormone releasing hormone in the rat hypothalamus. Endocrinology *100*, 1068–1079 (1977)

Barrack, E. R., Hawkins, E. F., Allen, S. L., Hicks, L. L., Coffey, D. S.: Concepts related to salt resistant estradiol receptors in rat uterine nuclei: nuclear matrix. Biochem. Biophys. Res. Commun. *79*, 829–836 (1977)

Barry, J., Gorski, J.: Uterine ribonucleic acid polymerase: Effect of estrogen on nucleotide incorporation into 3' chain termini. Biochemistry *10*, 2384–2390 (1971)

Baulieu, E.-E., Raynaud, J.-P.: A proportion graph method for measuring binding systems. Eur. J. Biochem. *13*, 293–304 (1970)

Baulieu, E.-E., Alberga, A., Jung, I., Lebeau, M.-C., Mercier-Bodard, C., Milgrom, E., Raynaud, J.-P., Raynaud-Jammet, C., Rochefort, H., Truong, H., Robel, P.: Metabolism and protein binding of sex steroids in target organs: An approach to the mechanism of hormone action. Rec. Prog. Horm. Res. *27*, 351–419 (1971)

Baulieu, E.-E., Alberga, A., Raynaud-Jammet, C., Wira, C. R.: New look at the very early steps of oestrogen action in uterus. Nature New Biol. *236*, 236–239 (1972a)

Baulieu, E. E., Wira, C. R., Milgrom, E., Raynaud-Jammet, C.: Ribonucleic acid synthesis and oestradiol action in the uterus. In: Gene transcription in reproductive tissue. Karolinska Symposia on Research Methods in Reproductive Endocrinology, 5th Symposium (May 29–31, 1972), pp. 396–419. Stockholm: Karolinska Institutet 1972b

Baulieu, E. E., Atger, M., Best-Belpomme, M., Corvol, P., Courvalin, J. C., Mester, J., Milgrom, E., Robel, P., Rochefort, H., Decatalogne, D.: Steroid hormone receptors. Vitam. Horm. *33*, 649–736 (1975)

Baxter, J. D., Rousseau, G. G., Higgins, S. J., Tomkins, G. M.: Mechanism of glucocorticoid hormone action and of regulation of gene expression in cultured mammalian cells. In: The biochemistry of gene expression in higher organisms. Pollack, J. K., Lee, J. W. (eds.), pp. 206–224. Sydney: Australia and New Zealand Book 1973

Beato, M., Kalimi, M., Feigelson, P.: Correlation between glucocorticoid binding to specific liver cytosol receptors and enzyme induction in vivo. Biochem. Biophys. Res. Commun. *47*, 1464–1472 (1972)

Beers, P. C., Rosner, W.: The binding of estrogen in the liver of the rat: demonstration and endocrine influences. J. Steroid Biochem. *8*, 251–258 (1977)

Belleau, B.: A molecular theory of drug action based on induced conformational perturbations of receptors. J. Med. Chem. *7*, 776–784 (1964)

Ben-Jonathan, N., Mical, R. S., Porter, J. C.: Superfusion of hemipituitaries with portal blood. I. LRF secretion in castrated and diestrous rats. Endocrinology *93*, 497–503 (1973)

Bennett, G. W., Edwardson, J. A., Holland, D., Jeffcoate, S. L., White, N.: Release of immunoreactive luteinizing hormone-releasing hormone and thyrotrophin-releasing hormone from hypothalamic synaptosomes. Nature *257*, 323–325 (1975)

Berezney, R., Coffey, D. S.: Nuclear matrix. Isolation and characterization of a framework structure from rat liver nuclei. J. Cell. Biol. *73*, 616–637 (1977)

Bergink, E. W., Kloosterboer, H. J., Gruber, M., Ab, G.: Estrogen-induced phosphoprotein synthesis in roosters: kinetics of induction. Biochim. Biophys. Acta *294*, 497–506 (1973)

Berthold, A. A.: Transplantation der Hoden. Arch. Anat. Physiol. Wiss. Med. *16*, 42–46 (1849)

Best-Belpomme, M., Fries, J., Erdos, T.: Interactions entre l'oestradiol et des sites recepteurs uterins: donnees clinetique et d'equilibre. Eur. J. Biochem. *17*, 425–432 (1970)

Billing, R. J., Barbiroli, B., Smellie, R. M. S.: The early effects of oestradiol-17β on ribonucleic acid synthesis in rat uterus. Biochem. J. *109*, 705–706 (1968)

Blaha, G. C.: Deciduomata in ovariectomized golden hamsters with locally implanted progestins. Anat. Rec. *178*, 311 (1974)

de Boer, W., de Vries, J., Mulder, E., van der Molen, H. J.: Comparative study of nuclear binding sites for oestradiol in rat testicular and uterine tissue. Determination of low amounts of specific binding site by an [^{3}H] oestradiol-exchange method. Biochem. J. *162*, 331–339 (1977)

Boeynaems, J. M., Dumont, J. E.: Binding of ligands to receptors: Theory. In: Eukaryotic cell function and growth: regulation by intracellular cyclic nucleotides. Dumont, J. E., Brown, B. L., Marshall, N. J. (eds.), pp. 19. New York: Plenum 1975

Bonne, C., Raynaud, J.-P.: Assay of androgen binding sites by exchange with methyltrienolone (R-1881). Steroids *27*, 497–507 (1976)

Borthwick, N. M., Smellie, R. M. S.: The effects of oestradiol-17β on the ribonucleic acid polymerases of immature rat uterus. Biochem. J. *147*, 91–101 (1975)

Bosley, C. G., Leavitt, W. W.: Specificity of progesterone action during the preovulatory period in the cyclic hamster. Fed. Proc. *31*, Abs., 257 (1972)

Boucek, M. M., Synderman, R.: Calcium influx requirement for human neutrophil chemotaxis: Inhibition by lanthanum chloride. Science *193*, 905–907 (1976)

Bouteille, M., Kalifat, S. R., Delarue, J.: Ultrastructural variations of nuclear bodies in human diseases. J. Ultrastruc. Res. *19*, 474–486 (1967)

Boylan, E. S., Wittliff, J. L.: Specific estrogen binding in rat mammary tumors induced by 7,12 dimethylbenz (a) anthracene. Cancer Res. *35*, 506–511 (1975)

Bradbury, J. T.: Direct action of estrogen on the ovary of the immature rat. Endocrinology *68*, 115–120 (1961)

Brecher, P. I., Vigersky, R., Wotiz, H. S.: An in vitro system for the binding of estradiol to rat uterine nuclei. Steroids *10*, 635–651 (1967)

Brenner, R. M., Resko, J. A., West, N. B.: Cyclic changes in oviductal morphology and residual cytoplasmic estradiol binding capacity induced by sequential estradiol-progesterone treatment of spayed rhesus monkeys. Endocrinology *95*, 1094–1104 (1974)

Bronson, F. H., Hamilton, T. H.: A comparison of nucleic acid synthesis in the mouse oviduct and uterus: Interactions between estradiol and progesterone. Biol. Reprod. *6*, 160–167 (1972)

Brown, M. S., Goldstein, J. L.: Receptor-mediated control of cholesterol metabolism. Science *191*, 150–154 (1976a)

Brown, M. S., Goldstein, J. L.: The LDL pathway in human fibroblasts: A system for the study of receptor-mediated adsorptive endocytosis. Trends Biochem. Sci. *1*, 193–195 (1976b)

Brown, M. S., Goldstein, J. L.: Analysis of a mutant strain of human fibroblasts with a defect in the internalization of receptor-bound low density lipoprotein. Cell *9*, 663–674 (1976c)

Bruce, J. O., Ramirez, V. D.: Site of action of the inhibitory effect of estrogen upon lactation. Neuroendocrinology *6*, 19–29 (1970)

Bruchovsky, N., Wilson, J. D.: The conversion of testosterone to 5α-androstan-17β-ol-3-one by rat prostate in vivo and in vitro. J. Biol. Chem. *243*, 2012–2021 (1968a)

Bruchovsky, N., Wilson, J. D.: The intranuclear binding of testosterone and 5α-androstan-17β-ol-3-one by rat prostate. J. Biol. Chem. *243*, 5953–5960 (1968b)

Brush, M. G., Taylor, R. W., King, R. J. B.: The uptake of [6,7,-^{3}H] oestradiol by the normal human reproductive tract. J. Endocrinol. *39*, 599–607 (1967)

Buller, R. E., O'Malley, B. W.: The biology and mechanism of steroid hormone receptor interaction with the eukaryotic nucleus. Biochem. Pharm. *25*, 1–12 (1976)

Buller, R. E., Toft, D. O., Schrader, W. T., O'Malley, B. W.: Progesterone binding components of chick oviduct. VIII. Receptor activation and hormonedependent binding to purified nuclei. J. Biol. Chem. *250*, 801–808 (1975a)

Buller, R. E., Schrader, W. T., O'Malley, B. W.: Progesterone-binding components of chick oviduct. IX. The kinetics of nuclear binding. J. Biol. Chem. *250*, 809–818 (1975b)

Buller, R. E., Schrader, W. T., O'Malley, B. W.: Steroids and the practical aspects of performing binding studies. J. Steroid Biochem. *7*, 321–326 (1976)

Bullock, L. P., Bardin, C. W., Ohno, S.: The androgen insensitive mouse: Absence of intranuclear androgen retention in the kidney. Biochem. Biophys. Res. Commun. *44*, 1537–1543 (1971)

Burke, C. W., Anderson, D. C.: Sex-hormone-binding globulin is an oestrogen amplifier. Nature *240*, 38–40 (1972)

Butcher, R. L., Collins, W. E., Fugo, N. W.: Plasma concentrations of LH, FSH, prolactin, progesterone and estradiol-17β throughout the 4-day estrous cycle of the rat. Endocrinology *94*, 1704–1708 (1974)

Callantine, M. R., Humphrey, R. R., Lee, S. L., Windsor, B. L., Schottin, N. H., O'Brien, O. P.: Action of an estrogen antagonist on reproductive mechanisms in the rat. Endocrinology *79*, 153–167 (1966)

Capony, F., Rochefort, H.: In vivo effect of anti-estrogens on the localization and replenishment of estrogen receptor. Mol. Cell. Endocrinol. *3*, 233–251 (1975)

Capony, F., Rochefort, H.: In vitro and in vivo interaction of [^{3}H-]dimethylstilbestrol with the estrogen receptor. Mol. Cell. Endocrinol. *8*, 47–64 (1977)

Castaneda, E., Liao, S.: The use of anti-steroid antibodies in the characterization of steroid receptors. J. Biol. Chem. *250*, 883–888 (1975)

Chader, G. J., Villee, C. A.: Uptake of oestradiol by the rabbit hypothalamus: Specificity of binding by nuclei in vitro. Biochem. J. *118*, 93–97 (1970a)

Chader, G. J., Villee, C. A.: Estrogen receptors in the hypothalamus. In: Influence of hormones on the nervous system. Proc. Int. Soc. Psychoneuroendocrinol. Ford, D. H. (ed.), pp. 17–24. Brooklyn: Karger (Basel) 1970b

Chamberlin, M. J.: Selectivity of transcription. Ann. Rev. Biochem. *43*, 721–775 (1974)

Chamness, G. C., McGuire, W. L.: Estrogen receptor in the rat uterus, physiological forms and artifacts. Biochemistry *11*, 2466–2472 (1972)

Chamness, G. C., McGuire, W. L.: Scatchard plots: Common errors in correction and interpretation. Steroids *26*, 538–542 (1975)

Chamness, G. C., Jennings, A. W., McGuire, W. L.: Oestrogen receptor binding is not restricted to target nuclei. Nature *241*, 458–460 (1973)

Chamness, G. C., Jennings, A. W., McGuire, W. L.: Estrogen receptor binding to isolated nuclei. A nonsaturable process. Biochemistry *13*, 327–331 (1974)

Chamness, G. C., Huff, K., McGuire, W. L.: Protamine-precipitated estrogen receptor: a solid-phase ligand exchange assay. Steroids *25*, 627–635, (1975a)

Chamness, G. C., Costlow, M. E., McGuire, W. L.: Estrogen receptor in rat liver and its dependence on prolactin. Steroids *26*, 363–371 (1975b)

Chan, L., Means, A. R., O'Malley, B. W.: Rates of steroid hormone induction of specific translatable messenger RNA's for ovalbumin and avidin. Proc. Natl. Acad. Sci. USA *70*, 1870–1874 (1973)

Chan, L., Jackson, R. L., O'Malley, B. W., Means, A. R.: Synthesis of very low density lipoproteins in the cockerel. J. Clin. Invest. *58*, 363–379 (1976)

Changeux, J.-P., Thiery, J., Tung, Y., Kittel, C.: On the cooperativity of biological membranes. Proc. Natl. Acad. Sci. USA *57*, 335–341 (1967)

Church, R. B., McCarthy, B. J.: Unstable nuclear RNA synthesis following estrogen stimulation. Biochem. Biophys. Acta *199*, 103–114 (1970)

Cidlowski, J. A., Munck, A.: Glucocorticoid binding in thymocyte nuclei: Physiologically distinct forms of nuclear interaction. Proc. 59th Annu. Endocrine Soc. Meet., Abs. *191*, p. 152 (1977)

Cidlowski, J. A., Muldoon, T. G.: Estrogenic regulation of cytoplasmic receptor populations in estrogen-responsive tissues of the rat. Endocrinology *95*, 1621–1629 (1974)

Cidlowski, J. A., Muldoon, T. G.: Sex-related differences in the regulation of cytoplasmic estrogen receptor levels in responsive tissues of the rat. Endocrinology *98*, 833–841 (1976)

Clark, A. J.: The mode of action of drugs on cells. Baltimore: Williams & Wilkins 1937

Clark, J. H., Gorski, J.: Estrogen receptors: An evaluation of cytoplasmic-nuclear interactions in a cell-free system and a method for assay. Biochem. Biophys. Acta *192*, 508–515 (1969)

Clark, J. H., Gorski, J.: Ontogeny of the estrogen receptor during early uterine development. Science *169*, 76–78 (1970)

Clark, J. H., McCormack, S. A.: Clomid or nafoxidine administered to neonatal rats causes reproductive tract abnormalities. Science *197*, 164–165 (1977)

Clark, J. H., Peck, E. J., Jr.: Nuclear retention of receptor-oestrogen complex and nuclear acceptor sites. Nature *260*, 635–637 (1976a)

Clark, J. H., Peck, E. J., Jr.: Steroid receptors: Biology, function and measurement. In: Hormone action and molecular endocrinology workshop. O'Malley, B. W., Schrader, W. T. (eds.), pp. 1–20. Houston: Dept. of Cell Biol. 1976b

Clark, J. H., Peck, E. J., Jr.: Steroid hormone receptors: Basic principles and measurement. In: Receptors and hormone action. O'Malley, B. W., Birnbaumer, L. (eds.), pp. 383–410. New York: Academic Press 1977

Clark, J. H., Anderson, J., Peck, E. J., Jr.: Receptor estrogen complex in the nuclear fraction of the rat uterus during the estrous cycle. Science *176*, 528 (1972a)

Clark, J. H., Campbell, P. S., Peck, E. J., Jr.: Receptor estrogen complex in the nuclear fraction of the pituitary and hypothalamus of male and female immature rats. Neuroendocrinology *77*, 218–228 (1972b)

Clark, J. H., Anderson, J. N., Peck, E. J., Jr.: Nuclear receptor estrogen complexes of rat uteri: Concentration-Time-Response parameters. In: Hormone receptors and reproduction. O'Malley, B. W., Means, A. R. (eds.), Vol. 36, pp. 15–59. Adv. Exptl. Med. & Biol., New York: Plenum Pub. Co. 1973a

Clark, J. H., Anderson, J. N., Peck, E. J., Jr.: Estrogen receptor-antiestrogen complex: Atypical binding by uterine nuclei and effects on uterine growth. Steroids *22*, 707–718 (1973b)

Clark, J. H., Anderson, J. N., Peck, E. J., Jr.: Oestrogen receptors and antagonism of steroid hormone action. Nature *251*, 446–448 (1974a)

Clark, J. H., Eriksson, H. A., Hardin, J. W.: Uterine receptor-estradiol complexes and their interaction with nuclear binding sites. J. Steroid Biochem. *7*, 1039–1043 (1976a)

Clark, J. H., Baulieu, E.-E., Baxter, J. D., de Crombrugghe, B., Jorgenson, E. C., Katzenellenbogen, B. S., Katzenellenbogen, J. A., Luebke, K., Moran, J., Rochefort, H. M., Sherman, M. R., Topert, M.: Intracellular receptors. In: Hormone and anti-hormone action at the target cell. Clark, J. H., Klee, W., Levitzki, A., Wolff, J. (eds.), pp. 147–169. Berlin: Life Sci. Res. Report No. 3, Dahlem Konferenzen Abakon Verlagsgesellschaft 1976b

Clark, J. H., Peck, E. J., Jr., Glasser, S. R.: Mechanism of action of sex steroid hormones in the female. In: Reproduction in Domestic Animals 3rd ed., Cole, H. H., Cupps, P. T. (eds.), pp. 143–174. New York: Academic Press 1977a

Clark, J. H., Hseuh, A., Peck, E. J., Jr.: Regulation of estrogen receptor replenishment by progesterone. Ann. N.Y. Acad. Sci. USA: Biochemical Actions of Progesterone and Progestins, Vol. 286, pp. 161–179. 1977b

Clark, J. H., Paszko, Z., Peck, E. J., Jr.: Nuclear binding and retention of the receptor estrogen complex: Relation to the agonistic and antagonistic properties of estriol. Endocrinology *100*, 91–96 (1977c)

Clark, J. H., Anderson, J. N., Hardin, J. W., Eriksson, H. E., Peck, E. J., Jr.: Methods for assessing the binding of steroid hormones in nuclei and chromatin. In: Methods in cell biology. Stein, G., Stein, J. (eds.), Vol. 17, pp. 335–348. New York: Academic Press 1978a

Clark, J. H., Hardin, J. W., Upchurch, S., Eriksson, H.: Heterogenity of estrogen binding sites in the cytosol of the rat uterus. J. Biol. Chem. *253*, 2630–2634 (1978b)

Clark, J. H., Peck, E. J., Jr., Hardin, J. W., Eriksson, H.: Biology and pharmacology of estrogen receptors: Relation to physiological response. In: Receptors and hormone action. O'Malley, B. W., Birnbaumer, L. (eds.), Vol. II, pp. 1–31. New York: Academic Press 1978c

Clark, J. H., Hardin, J. W., Padykula, H. A., Cardasis, C. A.: Role of estrogen receptor binding and transcriptional activity in the stimulation of hyperestrogenization and nuclear bodies. Proc. Natl. Acad. Sci. USA *75*, 2781–2784 (1978d)

Clark, J. H., Hardin, J. W., McCormack, S. A., Padykula, H. A.: Mechanism of action of estrogen antagonists: Relationship to estrogen receptor binding and hyperestrogenization. In: Hormones and Receptors in Breast Cancer. McGuire, W. L. (ed.), 107–133. New York: Raven Press 1978e

Clementi, F., Ceccarelli, B., Cerati, E., Delmonte, M. L., Felici, M., Motta, M., Pecile, A.: Subcellular localization of neurotransmitters and releasing factors in the rat median eminence. J. Endocrinol. *48*, 205–213 (1970)

Cochet, C., Chedin, M., Chambaz, E. M.: Characterization of steroid binding macromolecules in small biological samples by a rapid minigel filtration method. Anal. Biochem. *76*, 351–357 (1976)

Cole, P., MacMahon, B.: Oestrogen fractions during early reproductive life in the etiology of breast cancer. Lancet *1*, 604–606 (1969)

Concolino, G., Marocchi, A., Silverio, F. D., Conti, C.: Progestational therapy in human renal carcinoma and steroid receptors. J. Steroid Biochem. *7*, 923–927 (1976a)

Concolino, G., Marocchi, A., Concolino, F., Sciarra, F., DiSilverio, F., Conti, C.: Human kidney steroid receptors, J. Steroid Biochem. *7*, 831–836 (1976b)

Conn, P. M.: The chick oviduct nuclear membrane during estrogen-induced differentiation. Ph. D. Thesis, Baylor College of Medicine, Houston, Texas, 1976

Conn, P. M., O'Malley, B. W.: Lysozyme in isolated chick oviduct. Biochem. Biophys. Res. Commun. *64*, 740–746 (1975)

Cornish-Bowden, A., Eisenthal, R.: Statistical considerations in the estimation of enzyme kinetic parameters by the direct linear plot and other methods. Biochem. J. *139*, 721–730 (1974)

Corvol, P., Falk, R., Freifeld, M., Bardin, C. W.: In vitro studies of progesterone binding proteins in guinea pig uterus. Endocrinology *90*, 1464–1469 (1972)

Coulson, P. B., Pavlik, E. J.: Effects of estrogen and progesterone on cytoplasmic estrogen receptor and rates of protein synthesis in rat uterus. J. Steroid Biochem. *8*, 205–212 (1977)

Courralin, J.-C., Bouton, M.-M., Baulieu, E. E., Nuret, P., Chambon, P.: Effect of estradiol on rat uterus DNA-dependent RNA polymerases. Studies on solubilized enzymes. J. Biol. Chem. *251*, 4843–4849 (1976)

Cramer, W., Horning, E. S.: Experimental production by oestrin of pituitary tumors with hypopituitarism and of mammary cancer. Lancet *1*, 247–248 (1936)

Csapo, A. I., Wiest, W. G.: An examination of the quantitative relationship between progesterone and the maintenance of pregnancy. Endocrinology *85*, 735–746 (1969)

Daehnfeldt, J. L., Briand, P.: Determinations of high-affinity gestagen receptors in hormone-responsive and hormone-independent GR mouse mammary tumors by an exchange assay. In: Progesterone receptors in normal and neoplastic tissues. McGuire, W. L., Raynaud, J.-P., Baulieu, E. E. (eds.), pp. 59–69, New York: Raven Press 1977

Davies, I. J., Ryan, K. J.: The modulation of progesterone concentration in the myometrium of the pregnant rat by changes in cytoplasmic receptor protein activity. Endocrinology *92*, 394–401 (1973)

Davies, P., Thomas, P., Griffiths, K.: The influence of steroid receptor complexes on the stages of transcription of target tissue chromatin. J. Steroid Biochem. *7*, 993–1000 (1976)

DeAngelo, A. B., Gorski, J.: The role of RNA synthesis in the estrogen induction of a specific uterine protein. Proc. Natl. Acad. Sci. USA *66*, 693–700 (1970)

Deeley, R. G., Mullinix, K. P., Wetckam, W., Kronenberg, H. M., Meyers, M., Eldridge, J. D., Goldberger, R. F.: Vitellogenin synthesis in the avian liver. J. Biol. Chem. *250*, 9060–9066 (1975)

Deeley, R. G., Udell, D. S., Burns, A. T. H., Gordon, J. I., Goldberger, R. F.: Kinetics of avian vitellogenin messenger RNA induction. J. Biol. Chem. *252*, 7913–7915 (1977a)

Deeley, R. G., Gordon, J. I., Barns, A. T. H., Mullinix, K. P., Binastein, M., Goldberger, R. F.: Primary activation of the vitellogenin gene in the rooster. J. Biol. Chem. *252*, 8310–8319 (1977b)

De Hertogh, R., Ekka, E., Vanderheyden, I., Hoet, J. J.: Slowly exchangeable pool of estradiol in the rat uterus. J. Steroid Biochem. *4*, 313–320 (1973)

DeRobertis, E., Delraldi, A. P., Rodriquez, G., Gomez, C. J.: On the isolation of nerve endings and synaptic vesicles. J. Biophys. Biochem. Cytol. *9*, 229–235 (1961)

Dewhurst, C. J., De Koos, E. B., Haines, R. M.: Replacement hormone therapy in gonadal dysgenesis. Am. J. Obstet. Gynecol. *82*, 412–416 (1975)

Dorfman, R. I.: Anti-estrogenic compounds. In: Methods in hormone research, Dorfman, R. I. (ed.), Vol. II, pp. 113–126. New York: Academic Press 1962

Dunn, T. B., Green, A. W.: Cysts of the epididymus, cancer of the cervix, granular cell myoblastoma and other lesions after estrogen injection in newborn mice. J. Natl. Cancer Inst. *31*, 425–455 (1963)

Dupuy-Coin, A. M., Kalifat, S. R., Bouteille, M.: Nuclear bodies as proteinaceous structures containing ribonucleoproteins, J. Ultrastruct. Res. *38*, 174–183 (1972)

Dyer, R. D., Sarto, G. E., Col'as, A. E.: Binding of progesterone by human uterine cytosol. J. Clin. Endocrinol. Metab. *43*, 1211–1218 (1976)

Ehrlich, P.: Chemotherapeutics: Scientific principles, methods, and results. Lancet II, 445–451 (1913)

Eisenfeld, A. J.: Computer analysis of the distribution of [^{3}H]-estradiol. Biochem. Biophys. Acta *136*, 498–507 (1967)

Eisenfeld, A. J.: Hypothalamic estradiol binding molecules. Nature *224*, 1202–1203 (1969)

Eisenfeld, A. J.: [^{3}H]-estradiol: In vivo binding to macromolecules from the rat hypothalamus, anterior pituitary and uterus. Endocrinology *86*, 1313–1318 (1970)

Eisenfeld, A. J., Axelrod, J.: Selectivity of estrogen distribution in tissues. J. Pharmacol. Exp. Ther. *150*, 469–475 (1965)

Eisenfeld, A. J., Weinberger, R. A. M., Hasebacher, G., Halperm, K.: Estrogen receptor in the mammalian liver. Science *191*, 862–865 (1976)

Eisenfeld, A. J., Aten, R. F., Haselbacher, G. K., Halpern, K.: Specific macromolecular binding of estradiol in the mammalian liver supernate. Biochem. Pharmacol. *26*, 919 (1977)

Eisenthal, R., Cornish-Bowden, A.: The direct linear plot: A new graphical procedure for estimating enzyme kinetic parameters. Biochem. J. *139*, 715–720 (1974)

Ellis, D. J., Ringold, H. J.: The uterine estrogen receptor: A physicochemical study. In: The sex steroids. McKerns, K. W. (ed.), pp. 73–98. New York: Appelton-Century-Crofts 1971

Elsner, C. W., Illingworth, D. V., De Groot, K., Flickinger, G. L., Mikhail, G.: Cytosol and nuclear estrogen receptor in the genital tract of the rhesus monkey. J. Steroid Biochem. *8*, 151–155 (1977)

Emmens, C. W.: Antifertility agents. Ann. Rev. Pharmacol. *4*, 237–254 (1970)

Emmens, C. W., Carr, W. L.: Further studies of compounds exhibiting prolonged antioestrogenic and antifertility activity in the mouse. J. Reprod. Fert. *34*, 29–40 (1973)

Emmens, C. W., Miller, B. G.: Estrogens, progestogens and antiestrogens. Steroids *13*, 725–750 (1969)

Epifanova, O. I.: Mitotic cycles in estrogen-treated mice: A radiographic study. Exp. Cell. Res. *42*, 562–577 (1966)

Erdos, T.: Properties of a uterine oestradiol receptor. Biochem. Biophys. Res. Commun. *32*, 338–343 (1968)

Erdos, T., Fries, J.: The subunit-structure of the uterine estradiol-receptor. Biochem. Biophys. Res. Commun. *58*, 932–939 (1974)

Erdos, T., Bessada, R., Fries, J.: Binding of oestradiol to receptor substances present in extracts from calf uterus. FEBS Letters *5*, 161–164 (1969)

Erdos, T., Best-Belpomme, M., Bessada, R.: A rapid assay for binding estradiol to uterine receptor(s). Anal. Biochem. *37*, 244–252 (1970)

Eriksson, H., Upchurch, S., Hardin, J. W., Peck, E. J., Jr., Clark, J. H.: Heterogeneity of estrogen receptors in the cytosol and nuclear fractions of the rat uterus. Biochem. Biophys. Res. Commun. *81*, 1–7 (1978)

Evans, L. H., Hähnel, R.: Oestrogen receptors in human uterine tissue. J. Endocrinol. *50*, 209–229 (1971)

Faber, L. E., Sandmann, M. L., Stavely, H. E.: Progesterone-binding proteins of the rat and rabbit uterus. J. Biol. Chem. *247*, 5648–5649 (1972)

Faber, L. E., Sandmann, M. L., Stavely, H. E.: Progesterone and corticosterone binding in rabbit uterine cytosols. Endocrinology *93*, 74–80 (1973)

Falck, B.: Occurrence of cholesterol and formation of oestrogen in the infantile rat ovary. Acta Endocrinol. *12*, 115–122 (1953)

Falk, R. J., Bardin, C. W.: Uptake of tritiated progesterone by the uterus of the ovariectomized guinea pig. Endocrinology *86*, 1059–1063 (1970)

Fang, S., Liao, S.: Steroid and tissue specific retention of a 17β-hydroxy-5α-androstan-3-one-protein complex by the cell nuclei of ventral prostate. J. Biol. Chem. *246*, 16–24 (1971)

Fechner, R. E., Kaufman, R. H.: Endometrial adenocarcinoma. Cancer *34*, 444–452 (1974)

Feder, H. H., Marrone, B. L.: Progesterone: Its role in the central nervous system as a facilitator and inhibitor of sexual behavior and gonadotropin release. Annu. N. Y. Acad. Sci. USA *286*, 331–354 (1977)

Feder, H. H., Landau, I. T., Walker, W. A.: Anatomical and biochemical substrates of the action of estrogens and anti-estrogens on brain tissues that regulate female sex behavior of rodents. In: Endocrine control of sexual behavior. Beyer, C. (ed.), New York: Raven Press (in press) (1978)

Feherty, P., Robertson, D. M., Waynforth, H. B., Kellie, A. E.: Changes in the concentration of high affinity oestradiol receptors in rat uterine supernatant preparations during the oestrous cycle, pseudopregnancy, pregnancy, maturation and after ovariectomy. Biochem. J. *120*, 837–844 (1970)

Feil, P. D., Bardin, C. W.: Cytoplasmic and nuclear progesterone receptors in the guinea pig uterus. Endocrinology *97*, 1398–1407 (1975)

Feil, P. D., Glasser, S. R., Toft, D. O., O'Malley, B. W.: Progesterone binding in the mouse and rat uterus. Endocrinology *91*, 738–746 (1972)

Feil, P. D., Miljkovi'c, M., Bardin, C. W.: Medroxyprogesterone acetate: A steroid with potent progestational activity but low receptor affinity in the guinea pig uterus. Endocrinology *98*, 1508–1515 (1976)

Feldman, H. A.: Mathematical theory of complex ligand-binding sites at equilibrium. Anal. Biochem. *48*, 317–338 (1972)

Fels, E.: Die Keimdrüsenfunktion der Ratte nach prae- oder postnataler Injektion verschiedener Hormone. Arch. Gynäkol. *221*, 103–118 (1976)

Ferenczy, A.: Ultrastructural pathology of the uterus. In: Biology of the uterus. Wynn, R. M. (ed.), pp. 545–585. New York: Plenum Press 1977

Ferguson, E. R., Katzenellenbogen, B. S.: A comparative study of antiestrogen action: Temporal patterns of antagonism of estrogen stimulated uterine growth and effects on estrogen receptor levels. Endocrinology *100*, 1242–1251 (1977)

Fink, G., Smith, G. C., Tibballs, J., Lee, V. W. K.: LRF and CRF release in subcellular fractions of bovine median eminence. Nature New Biol. *239*, 57–59 (1972)

Finn, C. A.: The biology of decidual cells. In: Advances in reproductive physiology. Bishop, M. W. H. (ed.), pp. 1–27. New York: Academic Press 1971

Finn, C. A., Martin, L.: The role of oestrogen secreted before oestrus in the preparation of the uterus for implantation in the mouse. J. Endocrinol. *47*, 431–438 (1970)

Fishman, J.: The catechol estrogens. Neuroendocrinology *22*, 363–374 (1976)

Fishman, J., Fishman, J. H.: Competitive binding assay for estradiol receptor using immobilized antibody. J. Clin. Endocrinol. Metab. *39*, 603–606 (1974)

Floridi, A., Citro, G., Sega, E., Natali, P. G.: Quantitation of estradiol receptors employing a sensitive radioimmunossay. Immunol. Commun. *4*, 603–615 (1975)

Forsberg, J. G.: Estrogen, vaginal cancer and vaginal development. Am. J. Obstet. Gynecol. *113*, 83–87 (1972)

Forsberg, J. G.: Late effects in the vaginal and cervical epithelia after injections of diethylstilbestrol into neonatal mice. Am. J. Obstet. Gynecol. *121*, 101–104 (1975)

Fox, T. O.: Oestradiol receptors of neonatal mouse brain, Nature *258*, 441–444 (1975)

Fox, T. O., Johnston, C.: Estradiol receptors from mouse brain and uterus: Binding to DNA. Brain Res. *77*, 330–336 (1974)

Freeman, M. C., Dupke, K. C., Croteau, C. M.: Extinction of the estrogen induced daily surge for LH release in the rat: A role for the proestrus surge of progesterone. Endocrinology *99*, 223–229 (1976)

Freifeld, M. L., Feil, P. D., Bardin, C. W.: The in vivo regulation of the progesterone "receptor" in guinea pig uterus: Dependence on estrogen and progesterone. Steroids *23*, 93–103 (1974)

Funder, J. W., Feldman, D., Edelman, I. S.: The roles of plasma binding and receptor specificity in the mineralocorticoid action of aldosterone. Endocrinology *92*, 994–1004 (1973)

Furth, J., Ueda, G., Clifton, K. H.: Pathophysiology of pituitaries and their tumors; methodological advances. In: Methods in cancer research. Busch, H. (ed.), pp. 201–277. New York: Academic Press 1973

Gaddum, J. H.: Discussion in the chemical and physical bases of pharmacological action. Proc. R. Soc. Lond. Biol. J. *121*, 598–601 (1937)

Gallez, G., Heuson, J. C., Waelbroeck, C.: Growth stimulating effect of Nafoxidine on rat mammary tumor after ovariectomy. Eur. J. Cancer *9*, 699–700 (1973)

Gannon, F., Gorski, J.: Nuclear binding of the estrogen receptor in whole uteri. Steroids *28*, 669–697 (1976)

Garcia, M., Rochefort, H.: Androgens on the estrogen receptor. II-Correlation between nuclear translocation und uterine protein synthesis. Steroids *29*, 111–126 (1977)

Gardner, W. U., Pfeiffer, C. A., Trentin, J. J.: Hormonal factors in experimental carcinogenesis. In: The physiopathology of cancer. Humburger, F. (ed.)., pp. 152–237. New York: Harper & Row 1959

Gehring, U., Tomkins, G. M., Ohno, S.: Effect of the androgen-insensitivity mutation on a cytoplasmic receptor for dihydrotestosterone. Nature New Biol. *232*, 106–107 (1971)

Gellert, R. J., Lewis, J., P'etra, P. H.: Neonatal treatment with sex steroids: Relationship between the uterotropic response and the estrogen "receptor" in prepubertal rats. Endocrinology *100*, 520–528 (1977)

Giannopoulos, G.: Binding of testosterone to cytoplasmic components of the immature rat uterus. Biochem. Biophys. Res. Commun. *44*, 943–951 (1971)

Giannopoulos, G.: Binding of testosterone to uterine components of the immature rat. J. Biol. Chem. *248*, 1004–1010 (1973)

Giannopoulos, G., Gorski, J.: Estrogen receptors. Quantitative Studies on transfer of estradiol from cytoplasmic to nuclear binding sites. J. Biol. Chem. *246*, 2524–2529 (1971a)

Giannopoulos, G., Gorski, J.: Estrogen binding protein of the rat uterus. Different molecular forms associated with nuclear uptake of estradiol. J. Biol. Chem. *246*, 2530–2536 (1971b)

Giese, A. C.: Cell physiology. 4th ed. pp. 289–292. Philadelphia: Saunders 1973

Ginsburg, M. MacLusky, N. J., Morris, I. D., Thomas, P. J.: Cyclical fluctuation of oestradiol receptors in hypothalamus and pituitary. J. Physiol. (Lond.) *224*, 72–74 (1972)

Ginsburg, M., MacLusky, N. J., Morris, I. D., Thomas, P. J.: Physiological variation in abundance of oestrogen specific high-affinity binding sites in hypothalamus, pituitary and uterus of the rat. J. Endocrinol. *64*, 443–449 (1975)

Glasscock, R. F., Hoekstra, W. G.: Selective accumulation of tritium-labeled hexosterol by the reproductive organs of immature female goats and sheep. Biochem. J. *72*, 673–682 (1959)

Glasscock, R. F., Michael, R. P.: The localization of oestrogen in a neurological system in the brain of the female rat. J. Physiol. (Lond.) *163*, 38P–39P (1962)

Glasser, S. R., Clark, J. H.: A determinant role for progesterone in the development of uterine sensitivity to decidualization and ovo-implantation. Symp. Soc. Dev. Biol. *33*, 311–345 (1975)

Glasser, S. R., Spelsberg, T. C.: Differential modulation of estrogen induced activity of uterine nuclear RNA polymerases (I and II) by cycloheximide and actinomycin D. Endocrinology *92*, Suppl., (Abs. 88) (1973)

Glasser, S. R., Chytil, F., Spelsberg, T. C.: Early effects of oestradiol-17 β on the chromatin and activity of the deoxyribonucleic acid-dependent ribonucleic acid polymerases I and II of the rat uterus. Biochem. J. *130*, 947–957 (1972)

Goebelsmann, U., Midgley, A. R., Jr., Jaffe, R. B.: Regulation of human gonadotropins: VII. Daily individual urinary estrogens, pregnanediol and serum luteinizing and follicle stimulating hormone during the menstrual cycle. J. Clin. Endocrinol. *29*, 1222–1230 (1969)

Goldenberg, R. L., Vaitukaitus, J. L., Ross, G. T.: Estrogen and follicle stimulating hormone interactions on follicle growth in rats. Endocrinology *90*, 1492–1498 (1972)

Goldstein, J. L., Brown, M. S.: The LDL pathway in human fibroblasts: A receptor mediated mechanism for the regulation of cholesterol metabolism. Curr. Top. Cell Regul. *11*, 147–181 (1976)

Goldstein, J. L., Brown, M. S.: The low-density lipoprotein pathway and its relation to atherosclerosis. Ann. Rev. Biochem. *46*, 897–930 (1977)

Goldstein, J. L., Basu, S. K., Brunschede, G. Y.: Release of low density lipoprotein from its cell surface receptor by sulfated glycosaminoglycans. Cell *7*, 85–95 (1976)

Gomez, F., Bohnet, H. G., Friesen, H. G.: Changes in progesterone and estrogen receptors in the rat uterus during the estrous cycle and the puerperium. In: Progesterone receptors in normal and neoplastic tissues. McGuire, W. L., Raynaud, J. P., Baulieu, E. E. (eds.), pp. 245–259. New York: Raven Press 1977

Gorski, J.: Early estrogen effects on the activity of uterine ribonucleic acid polymerase. J. Biol. Chem. *239*, 889–892 (1964)

Gorski, R.: Modification of ovulatory mechanisms by postnatal administration of estrogen to the rat. Am. J. Physiol. *205*, 842–844 (1963)

Gorski, R. A., Clemens, L. G.: Action of gonadal steroids. In: Handbook of Neurochemistry. Lajtha, A. (ed.), Vol. IV, pp. 429–449. Control mechanisms in the nervous system. New York: Plenum Press 1970

Gorski, J., Gannon, F.: Current models of steroid hormone action: A critique. Ann. Rev. Physiol. *38*, 425–450 (1976)

Gorski, J., Raker, B.: Estrogen action in the uterus: The requisite for sustained estrogen binding in the nucleus. Gynecol. Oncol. *2*, 249–258 (1974)

Gorski, J., Toft, D., Shyamala, G., Smith, D., Notides, A.: Hormone receptors: Studies on the interaction of estrogen with the uterus. Rec. Prog. Horm. Res. *24*, 45–80 (1968)

Gorski, J., Sarff, M., Clark, J.: The regulation of uterine concentration of estrogen binding protein. In: Adv. in the Biosciences. Raspe, G. (ed.), Vol. 7, pp. 5–20. Vieweg: Pergamon Press 1971

Greely, G. H., Muldoon, T. G., Mahesh, V. B.: Correlative aspects of luteinizing hormone releasing hormone sensitivity and cytoplasmic estrogen receptor concentration in the anterior pituitary and hypothalamus of the cycling rat. Biol. Reprod. *13*, 505–512 (1975)

Greenblatt, R. B., Barfield, W. E., Jungck, E. C., Ray, A. W.: Induction of ovulation with MRL/41. J. Am. Med. Assoc. *178*, 101–104 (1961)

Greengard, O: The developmental formation of enzymes in rat liver. In: Biochemical actions of hormones. Litwack, G. (ed.), p. 53–87. New York: Academic Press 1970

Greengard, O: Steroids and the maturation of rat tissues. J. Steroid Biochem. *6*, 639–642 (1975)

Gschwendt, M.: Solubilization of the chromatin-bound estrogen receptor from chicken liver and fractionation in hydroxylapatite. Eur. J. Biochem. *67*, 411–419 (1976)

Gschwendt, M., Hamilton, T. H.: The transformation of the cytoplasmic oestradiol-receptor complex into the nuclear complex in a uterine cellfree system. Biochem. J. *128*, 611–616 (1972)

Gschwendt, M., Kittstein, W.: Specific binding of estradiol to the liver chromatin of estrogenized roosters. Biochim. Biophys. Acta *361*, 84–96 (1974)

Gurpide, E.: Tracer methods in hormone research. In: Monographs on Endocrinology. Gross, F., Labhart, A., Lipsett, M. B., Mann, T., Samuels, L. T., Zander, J. (eds.), Vol. 8, Berlin, Heidelberg, New York: Springer 1975

Gurpide, E., Tseng, L.: Factors controlling intracellular levels of estrogens in human endometrium. Gynecol. Oncol. *2*, 221–227 (1974)

Gurpide, E., Welch, M.: Dynamics of uptake of estrogens and androgens by human endometrium. J. Biol. Chem. *244*, 5159–5167 (1969)

Gurpide, E., Stolee, A., Tseng, L.: Quantitative studies of tissue uptake and disposition of hormones. In: Karolinska symposia on research methods in reproductive endocrinology, 3rd Symposia, pp. 247–278. In Vitro Methods in Reproductive Cell Biology. 1971

Gurpide, E., Gusberg, S. B., Tseng, L.: Estradiol binding and metabolism in human endometrial hyperplasia and adenocarcinoma. J. Steroid Biochem. *7*, 891–896 (1976)

Gusberg, S. B.: Hormone dependence of endometrial cancer. Obstet. Gynecol. *30*, 287–292 (1967)

Gusberg, S. B., Kardon, P.: Proliferative endometrial response to thecagranulosa cell tumors. Am. J. Obstet. Gynecol. *111*, 633–641 (1971)

Ham, E. A., Cirillo, V. J., Zanetti, M. E., Kuehl, F. A.: Estrogen directed synthesis of specific prostaglandins in uterus. Proc. Natl. Acad. Sci. USA *72*, 1420–1424 (1975)

Hamilton, T. H.: Isotopic studies on estrogen-induced accelerations of ribonucleic acid and protein synthesis. Proc. Natl. Acad. Sci. USA *49*, 373–379 (1963)

Hamilton, T. H.: Sequences of RNA and protein synthesis during early estrogen action. Proc. Natl. Acad. Sci. USA *51*, 83–89 (1964)

Hamilton, T. H.: Control by estrogen of genetic transcription and translation. Science *161*, 649–660 (1968)

Hamilton, T. H., Widnell, C. C., Tata, J. R.: Sequential stimulations by oestrogen of nuclear RNA synthesis and DNA-dependent RNA polymerase activities in rat uterus. Biochim. Biophys. Acta *108*, 168–172 (1965)

Hamilton, T. H., Widnell, C. C., Tata, J. R.: Synthesis of ribonucleic acid during early estrogen action. J. Biol. Chem. *243*, 408–417 (1968a)

Hamilton, T. H., Teng, C. S., Means, A. R.: Early estrogen action: Nuclear synthesis and accumulation of protein correlated with enhancement of two DNA-dependent RNA polymerase activities. Proc. Natl. Acad. Sci. USA *59*, 1265–1272 (1968b)

Hansen, P. H., Johnson, A., Schrader, W. T., O'Malley, B. W.: Kinetics of progesterone binding to the chick oviduct receptor protein. J. Steroid Biochem. 7, 723–732 (1976)

Hardin, J. W., Clark, J. H., Glasser, S. R., Peck, E. J., Jr.: Estrogen receptor binding by uterine nuclei: Relationship to endogenous nuclear RNA polymerase activity. Biochemistry *15*, 1370–1374 (1976)

Harman, S. M., Louvet, J.-P., Ross, G. T.: Interaction of estrogen and gonadotrophin on follicular atresia. Endocrinology *96*, 1145–1152 (1975)

Harris, G. S.: Nature of oestrogen specific binding sites in the nuclei of mouse uteri. Nature New Biol. *231*, 246–248 (1971)

Harris, S. E., Rosen, J. M., Means, A. R., O'Malley, B. W.: Use of a specific probe for ovalbumin messenger RNA to quantitiate estrogen-induced gene transcripts. Biochemistry *14*, 2072–2081 (1975)

Hart, H. E.: Determination of equilibrium constants and maximum binding capacities in complex in vitro systems. Bull. Math. Biophys. *27*, 87–98 (1965)

Hashimoto, I., Henricks, D. M., Anderson, L. L., Melampy, R. M.: Progesterone and pregn-4-en-20α-ol-3-one in ovarian venous blood during various reproductive states in the rat. Endocrinology *82*, 333–341 (1968)

Haukkamaa, M., Karjalainen, O., Luukkainen, T.: In vitro binding of progesterone by the human endometrium during the menstrual cycle and by hyperplastic, atrophic, and carcinomatous endometrium. Am. J. Obstet. Gynecol. *111*, 205–210 (1971)

Herbst, A. L., Ulfelder, H., Poskanzer, D. C.: Adenocarcinoma of the vagina; association of maternal stilbestrol therapy with tumor appearance in young women. N. Engl. J. Med. *284*, 878–881 (1971)

Herranen, A., Mueller, G. C.: Effect of estradiol on the metabolism of serine-3-[C^{14}] in surviving uterine segments. J. Biol. Chem. *223*, 369–375 (1956)

Hertz, R.: The estrogen-cancer hypothesis. Cancer *38*, 534–540 (1974)

Higgins, S. J., Rousseau, G. G., Baxter, J. D., Tomkins, G. M.: Nature of nuclear acceptor sites for glucocorticoid- and estrogen-receptor complexes. J. Biol. Chem. *248*, 5873–5879 (1973)

Hilliard, J., Scaramuzzi, R. J., Penardi, R., Sawyer, C. H.: Serum progesterone levels in hysterectomized pseudopregnant rabbits. Proc. Soc. Exp. Biol. Med. *145*, 151–153 (1974)

Hillyard, L. A., Entenman, C., Chaikoff, I. L.: Concentration and composition of serum lipoproteins of cholesterol-fed and stilbestrol-injected birds. J. Biol. Chem. *223*, 359–368 (1956)

Hisaw, F. L., Jr.: Comparative effectiveness of estrogens on fluid imbibition and growth of the rat's uterus. Endocrinology *64*, 276–289 (1959)

Hobkirk, R., Nilsen, M.: Observations on the occurrence of six estrogen fractions in human pregnancy urine. J. Clin. Endocrinol. Metab. *22*, 134–141 (1962)

Hoff, J. D., Lasley, B. L., Wang, C. F., Yen, S. S. C.: The two pools of pituitary gonadotropin: Regulation during the menstrual cycle. J. Clin. Endocrinol. Metab. *44*, 302–312 (1977)

Hohlweg, W.: Veränderungen des Hypophysenvorderlappens und des Ovariums nach Behandlung mit großen Dosen von Follikelhormon. Klin. Wochenschr. *13*, 92–95 (1934)

Hollenberg, M. D., Cuatrecasas, P.: Principles and techniques for the study of plasma membrane receptors related to hormone action. In: Methods in cancer research. Busch, H. (ed.), Vol. XII. pp. 317–366. New York: Academic Press 1976

Holtkamp, D. E., Greslin, J. G., Root, C. A., Lerner, L. J.: Gonadotropin inhibiting and antifecundity effects of chloramiphene. Proc. Soc. Exp. Biol. Med. *105*, 197–198 (1960)

Hori, T., Ide, M., Miyake, T.: Ovarian estrogen secretion during the estrous cycle and under the influence of exogenous gonadotropins in rats. Endocrinol. Jpn. *15*, 215–222 (1968)

Horwitz, K. B., McGuire, W. L.: Estrogen and progesterone: Their relationship in hormone-dependent breast cancer. In: Progesterone receptors in normal and neoplastic tissues. McGuire, W. L., Raynaud, J.-P., Baulieu, E.-E. (eds.), pp. 103–124. New York: Raven Press 1977

Horwitz, K. B., McGuire, W. L.: Estrogen control of progesterone receptor in human breast cancer. J. Biol. Chem. *253*, 2223–2228 (1978)

Horwitz, K. B., McGuire, W. L., Pearson, O. H., Segaloff, A.: Predicting response to endocrine therapy in human breast cancer: A hypothesis. Science *189*, 726–727 (1975)

Howard, P. D., Wiest, W. G.: Progesterone metabolism by uterine tissue by pregnant rats. Steroids *19*, 35–45 (1972)

Hsueh, A. J. W., Peck, E. J., Jr., Clark, J. H.: Oestrogen receptors in the mammary gland of the lactating rat. J. Endocrinol. *58*, 503–511 (1973)

Hsueh, A. J., Peck, E. J., Jr., Clark, J. H.: Receptor progesterone complex in the nuclear fraction of the rat uterus: Demonstration by ^{3}H-progesterone exchange. Steroids *24*, 599–611 (1974)

Hsueh, A. J., Peck, E. J., Jr., Clark, J. H.: Progesterone antagonism of the oestrogen receptor and oestrogen-induced uterine growth. Nature *254*, 337–339 (1975)

Hsueh, A. J., Peck, E. J., Jr., Clark, J. H.: Control of uterine estrogen receptor levels by progesterone. Endocrinology *98*, 438–444 (1976)

Huggins, C.: Two principles in endocrine therapy of cancers: Hormone deprival and hormone interference. Cancer Res. *25*, 1163–1175 (1965)

Huggins, C., Jensen, E. V.: The depression of estrone-induced uterine growth by phenolic estrogens with oxygenated functions at positions 6 or 16; The impeded estrogens. J. Exptl. Med. *102*, 335–346 (1955)

Huggins, C., Yang, N. C.: Induction and extinction of mammary cancer. Science *137*, 257–262 (1962)

Huggins, C., Moon, R. C., Morii, S.: Extinction of experimental mammary cancer. I. Estradiol and progesterone. Proc. Natl. Acad. Sci. USA *48*, 379–386 (1962)

Hunt, M. E., Muldoon, T. G.: Factors controlling estrogen receptor levels in normal mouse mammary tissue. J. Steroid Biochem. *8*, 181–186 (1977)

Illingworth, D. V., Elsner, C., De Groot, K., Flickinger, G. L., Mikhail, G.: A specific progesterone receptor of myometrial cytosol from the rhesus monkey. J. Steroid Biochem. *8*, 157–160 (1977)

Iramain, C. A., Danzo, B. J., Strott, C. A., Toft, D. O.: Progesterone binding in the hypothalamus and hypophysis of female guinea pigs and rabbits (Abs.). 4th Int. Congr., Int. Soc. Psychoneuroendocrinol. 1973

Ishii, S.: Association of luteinizing hormone-releasing factor with granules separated from equine hypophysial stalk. Endocrinology *86*, 207–216 (1970)

Ishii, D. N., Aronow, L.: In vitro degradation and stabilization of the glucocorticoid binding component from mouse fibroblasts. J. Steroid Biochem. *4*, 593–603 (1973)

Jackson, V., Chalkley, R.: The binding of estradiol-17β to the bovine endometrial nuclear membrane. J. Biol. Chem. *249*, 1615–1626 (1974)

Jackson, R. L., Dockerty, M. B.: The Stein-Leventhal syndrone; analysis of 43 cases with special reference to association with endometrial carcinoma. Am. J. Obstet. Gynecol. *67*, 161–173 (1957)

Jacobson, H. I., Keyes, P. L., Bullock, D. W.: Regulation of uterine estrogen receptor by luteal progesterone in the pseudopregnant rabbit. Biol. Reprod. *7*, Abs., 108 (1972)

Jaffe, R. C., Socher, S. H., O'Malley, B. W.: An analysis of the binding of the chick oviduct progesterone receptor to chromatin. Biochim. Biophys. Acta *399*, 403–419 (1975)

Janne, O., Kontula, K., Luukkainen, T., Vihko, R.: Oestrogen-induced progesterone receptor in human uterus. J. Steroid Biochem. *6*, 501–509 (1975)

Janne, O., Kontula, K., Vihko, R.: Progestin receptors in human tissues: concentrations and binding kinetics. J. Steroid Biochem. *7*, 1061–1068 (1976)

Jensen, E. V., DeSombre, E. R.: Mechanism of action of the female sex hormones. Annu. Rev. Biochem. *41*, 203–230 (1972)

Jensen, E. V., DeSombre, E. R.: Estrogen-receptor interaction. Science *182*, 126–134 (1973)

Jensen, E. V., Jacobson, H. I.: Basis guides to the mechanism of estrogen action. Rec. Prog. Horm. Res. *18*, 387–414 (1962)

Jensen, E. V., Jacobson, H. I., Flesher, J. W., Saha, N. N., Gupta, G., Smith, S., Colucci, V., Shiplacoff, D., Neumann, H. G., DeSombre, E. R., Jungblut, P. W.: Estrogen receptors in target tissues. In: Steroid dynamics. Pincus, G., Nakao, T., Tait, J. F. (eds.), pp. 133–157. New York: Academic Press 1966

Jensen, E. V., DeSombre, E. R., Hurst, D. J., Kawashima, T., Jungblut, P. W.: IV. Mode d'action des hormones intervenant dars la reproduction. Arch. Anat. Microsc. Morphol. Exp. *56*, Suppl. 3–4, 547–569 (1967)

Jensen, E. V., Suzuki, T., Kawashima, T., Stumpf, W. E., Jungblut, P., DeSombre, E. R.: A two-step mechanism for the interaction of estradiol with rat uterus. Proc. Natl. Acad. Sci. USA *59*, 632–638 (1968)

Jensen, E. V., Suzuki, T., Numata, M., Smith, S., DeSombre, E. R.: Estrogen-binding substances of target tissues. Steroids *13*, 417–427 (1969)

Jensen, E. V., Numata, M., Brecher, P. I., DeSombre, E. R.: Hormone receptor interaction as a guide to biochemical mechanism. In: The biochemistry of steroid hormone action. Smellie, R. M. S. (ed.), pp. 133–159. London: Academic Press 1971

Jensen, E. V., Block, G. E., Smith, S., Kyser, K., DeSombre, E. R.: Estrogen receptors and hormone dependency. In: Estrogen target tissues and neoplasia. Dao, T. L. (ed.), pp. 23–57. Chicago: University of Chicago Press 1972

Jensen, E. V., Mohla, S., Gorell, T. A., DeSombre, E. R.: The role of estrophilin in estrogen action. Vitam. Horm. *32*, 89–127 (1974)
Jensen, E. V., Smith, S., DeSombre, E. R.: Hormone dependency in breast cancer. J. Steroid Biochem. *7*, 911–917 (1976)
Jordan, V. C., Dowse, L. J.: Tamoxifen as an anti-tumor agent: effect on oestrogen binding. J. Endocrinol. *68*, 297–303 (1976)
Jordan, V. C., Jaspan, T.: Tamoxifen as an anti-tumor agent: Oestrogen binding as a predictive test for tumor response. J. Endocrinol. *68*, 453–460 (1976)
Jordan, V. C., Koerner, S.: Tamoxifen as an anti-tumor agent: Role of estradiol and prolactin. J. Endocrinol. *68*, 305–311 (1976)
Jordan, V. C., Dix, C. J., Rowsby, L., Prestwich, G.: Studies on the mechanism of action of the nonsteroidal antioestrogen tamoxifen (I.C.I. 46, 474) in the rat. Mol. Cell Endocrinol. *7*, 177–192 (1977)
Jost, J. P., Pehling, G., Baca, O. G.: Rate of synthesis of β-lipovitellin in the liver of immature chicks treated with 17β-estradiol. Biochem. Biophys. Res. Commun. *62*, 957–965 (1975)
Juliano, J. V., Stancel, G. M.: Estrogen receptors in the rat uterus. Retention of hormone-receptor complexes. Biochemistry *15*, 916–920 (1976)
Kahwanago, I., Heinrichs, W. L., Hermann, W. L.: Isolation of oestradiol receptors from bovine hypothalamus and anterior pituitary gland. Nature *223*, 313–314 (1969)
Kahwanago, I., Heinrichs, W. L., Herrmann, W. L.: Oestradiol "receptors" in the hypothalamus and anterior pituitary gland. Inhibition of estradiol binding by SH group blocking agents and clomiphene citrate. Endocrinology *86*, 1319–1326 (1970)
Kalimi, M., Beato, M., Feigelson, P.: Interaction of glucocorticoids with rat liver nuclei I. role of the cytosol proteins. Biochemistry *12*, 3365–3371 (1973)
Kalimi, M., Lozman, P., Feigelson, P.: The "activated" hepatic glucocorticoid receptor complex. Its generation and properties. J. Biol. Chem. *250*, 1080–1086 (1975)
Kalimi, M., Tsai, S. Y., Tsai, M.-J., Clark, J. H., O'Malley, B. W.: Effect of estrogen on gene expression in the chick oviduct: Correlation between nuclear-bound estrogen receptor and chromatin initiation sites for transcription. J. Biol. Chem. *251*, 516–523 (1976)
Kalman, S. M., Lowenstein, J. M.: The effect of estrogens on the uptake of albumin and electrolytes by the rat uterus. J. Pharmacol. Exp. Ther. *122*, 163–167 (1958)
Kalra, S.: Tissue levels of luteinizing hormone-releasing hormone in the preoptic area and hypothalamus, and serum concentrations of gonadotropins following anterior hypothalamic differentiation and estrogen treatment of the female rat. Endocrinology *99*, 101–107 (1976)
Kalra, S. P., Prasad, M. R. N.: Effect of clomiphene on fertility in female rats. Indian J. Exp. Biol. *5*, 1–8 (1967)
Kang, Y.-H., Anderson, W. A., DeSombre, E. R.: Modulation of uterine morphology and growth by estradiol-17β and an estrogen antagonist. J. Cell Biol. *64*, 682–691 (1975)
Karavolas, H. J., Herf, S. M.: Conversion of progesterone by rat medial basal hypothalamic tissue to 5α-pregnane-3, 20-dione. Endocrinology *89*, 940–942 (1971)
Kato, J.: Localization of oestradiol receptors in the rat hypothalamus. Acta Endocrinol. (Kbh) *72*, 663–670 (1973)
Kato, J., Villee, C. A.: Preferential uptake of estradiol by the anterior hypothalamus of the rat. Endocrinology *80*, 567–575 (1967a)
Kato, J., Villee, C. A.: Factors affecting the uptake of estradiol-6, 7-[^{3}H] by the hypophysis and hypothalamus. Endocrinology *80*, 1133–1138 (1967b)
Kato, J., Onuouchi, T.: Specific progesterone receptors in the hypothalamus and anterior hypophysis of the rat. Endocrinology *101*, 920–928 (1977)
Kato, J., Suqimara, N., Kobayashi, T.: Changing patterns of the uptake of estradiol by the anterior hypothalamus, the median eminences and the hypophysis in the developing female rat. In: Hormones in development. Hamburgh, M., Barrington, E. J. W. (eds.), pp. 689–703. New York: Appleton Century-Crofts 1971a
Kato, J., Atsumi, Y., Inaba, M.: Development of oestrogen receptor in rat hypothalamus. J. Biochem. *70*, 1051–1053 (1971b)
Kato, J., Atsumi, Y., Inaba, M.: Estradiol receptors in female rat hypothalamus in the developmental stages and during pubescence. Endocrinology *94*, 309–317 (1974)
Katzenellenbogen, B. S.: Synthesis and inducibility of the uterine estrogen-induced protein, IP, During the rat estrous cycle: Clues to uterine estrogen sensitivity. Endocrinology *96*, 289–297 (1975)

Katzenellenbogen, B. S., Ferguson, E. R.: Antiestrogen action in the uterus: Biological ineffectiveness of nuclear bound estradiol after antiestrogen. Endocrinology *97*, 1–12 (1975)

Katzenellenbogen, B. S., Gorski, J.: Estrogen action in vitro: Induction of the synthesis of a specific uterine protein. J. Biol. Chem. *247*, 1299–1305 (1972)

Katzenellenbogen, B. S., Gorski, J.: Estrogen actions on synthesis of macromolecules in target cells. In: Biochemical actions of hormones. Litwack, G. (ed.), pp. 187–243. New York: Academic Press 1975

Katzenellenbogen, B. S., Greger, N. G.: Ontogeny of uterine responsiveness to estrogen during early development in the rat. Mol. Cell Endocrinol. *2*, 31–42 (1974)

Katzenellenbogen, B. S., Katzenellenbogen, J. A.: Antiestrogens: Studies using an in vitro estrogen-responsive uterine system. Biochem. Biophys. Res. Commun. *50*, 1152–1159 (1973)

Katzenellenbogen, B. S., Ferguson, E. R., Lan, N. C.: Fundamental differences in the action of estrogen and antistrogens on the uterus: Comparison between compounds with similar duration of action. Endocrinology *10*, 1252–1259 (1977)

Katzenellenbogen, J. A., Johnson, H. J., Carlson, K. E.: Studies on the uterine, cytoplasmic estrogen binding protein. Thermal stability and ligand dissociation rate. An assay of empty and filled sites by exchange. Biochemistry *12*, 4092–4099 (1973)

Kaye, A.: The ontogeny of estrogen receptors. In: Biochemical actions of hormones. Litwack, G. (ed.). (in press) (1978)

Keefer, D. A., Stumpf, W. E., Sar, M.: Estrogen-topographical localization of estrogen-concentrating cells in the rat spinal cord following [^{3}H]-estradiol administration. Proc. Soc. Exp. Biol. Med. *143*, 414–417 (1973)

Keller, N., Richardson, U. I., Yates, F. E.: Protein binding and the biological activity of corticosteroids: In vivo induction of hepatic and pancreatic alanine aminotransferases by corticosteroids in normal and estrogen-treated rats. Endocrinology *84*, 49–62 (1969)

Kennedy, T. G., Armstrong, D. T.: Loss of uterine luminal fluid in the rat: Relative importance of changing peripheral levels of estrogen and progesterone. Endocrinology *97*, 1379–1385 (1975)

Keran, E. E., Barker, K. L.: Regulation of glucose-6-phosphate dehydrogenase activity in uterine tissue in organ culture. Endocrinology *99*, 1386–1397 (1976)

Kim, U.: Pituitary function and hormonal therapy of experimental breast cancer. Cancer Res. *25*, 1146–1161 (1965)

Kimball, F. A., Hansel, W.: Estogen cytosol binding proteins in bovine endometrium and corpus luteum. Biol. Reprod. *11*, 566–577 (1974)

King, R. J. B., Gordon, J.: The localization of [^{3}H]-oestradiol-17β in rat uterus. J. Endocrinol. *34*, 431–437 (1966)

King, R. J. B., Gordon, J.: Involvement of DNA in the acceptor mechanism for uterine oestradiol receptor. Nature New Biol. *240*, 185–187 (1972)

King, R. J. B., Mainwaring, W. I. P.: Steroid-cell interactions, Baltimore Md., University Park Press (1974)

King, R. J. B., Gordon, J., Inman, D. R.: The intracellular localization of oestrogen in rat tissues. J. Endocrinol. *32*, 9–15 (1965a)

King, R. J. B., Gordon, J., Martin, L.: The association of (6,7-$^3H_2$) oestradiol with nuclear chromatin. Biochem. J. *97*, 28P (1965b)

King, R. J. B., Cowan, D. M., Inman, D. R.: The uptake of [^{3}H]-oestradiol by dimethyl-benzanthracene induced rat mammary tumours. J. Endocrinol. *32*, 83–90 (1965c)

King, R. J. B., Gordon, J., Cowan, D. M., Inman, D. R.: The intranuclear localization of [^{3}H]-oestradiol in dimethylbenzanthracene induced rat mammary adenocarcinoma and other tissues. J. Endocrinol. *36*, 139–150 (1966)

King, R. J. B., Gordon, J., Steggles, A. W.: The properties of a nuclear acidic protein fraction that binds [^{3}H-]oestradiol-17β. Biochem. J. *114*, 649–657 (1969)

King, R. J. B., Beard, V., Gordon, J., Pauley, A. S., Smith, J. A., Steggles, A. W., Vertes, M.: Studies on estradiol-binding in mammalian tissues. Adv. Biosci. *7*, 21–44 (1971)

Kistner, R. W.: Endometrial alterations associated with estrogen and estrogen-progestin combinations. In: The uterus. Norris, H. J., Hertig, A. T., Abel, M. R. (eds.), pp. 227–254. Baltimore Md. Williams & Wilkins 1973

Kistner, R. W., Smith, O. W.: Observations on the use of non-steroidal estrogen antagonist: MER-25. II. Effects in endometrial hyperplasia and Stein-Leventhal Syndrome. Fertil. Steril. *12*, 121–132 (1961)

Knobil, E.: On the control of gonadotropin secretion in the Rhesus monkey. Recent Prog. Horm. Res. *30*, 1–46 (1974)

Knowler, J. T., Smellie, R. M. S.: The synthesis of ribonucleic acid in immature rat uterus responding to oestradiol-17β. Biochem. J. *125*, 605–614 (1971)

Kobayashi, R. M., Lu, K. H., Moore, R. Y., Yen, S. C. C.: Regional distribution of hypothalamic luteinizing hormone-releasing hormone in proestrous rats: Effects of ovariectomy and estrogen replacement. Endocrinology *102*, 98–105 (1978)

Kohler, P. O., Grimley, P. M., O'Malley, B. W.: Estrogen-induced cytodifferentiation of the ovalbumin - secreting glands of the chick oviduct. J. Cell Biol. *40*, 8–27 (1969)

Koligian, K. B., Stormshak, F.: Nuclear and cytoplasmic estrogen receptors in ovine endometrium during the estrous cycle. Endocrinology *101*, 524–533 (1977)

Kontula, K., Janne, O., Janne, J., Vihko, R.: Partial purification and characterization of progesterone-binding protein from pregnant guinea pig uterus. Biochem. Biophys. Res. Commun. *47*, 596–603 (1972)

Kontula, K., Janne, O., Rajakoski, E., Tanhuanp'a'a, E., Vihko, R.: Ligand specificity of progesterone-binding proteins in guinea pig and sheep. J. Steroid Biochem. *5*, 39–44 (1974a)

Kontula, K., Janne, O., Luukkainen, T., Vihko, R.: Progesterone-binding protein in human myometrium. Influence of metal ions on binding. J. Clin. Endocrinol. Metab. *38*, 500–503 (1974b)

Kontula, K., Janne, O., Vihko, R., de Jager, E., de Visser, J., Zeelen, F.: Progesterone-binding proteins: in vitro binding and biological activity of different steroidal ligands. Acta Endocrinol. (Kbh) *78*, 574–592 (1975)

Korach, K. S., Muldoon, T. G.: Studies on the nature of the hypothalamic estradiol-concentrating mechanism in the male and female rat. Endocrinology *94*, 785–793 (1974)

Korach, K. S., Muldoon, T. G.: Inhibition of anterior pituitary estrogen-receptor complex formation by low-affinity interaction with 5α-dihydrotestosterone. Endocrinology *97*, 231–236 (1975)

Korenman, S. G., Rao, B. R.: Reversible disaggregation of the cytosol estrogen binding protein of uterine cytosol. Proc. Natl. Acad. Sci. USA *61*, 1028–1033 (1968)

Korenman, S. G., Perrin, L. E., McCallum, T. P.: A radio-ligand binding assay system for estradiol measurement in human plasma. J. Clin. Endocrinol. *29*, 879–883 (1969)

Korenman, S. G., Sanborn, B. M., Bhalla, R. C.: Adenyl cyclase and the cyclic AMP responsive systems in the uterus. In: Receptors for reproductive hormones. O'Malley, B. W., Means, A. R. (eds.), pp. 241–262. New York: Plenum Press 1973

Korsten, C. B., Persijn, J. P.: A simple assay for specific estrogen binding capacity in human mammary tumours. Z. Klin. Chem. Klin. Biochem. *10*, 502–508 (1972)

Koshland, D. E., Jr.: Biological specificity in protein small molecule interactions. Biochem. Pharmacol. *8*, Abs., 57 (1961)

Koshland, D. E., Jr., Neet, K. E.: The catalytic and regulatory properties of enzymes. Ann. Rev. Biochem. *27*, 359–410 (1968)

Koshland, D. E., Jr., Nemethy, G., Filmer, D.: Comparison of experimental binding data and theoretical models in proteins containing subunits. Biochemistry *5*, 365–385 (1966)

Krishan, A., Uzman, B. G., Hedley-Whyte, E. T.: Nuclear bodies: A component of cell nuclei in hamster tissues and human tumors. J. Ultrastruct. Res. *19*, 563–572 (1967)

Kuehl, F. A., Ham, E. A., Zanetti, M. E., Sanford, C., Nicol, S. E., Goldberg, N. D.: Estrogen-related increases in uterine guanosine 3':5'-cyclic monophosphate levels. Proc. Natl. Acad. Sci. USA *71*, 1866–1870 (1974)

Kuhn, R. W., Schrader, W. T., Smith, R. G., O'Malley, B. W.: Progesterone binding components of chick oviduct. X. Purification by affinity chromatography. J. Biol. Chem. *250*, 4220–4228 (1975)

Lan, N. C., Katzenellenbogen, B. S.: Temporal relationships between hormone receptor binding and biological responses in the uterus: Studies with short- and long-acting derivatives of estriol. Endocrinology *98*, 220–227 (1976)

Langley, J. N.: On the reaction of cells and of nerve-endings to certain poisons chiefly as regards the reaction of striated muscle to nicotine and to curari. J. Physiol. *33*, 374–413 (1905)

Larson, J. A.: Estrogens and endometrial carcinoma. Am. J. Obstet. Gynecol. *3*, 551–573 (1954)

Latham, K. R., Ring, J. C., Baxter, J. D.: Solubilized nuclear "receptors" for thyroid hormones: Physical characteristics and binding properties, evidence for multiple forms. J. Biol. Chem. *251*, 7388–7399 (1976)

Laumas, K. R., Farooq, A.: The uptake in vivo of [1,2-^{3}H]progesterone by the brain and genital tract of the rat. J. Endocrinol. *36*, 95–96 (1966)

Lazier, C.: [3H] estradiol binding by chick liver nuclear extracts: Mechanism of increase in binding following estradiol injection. Steroids *26*, 281–298 (1976)

Leavitt, W. W., Blaha, G. C.: An estrogen-stimulated, progesterone-binding system in the hamster uterus and vagina. Steroids *19*, 263–274 (1972)

Leavitt, W. W., Grossman, C. J.: Characterization of binding components for progesterone and 5 alpha-pregnane-3, 20-dione in the hamster uterus. Proc. Natl. Acad. Sci. USA *71*, 4341–4345 (1974)

Leavitt, W. W., Toft, D. O., Strott, C. A., O'Malley, B. W.: A specific progesterone receptor in the hamster uterus: Physiologic properties and regulation during the estrous cycle. Endocrinology *94*, 1041–1053 (1974)

Leavitt, W. W., Chen, T. J., Do, Y. S., Carlton, B. D., Allen, T. C.: Biology of progesterone receptors. In: Receptors and hormone action. O'Malley, B. W., Birnbaumer, L. (eds.), pp. 157–188. New York: Academic Press 1978

Lebeau, M. C., Massol, N., Baulieu, E. E.: An insoluble receptor for oestrogens in the "residual" nuclear proteins of chick liver. Eur. J. Biochem. *36*, 294–300 (1973)

Lebeau, M. C., Massol, N., Baulieu, E. E.: Extraction, partial purification and characterization of "the insoluble estrogen receptor" from chick liver nuclei. FEBS Letters *43*, 107–111 (1974)

Leclercq, G., Hulin, N., Heuson, J. C.: Interaction of activated estradiol receptor complex and chromatin in isolated uterine nuclei. Eur. J. Cancer *9*, 681–685 (1973)

Leclercq, G. Heuson, J. C., Deboel, M. C., Legros, N., Longeval, E., Mattheiem, W. H.: Estrogen and progesterone receptor in human breast cancer. In: Progesterone receptors in normal and neoplastic tissues. McGuire, W. L., Raynaud, J.-P., Baulieu, E.-E. (eds.), pp. 141–153. New York: Raven Press 1977

Lee, C., Keyes, P. L., Jacobsen, H. I.: Estrogen receptor in the rabbit corpus luteum. Science *173*, 1032–1033 (1971)

Legan, S. J., Karsch, F. J.: A daily signal for the LH surge in the rat. Endocrinology *96*, 57–62 (1975)

Legan, S. J., Gay, V. L., Midgley, A. R.: LH release following steroid administration in ovariectomized rats. Endocrinology *93*, 781–785 (1973)

Legan, S. J., Coon, G. A., Karsh, F. J.: Role of estrogen as initiator of daily LH surges in the ovariectomized rat. Endocrinology *96*, 50–56 (1975)

Lemon, H. M., Miller, D. M., Foley, J. F.: Competition between steroids for hormonal receptor. Natl. Cancer Inst. Monogr. *34*, 77–83 (1971)

Lerner, L. J.: Hormone antagonists: Inhibitors of specific activities of estrogen and androgen. Recent Prog. Horm. Res. *20*, 435–490 (1964)

Leung, B. S., Sasaki, G. H.: Prolactin and progesterone effect on specific estradiol binding in uterine and mammary tissues in vitro. Biochem. Biophys. Res. Commun. *55*, 1180–1187 (1973)

Leung, B. S., Jack, W. M., Reiney, C. G.: Estrogen receptor in mammary glands and uterus of rats during pregnancy, lactation and involution. J. Steroid Biochem. *7*, 89–95 (1976)

Levitzki, A.: Membrane receptor function. In: Hormone and antihormone action at the target cell. Clark, J. H., Klee, W., Levitzki, A., Wolff, J. (eds.), pp. 79–86. Berlin: Life Sci. Res. Report *3*., Daklem Konferenzen Abakon Verlagsgesellschaft 1976

Levy, W. B., Redburn, D. A., Cotman, C. W.: Stimulus-coupled secretion of GABA from synaptosomes. Science *181*, 676–678 (1973)

Li, J. J., Talley, D. J., Li, S. A., Villee, C. A.: Receptor characteristics of specific estrogen binding in the renal adenocarcinoma of the golden hamster. Cancer Res. *36*, 1127–1132 (1976)

Liang, T., Liao, S.: Association of the uterine 17beta-estradiol-receptor complex with ribonucleoprotein in vitro and in vivo. J. Biol. Chem. *249*, 4671–4678 (1974)

Liang, T., Liao, S.: A very rapid effect of androgen or initiation of protein synthesis in prostate. Proc. Natl. Acad. Sci. USA *72*, 706–709 (1975)

Liao, S., Liang, T., Tymoczko, J. L.: Ribonucleoprotein binding of steroid-"receptor" complexes. Nature New Biol. *241*, 211–213 (1973)

Liarakos, C. D., Rosen, J. M., O'Malley, B. W.: Effect of estrogen on gene expression in the chick oviduct. II. Transcription of chick tritiated unique deoxyribonucleic acid as measured by hybridization in ribonucleic acid excess. Biochemistry *12*, 2809–2816 (1973)

Lieberburg, I., McEwen, B. S.: Brain cell nuclear retention of testosterone metabolites, 5α-dihydrotestosterone and estradiol-17β in adult rats. Endocrinology *100*, 588–597 (1977)

Linkie, D. M.: Estrogen receptors in different target tissues: Similarities of form – dissimilarities of transformation. Endocrinology *101*, 1862–1870 (1977)

Linkie, D. M., Siiteri, P. K.: A re-examination of the interaction of estradiol with target cell receptors. J. Steroid Biochem. *9*, 1071–1078 (1978)
Lippe, B. M., Szego, C. M.: Participation of adrenocortical hyperactivity in the suppressive effect of systemic actinomycin-D on uterine stimulation by oestrogen. Nature *207*, 272–274 (1965)
Lippman, M., Bolan, G., Monaco, M., Pinkus, L., Engel, L.: Model systems for the study of estrogen action in tissue culture. J. Steroid Biochem. *7*, 1045–1051 (1976)
Lippman, M., Huff, K., Bolan, G.: Progesterone and glucocoerticoid interactions with receptor in breast cancer cells in long-term tissue culture. Ann. N.Y. Acad. Sci. *286*, 101–115 (1977a)
Lippman, M., Monaco, M. E., Bolan, G.: Effects of estrone, estradiol and estriol on hormone-responsive human breast cancer in long-term tissue culture. Cancer Res. *37*, 1901–1907 (1977b)
Lobl, R. T., Maenza, R. M.: Androgenization: Alterations in uterine growth and morphology. Biol. Reprod. *13*, 255–264 (1975)
Loeb, L., Kountz, W. B.: The effect of injection of follicular extract on the sex organs in the guinea pig and the interaction between the follicular substances and substances given off by the corpus luteum. Am. J. Physiol. *84*, 283–306 (1928)
Louvet, J. P., Harman, S. M., Schreiber, J. R., Ross, G. T.: Evidence for a role of androgens in follicular maturation. Endocrinology *97*, 366–372 (1975)
Lucas, W. E.: Casual relationships between endocrine-metabolic variables in patients with endometrial carcinoma. Obstet. Gynecol. Surv. *29*, 507–528 (1974)
Luck, D., Hamilton, T. H.: Early estrogen action: Stimulation of the metabolism of high molecular weight and ribosomal RNA's. Proc. Natl. Acad. Sci. USA *69*, 157–161 (1972)
Luck, D. N., Gschwendt, M., Hamilton, T. H.: Oestrogenic stimulation of macromolecular synthesis is correlated with nuclear hormone-receptor complex in neonatal rat uterus. Nature New Biol. *245*, 24–25 (1973)
Luskey, K. L., Brown, M. S., Goldstein, J. L.: Stimulation of the synthesis of very low density lipoproteins in the rooster liver by estradiol. J. Biol. Chem. *249*, 5939–5947 (1974)
Luttage, W. G., Wallis, C. J.: In vitro accumulation and saturation of [^{3}H-]progestins in selected brain regions and in the adenohypophysis, uterus and pineal of the female rat. Steroids *22*, 493–502 (1973)
Luu Thi, M. T., Baulieu, E. E., Milgrom, E.: Comparison of the characteristics and of the hormonal control of endometrial and myometrial progesterone receptors. J. Endocrinol. *66*, 349–356 (1975)

McCormack, G. M., Moon, R. C.: Effect of increasing doses of estrogen and progesterone on mammary carcinogenesis in the rat. Eur. J. Cancer *9*, 483–486 (1973)
McCormack, S. A., Glasser, S. R.: Ontogeny and regulation of a rat placental estrogen receptor. Endocrinology *102*, 273–280 (1978)
McCormack, S. A., Clark, J. H.: Clomid administration to pregnant rats causes abnormalities of the reproductive tract in the offspring and mothers. Science *204*, 629–631 (1979)
McEwen, B. S.: Gonadal steroid receptors in neuroendocrine tissues. In: Receptors for hormones. O'Malley, B. W., Birnbaumer, L. (eds.), Vol. II, pp. 353–400. New York: Academic Press 1978
McEwen, B. S., Pfaff, D. W.: Factors influencing sex hormone uptake by rat brain regions: I. Effects of neonatal treatment, hypophysectomy and competing steroid on estradiol uptake. Brain Res. *21*, 1–16 (1970)
McEwen, B. S., Zigmond, R. E., Gerlach, J. L.: Sites of steroid binding and action in the brain. In: Structures and function of nervous tissue. Bourne, G. H. (ed.), Vol. 5, pp. 205–291. New York: Academic Press 1972
McEwen, B. S., Denef, C. J., Gerlach, J. L., Plapinger, L.: Chemical studies of the brain as a steroid hormone target tissue. In: The neuroscience: Third study program. Schmitt, F. O., Worden, F. G. (eds.), pp. 599–620. Cambridge, Mass: MIT Press 1974
McEwen, B. S., Plapinger, L., Chaptal, C., Gerlach, J., Wallach, G.: Role of fetoneonatal estrogen binding proteins in the associations of estrogen with neonatal brain cell nuclear receptors. Brain Res. *96*, 400–406 (1975)
McEwen, B. S., De Kloet, R., Wallach, G.: Interactions in vivo of corticoids and progesterone with cell nuclei and soluble macromolecules from rat brain regions and pituitary. Brain Res. *105*, 129–135 (1976)
McGhee, J. D., von Hippel, P. H.: Theoretical aspects of DNA-protein interactions: Co-operative and non-cooperative binding of large ligands to a one-dimensional homogeneous lattice. J. Mol. Biol. *86*, 469–489 (1974)
McGuire, J. L., Lisk, R. D.: Localization of estrogen receptors in the rat hypothalamus. Neuroendocrinology *4*, 289–295 (1969)

McGuire, W. L., Huff, K., Chamness, G. C.: Temperature-dependent binding of estrogen receptor to chromatin. Biochemistry *11*, 4562–4565 (1973)

McGuire, W. L., Carbone, P. P., Vollmer, E. P.: Estrogen receptors in human breast cancer. New York: Raven Press 1975

McKelvy, J. F., LeBlanc, P., Laudes, C., Perrie, S., Grimm-Jorgensen, Y., Kordon, C.: The use of bacitracin as an inhibitor of the degradation of thyrotropin releasing factor and luteinizing hormone releasing factor. Biochem. Biophys. Res. Commun. *73*, 507–515 (1976)

MacLaughlin, D. T., Richardson, G. S.: Progesterone binding by normal and abnormal human endometrium. J. Clin. Endocrinol. Metab. *42*, 667–678 (1976)

MacLeod, J., Reynolds, S. R. M.: Vascular, metabolic, and motility responses of uterine tissue following administration of oestrin. Proc. Soc. Exp. Biol. Med. *37*, 666–668 (1938)

MacLuskey, N. J., McEwen, B. S.: Oestrogen modulates progestin receptor concentrations in some rat brain regions but not in others. Nature *274*, 276–278 (1978)

Mahesh, V. B., Muldoon, T. G., Eldridge, J. C., Korach, K. S.: The role of steroid hormones in the regulation of gonadotropin secretion. J. Steroid Biochem. *6*, 1025–1036 (1975)

Mainwaring, W. I. P., Peterken, B. M.: A reconstituted cell-free system for the specific transfer of steroid-receptor complexes into nuclear chromatin isolated from rat ventral prostate gland. Biochem. J. *125*, 285–295 (1971)

Major, J. S., Green, B., Heald, P. J.: Interactions of oestradiol-17beta and tamoxifen in the uterus of the pregnant rat. J. Endocrinol. *71*, 315–324 (1976)

Markaverich, B. M., Clark, J. H., Hardin, J. W.: RNA transcription and uterine growth: differential effects of estradiol, estriol and Nafoxidine on chromatin RNA initiation sites. Biochemistry *17*, 3146–3152 (1978)

Markee, J. E.: Rhythmic vascular uterine changes. Am. J. Physiol. *100*, 32–39 (1932)

Markland, F. S., Jr., Hutchens, T. W.: Characterization of the progestin receptor from lactating mammary glands of the goat. In: Progesterone receptors in normal and neoplastic tissues. McGuire, W. L., Raynaud, J.-P., Baulieu, E.-E. (eds.), pp. 23–38. New York: Raven Press 1977

Martin, L.: Dimethylstilbestrol and 16-oxo-estradiol: Antiestrogens or estrogens. Steroids *13*, 1–10 (1969)

Martin, L., Finn, C. A.: Duratin of progesterone treatment required for a stromal response to oestradiol 17β in the uterus of the mouse. J. Endocrinol. *44*, 279–280 (1969)

Martin, L., Finn, C. A.: Oestrogen-gestagen interactions on mitosis in target tissues. In: Basic actions of sex steroids on target organs. Hubinont, P. O., LeRoy, F., Galand, P., (eds.), pp. 172–188. Basel: S. Karger (1971)

Martin, L., Pollard, J. W., Fagg, B.: Oestriol, oestradiol-17β and the proliferation and death of uterine cells. J. Endocrinol. *69*, 103–115 (1976)

Martucci, C., Fishman, J.: Direction of estradiol metabolism as a control of its hormonal action-uterotropic activity of estradiol metabolites. Endocrinology *101*, 1709–1715 (1977)

Maurer, H. R., Chalkley, G. R.: Some properties of a nuclear binding site of estradiol. J. Mol. Biol. *27*, 431–441 (1967)

Maurer, R. A., Woolley, D. E.: Demonstration of nuclear [^{3}H]-estradiol binding in hypothalamus and amygdala of female, androgenized female and male rats. Neuroendocrinology *16*, 137–147 (1974)

Maurer, R. A., Woolley, D. E.: [^{3}H]-estradiol distribution in female, androgenized female and male rats at 100 and 200 days of age. Endocrinology *96*, 655–765 (1975)

Mayol, R. F., Thayer, S. A.: Synthesis of estrogen-specific proteins in the uterus of the immature rat. Biochemistry *9*, 2484–2489 (1970)

Means, A. R., Hamilton, T. H.: Evidence for depression of nuclear protein synthesis and concomitant stimulation of nuclear RNA synthesis during early estrogen action. Proc. Natl. Acad. Sci. USA *56*, 686–693 (1966)

Means, A. R., O'Malley, B. W.: Protein biosynthesis on chick oviduct polyribosomes. II. Regulation by progesterone. Biochemistry *10*, 1570–1576 (1971)

Means, A. R., Abrass, I. B., O'Malley, B. W.: Protein biosynthesis on chick oviduct polyribosomes. I. Changes during estrogen-mediated tissue differentiation. Biochemistry *10*, 1561–1570 (1971)

Means, A. R., Comstock, J. P., Rosenfeld, G. C., O'Malley, B. W.: Ovalbumin messenger RNA of chick oviduct: Partial characterization, estrogen dependence and translation in vitro. Proc. Natl. Acad. Sci. USA *69*, 1146–1150 (1972)

Medina, D., Iramian, C. A., Clark, J. H.: Hormone dependence and estradiol receptors in the D series of mammary nodule outgrowth lines and tumors. Cancer Res. *35*, 2355–2360 (1975)

Meiri, U., Rahaminoff, R.: Neuromuscular transmission: Inhibition by manganese ions. Science *176*, 308–309 (1972)

Mester, J., Baulieu, E. E.: Nuclear estrogen receptor of chick liver. Biochim. Biophys. Acta *261*, 236–244 (1972)

Mester, J., Baulieu, E. E.: Dynamics of oestrogen-receptor distribution between the cytosol and nuclear fractions of immature rat uterus after oestradiol administration. Biochem. J. *146*, 617–623 (1975)

Mester, J., Robertson, D. M., Feherty, P., Kellie, A. E.: Determination of high-affinity oestrogen receptor sites in uterine supernatant preparations. Biochem. J. *120*, 831–836 (1970)

Mester, J., Martel, D., Psychoyos, A., Baulieu, E. E.: Hormonal control of oestrogen receptor in uterus and receptivity for ovoimplantation in the rat. Nature *250*, 776–778 (1974)

Metzger, D. A., Schrader, W. T., Clark, J. H.: Various forms of the estrogen receptor in uterine cytosol. (submitted)

Michael, R. P.: Action of hormones on the cat brain. In: Brain and behavior. Gorski, R. A., Whalen, R. E. (eds.), Vol. III, pp. 82–98. Los Angeles: UCLA Press 1964

Michael, R. P.: Oestrogens in the central nervous system. Br. Med. Bull. *21*, 87–90 (1965)

Michel, G., Jung, I., Baulieu, E. E., Aussel, C., Uriel, J.: Two high affinity estrogen binding proteins of different specificity in the immature rat uterus cytosol. Steroids *24*, 437–449 (1974)

Middlebrook, J. L., Wong, M. D., Ishii, D. N., Aronow, L.: Subcellular distribution of glucocorticoid receptors in mouse fibroblasts. Biochemistry *14*, 180–186 (1975)

Miledi, R.: Lanthanum ions abolish "calcium response" of nerve terminals. Nature *229*, 410–411 (1971)

Milgrom, E., Baulieu, E. E.: Progesterone in uterus and plasma. I. Binding in rat uterus 105,000 g supernatant. Endocrinology *87*, 276–286 (1970a)

Milgrom, E., Baulieu, E. E.: Progesterone in the uterus and the plasma. II. The role of hormone availability and metabolism on selective binding to uterus protein. Biochem. Biophys. Res. Commun. *40*, 723–730 (1970b)

Milgrom, E., Atger, M., Baulieu, E. E.: Progesterone in uterus and plasma. IV. Progesterone receptors in guinea pig uterus cytosol. Steroids *16*, 741–764 (1970)

Milgrom, E., Atger, M., Baulieu, E.-E.: L'entree des oestrogens dans les cellules uterines dependelle d'une proteine? C. R. Acad. Sci. Paris *274*, 2771–2774 (1972a)

Milgrom, E., Perrot, M., Atger, M., Baulieu, E. E.: Progesterone in uterus and plasma. V. An assay of the progesterone cytosol receptor of the guinea pig uterus. Endocrinology *90*, 1064–1070 (1972b)

Milgrom, E., Thi, L., Atger, M., Baulieu, E. E.: Mechanisms regulating the concentration and the conformation of progesterone receptor(s) in the uterus. J. Biol. Chem. *248*, 6366–6377 (1973a)

Milgrom, E., Atger, M., Baulieu, E. E.: Studies on estrogen entry into uterine cells and on estradiol-receptor complex attachment to the nucleus is the entry of estrogen into uterine cells a protein-mediated process? Biochim. Biopys. Acta *320*, 267–283 (1973b)

Milgrom, E., Vu Hai, M. T., Logeat, F.: Use of [^{3}H-]R5020 for the assay of cytosol and nuclear progesterone receptor in the rat uterus. In: Progesterone receptors in normal and neoplastic tissues. McGuire, W. L., Raynaud, J.-P., Baulieu, E.-E. (eds.), pp. 261–270. New York: Raven Press 1977

Milin, B., Roy, A. K.: Androgen receptor in rat liver: Cytosol receptor deficiency in pseudohermaphrodite male rats. Nature New Biol. *242*, 248–250 (1973)

Miller, B. G.: The relative potencies of oestriol, oestradiol and oestrone in the uterus and vagina of the mouse. J. Endocrinol. *43*, 563–570 (1969)

Miller, B. G., Emmens, C. W.: The oestrogenic potency in the mouse of several substances closely related to diethylstilboestrol and mesohexoestrol. J. Endocrinol. *45*, 9–15 (1969)

Miller, J. B., Keyes, P. L.: A mechanism for regression of the rabbit corpus luteum; uterine-induced loss of luteal responsiveness to 17β estradiol. Biol. Reprod. *15*, 511–518 (1976)

Kawashima, M., Kamiyoshi, M., Tanaka, K.: A cytoplasmic progesterone receptor in hen pituitary and hypothalamic tissues. Endocrinology *102*, 1207–1213 (1978)

Moghissi, K. S., Syner, F. N., Evans, T. N.: A composite picture of the menstrual cycle. Am. J. Obstet. Gynecol. *114*, 405–418 (1972)

Moulton, B. C., Barker, K. L.: Synthesis and degradation of glucose-6-phosphate dehydrogenase in the rat uterus. Endocrinology *89*, 1131–1136 (1971)

Mowles, T. F., Ashkanazy, B., Mix, E., Jr., Sheppard, H.: Hypothalamic and hypophyseal estradiol binding complexes. Endocrinology *89*, 484–491 (1971)

Muechler, E. K., Flickinger, G. L., Mangan, C. E., Mikhail, G.: Estradiol binding by human endometrial tissue. Gynecol. Oncol. *3*, 244–250 (1975)

Mueller, G. C.: Incorporation of glycine-2-[C^{14}] into protein by surviving uteri from α-estradiol-treated rats. J. Biol. Chem *204*, 77–90 (1953)

Mueller, G. C., Herranen, A. M., Jervell, K.: Studies on the mechanism of action of estrogens. Recent Prog. Horm. Res. *14*, 95–139 (1958)

Muggia, F. M., Cassileth, P. A., Ochoa, M., Flatow, F. A., Gelhorn, A., Hyman, G. A.: Treatment of breast cancer with medroxyprogesterone acetate. Ann. Intern. Med. 46, 328–337 (1968)

Mulder, A. H.: On the subcellular localization of corticotropin-releasing factor (CRF) in the rat median eminence. In: Progress in brain research. De Weid, D., Weijnen, J. A. W. M. (eds.), Vol 32, pp. 33–41. New York: Elsevier 1970

Muldoon, T. G.: Mouse mammary tissue estrogen receptors: Ontogeny and molecular heterogeneity. In: Ontogeny of receptors and molecular mechanism of reproductive hormone action. Hamilton, T. H., Clark, J. H., Sadler, W. (eds.), New York: Raven Press 1978

Muller, R. E., Traish, A. M., Wotiz, H. H.: Interaction of receptor estradiol complex with uterine nuclei. J. Biol. Chem. *252*, 8206–8211 (1977)

Mullinix, K. P., Wetekam, W., Deeley, R. G., Gordon, J. I., Meyers, M., Kent, K. A., Goldberger, R. F.: Induction of vitellogenin synthesis by estrogen in the avian liver: Relationship between the level of vitellogenin mRNA and vitellogenin synthesis. Proc. Natl. Acad. Sci. USA *73*, 1442–1451 (1976)

Munck, A.: General aspects of steroid hormone-receptor interactions. In: Receptors and mechanism of action of steroid hormones. Pasqualini, J. R. (ed.), pp. 1–40. New York: Marcel Dekker 1976

Munck, A., Foley, R.: Kinetics of glucocorticoid receptor complexes in rat thymus cells. J. Steroid Biochem. *7*, 1117–1122 (1976)

Munck, A., Wira, C.: Glucocorticoid receptors in rat thymus cells. Adv. Biol. Sci. *7*, 301–327 (1971)

Munck, A., Wira, C., Young, D. A., Mosher, K. M., Hallahan, C., Bell, P. A.: Glucocorticoid-receptor complexes and the earliest steps in the action of glucocorticoids on thymus cells. J. Steroid Biochem. *3*, 567–578 (1972)

Murphy, B. E. P.: Clinical evaluation of urinary cortisol determinations by competitive protein-binding radioassay. J. Clin. Endocrinol. *28*, 343–349 (1968)

Musliner, T. A., Chader, G. J., Villee, C. A.: Studies on estradiol receptors of the rat uterus. Nuclear uptake in vitro. Biochemistry *9*, 4448–4452 (1970)

Naftolin, F., Ryan, K. J., Petro, Z.: Aromitization and androstenedione by the anterior hypothalamus of adult male and female rats. Endocrinology *90*, 295–298 (1972)

Naftolin, F., Morishita, H., Davies, I. J., Todd, R., Ryan, K. J., Fishman, J.: 2-hydroxyesterone induced rise in serum luteinizing hormone in the immature male rat. Biochem. Biophys. Res. Commun. *64*, 905–913 (1975a)

Naftolin, F., Ryan, K. J., Davies, I. J., Reddy, V. V., Flores, F., Petro, Z., Kuhn, M., White, R. J., Takaoka, Y., Wolin, L.: The formation of estrogens by central neuroendocrine tissues. Recent Prog. Horm. Res. *31*, 295–319 (1975b)

Nakai, T., Sakamota, S., Kigawa, T., Shigematsu, A.: Nuclear concentration of [^{3}H]-estradiol in target tissues of a fetal rat demonstrated by dry mount autoradiography. Endocrinol. Jpn. *19*, 47–52 (1972)

Nakai, T., Plant, M., Hess, D. L., Keogh, E. J., Knobil. E.: On the sites of the negative and positive feedback actions of estradiol in the control of gonadotropin secretion in the rhesus monkey. Endocrinology *102*, 1008–1014 (1978)

Nicholson, R. I., Golder, M. P.: The effect of synthetic anti-estrogens on the growth and biochemistry of rat mammary tumors. Eur. J. Cancer *11*, 571–579 (1975)

Nicholson, R. I., Golder, M. P., Davies, P., Griffiths, K.: Effects of oestradiol and tamoxifen on total and accessible cytoplasmic oestradiol receptors in DMBA-induced rat mammary tumours. Eur. J. Cancer *12*, 711–717 (1976)

Nielsen, S., Notides, A. C.: Transformation of the rat uterine estrogen receptor after partial purification. Biochim. Biophys. Acta *381*, 377–383 (1975)

Normann, J. C.: Neurosecretion by exocytosis. Int. Rev. Cytol. *46*, 1–77 (1976)

Norris, J. S., Kohler, P. O.: Estrogen accumulation by porcine granulosa cells in culture. Endocr. Res. Commun. *1*, 133–143 (1974)

Noteboom, W. D., Gorski, J.: Stereospecific binding of estrogens in the rat uterus. Arch. Biochem. Biophys. *111*, 559–568 (1965)

Notides, A. C.: Conformational forms of the estrogen receptor. In: Receptors and Hormone Action. O'Malley, B. W., Birnbaumer, L. (eds.), pp. 33–61. New York: Academic Press 1978

Notides, A. C., Nielsen, S.: The molecular mechanism of the in vitro 4S to 5S transformation of the uterine estrogen receptor. J. Biol. Chem. *249*, 1866–1873 (1974)

Notides, A. C., Nielsen, S.: A molecular and kinetic analysis of estrogen receptor transformation. J. Steroid Biochem. *6*, 483–486 (1975)

Notides, A. C., Hamilton, D. E., Auer, H. E.: A kinetic analysis of the estrogen receptor transformation. J. Biol. Chem. *250*, 3945–3950 (1975)

Nunez, E., Engelmann, F., Benassayag, C., Jayle, M. G.: Identification et purification preliminaire de la foeto-proteine liant les oestrogens dans le serum de rats nouveau-nes. C. R. Acad. Sci. Paris *273*, 831–834 (1971)

Nutting, E. F., Meyer, R. K.: Effect of oestrone on delay of nidation, implantation and foetal survival in ovariectomized rats. J. Endocrinol. *29*, 235–242 (1964)

Ogren, L., Woolley, D.: Brain and pituitary nuclear [^{3}H]-estradiol binding after saturating doses in male and female rats. Reduction by endogenous hormones. Neuroendocrinology (in press) (1978)

Ogren, L., Vertes, M., Woolley, D.: In vivo nuclear [^{3}H]-estradiol binding in brain areas of the rat: reduction by endogenous and exogenous androgens. Neuroendocrinology *21*, 350–365 (1976)

Ohta, T., Kimura, M.: Functional organization of genetic material as a product of molecular evolution. Nature *233*, 118–119 (1971)

Ojeda, S. R., Wheaton, J. E., Jameson, H. E., McCann, S. M.: The onset of puberty in the female rat: Plasma prolactin, gonadotropins, luteinizing hormone-releasing-hormone (LHRH), and hypothalamic LHRH content. Endocrinology *98*, 630–638 (1976)

Oka, T., Schimke, R. T.: Progesterone antagonism of estrogen-induced cytodifferentiation in chick oviduct. Science *163*, 83–84 (1969)

O'Malley, B. W., Means, A. R.: Female steroid hormones and target cell nuclei. Science *183*, 610–620 (1974)

O'Malley, B. W., Schrader, W. T.: Progesterone receptor components: Identification of subunits binding to the target-cell genome. J. Steroid Biochem. *3*, 617–629 (1972)

O'Malley, B. W., Strott, C. A.: The mechanism of action of progesterone. In: Handbook of Physiology. Greep, R. O. (ed.), Vol. II, Part 1, pp. 591–602. Washington D. C.: American Physiological Society 1973

O'Malley, B. W., McGuire, W. L., Kohler, P. O., Korenman, S. G.: Studies on the mechanism of steroid hormone regulation of synthesis of specific proteins. Recent Prog. Horm. Res. *25*, 105–160 (1969)

O'Malley, B. W., Sherman, M. R., Toft, D. O.: Progesterone "receptors" in the cytoplasm and nucleus of chick oviduct target tissue. Proc. Natl. Acad. Sci. *67*, 501–508 (1970)

O'Malley, B. W., Toft, D. O., Sherman, M. R.: Progesterone binding components of chick oviduct. II. Nuclear components. J. Biol. Chem. *246*, 1117–1122 (1971)

O'Malley, B. W., Spelsberg, T. C., Schrader, W. T., Chytil, F., Steggles, A. W.: Mechanisms of interaction of a hormone receptor complex with the genome of a eukaryotic target cell. Nature *235*, 141–144 (1972)

O'Malley, B. W., Schrader, W. T., Spelsberg, T. C.: Hormone receptor interactions with the genome of eukaryotic target cells. Adv. Exptl. Med. Biol. *36*, 174–196 (1973)

Oppenheimer, J. H.: Role of plasma proteins in the binding, distribution and metabolism of the thyroid hormones. N. Engl. J. Med. *278*, 1153–1162 (1968)

Oppenheimer, J. H., Schwartz, H. L., Surks, M. I., Koerner, D., Dillman, W. H.: Nuclear receptors and the initiation of thyroid hormone action. Recent Prog. Horm. Res. *32*, 529–565 (1976)

Overath, P., Trauble, H.: Phase transitions in cells, membranes, and lipids of *Escherichia coli*. Detection by fluorescent probes, light scattering, and dilatometry. Biochemistry *12*, 2625–2634 (1973)

Palmiter, R. D.: I. Independent regulation of ovalbumin, conalbumin, ovomucoid and lysozyme induction. J. Biol. Chem. *247*, 6450–6461 (1972)

Palmiter, R. D., Moore, P. B., Mulvihill, E. R.: A significant lag in the induction of ovalbumin messenger RNA by steroid hormones: A receptor translocation hypothesis. Cell *8*, 557–572 (1976)

Panko, W. B., Macleod, R. M.: Uncharged nuclear receptors for estrogen in breast cancers. Cancer Res. *38*, 1948–1951 (1978)

Panko, W. B., Watson, C., Clark, J. H.: Heterogeneity of estradiol binding sites in human breast cancer. (submitted)

Parker, C. R., Jr., Costoff, A., Muldoon, T. G., Mahesh, V. B.: Actions of pregnant mare serum gonadotropin in the immature female rat: correlative changes in blood steroids, gonadotropins and cytoplasmic estradiol receptors of anterior pituitary and hypothalamus. Endocrinology *98*, 129–138 (1976)

Pasqualini, J. R., Sumida, C., Gelly, C., Nguyen, B. L.: Specific [^{3}H] estradiol binding in the fetal uterus and testis of guinea pig. Quantitative evolution of [^{3}H] estradiol receptors in the different fetal tissues (kidney, lung, uterus and testis) during fetal development. J. Steroid Biochem. *7*, 1031–1038 (1976)

Paton, W. D. M.: A theory of drug action based on the rate of drug-receptor combination. Proc. R. Soc. Lond. [Biol.] *154*, 21–69 (1961)

Paton, W. D. M., Rang, H. P.: A kinetic approach to the mechanism of drug action. In: Advances in drug research. Harper, N. J., Simmonds, A. B. (eds.), pp. 57–80. New York: Academic Press 1966

Paup, D. C., Mennin, S. P., Gorski, R. A.: Androgen- and estrogen-induced copulatory behavior and inhibition of luteinizing hormone (LH) secretion in the male rat. Horm. Behav. *6*, 35–46 (1975)

Pavlik, E. J., Coulson, P. B.: Hydroxylapatite "batch" assay for estrogen receptors: increased sensitivity over present receptor assays. J. Steroid Biochem. *7*, 357–368 (1976a)

Pavlik, E. J., Coulson, P. B.: Modulation of estrogen receptors in four different target tissues: Differential effects of estrogen vs progesterone. J. Steroid Biochem. *7*, 369–376 (1976b)

Payne, R. W., Runser, R. H.: The influence of estrogen and androgen on the ovarian responses of hypophysectomized immature rats to gonadotrophs. Endocrinology *62*, 313–321 (1958)

Pearlman, W. H., DeHertogh, R., Lauman, D. R., Brueggemann, J. A., Pearlman, M. R. J.: Metabolism and tissue uptake of steroid sex hormones in patients with advanced carcinoma of the breast and in normal rats. In: Steroid dynamics. Pincus, G., Nakao, T., Tati, J. F. (eds.), pp. 159–176. New York: Academic Press 1966

Peck, E. J., Jr., Clark, J. H.: Effect of ionic strength on charcoal absorption assays of receptor estradiol complexes. Endocrinology *101*, 1034–1043 (1977)

Peck, E. J., Jr., Burgner, J., Clark, J. H.: Estrophilic binding sites of the uterus. Relation to uptake and retention of estradiol in vitro. Biochemistry *12*, 4596–4603 (1973a)

Peck, E. J., Jr., Delibero, J., Richards, R., Clark, J. H.: Instability of the uterine estrogen receptor under in vitro conditions. Biochemistry *12*, 4603–4608 (1973b)

Peck, E. J., Jr., Miller, A. L., Kelner, K. L.: Estrogen receptors in the activation of RNA polymerases by estrogens in the central nervous system. In: Ontogeny of receptors and molecular mechanisms of reproductive hormone action. Hamilton, T. H., Clark, J. H. (eds.), New York: Raven Press 1978

Peterson, R. P., Spaziani, E.: Autoradiographic localization of albumin-[^{131}I] accumulation in the estrogen-stimulated rat uterus. Endocrinology *89*, 1280–1286 (1971)

Petra, P. H., Schiller, H. S.: Sex steroid binding protein in the plasma of Macaca Nemestrina. J. Steroid Biochem. *8*, 655–661 (1977)

Pfaff, D. W.: Nature of sex hormone effects on rat sex behaviour: Specificity of effects of individual patterns of response. J. Comp. Physiol. Psychol. *73*, 349–358 (1970a)

Pfaff, D. W.: Synergistic and antagonistic effects of sex hormones on female rat sexual behaviour. Am. Zool. *10*, Abs., 478 (1970b)

Pfaff, D. W., Keiner, M.: Estradiol-concentrating cells in the rat amygdala as part of a limbic hypothalamic-hormone sensitive system. In: The neurobiology of the amygdala. Eleftheriou, B. (ed.), pp. 775–785. New York: Plenum Press 1972

Pfaff, D. W., Keiner, M.: Atlas of estradiol concentrating cells in the central nervous system of the female rat. J. Comp. Neurol. *151*, 121–158 (1973)

Philibert, D., Raynaud, J. P.: Binding of progesterone and R5020, a highly potent progestin, to human endometrium and myometrium. Contraception *10*, 457–466 (1974)

Philibert, D., Raynaud, J. P.: Cytoplasmic progestin receptors in mouse uterus. In: Progesterone receptors in normal and neoplastic tissue. McGuire, W. L., Raynaud, J. P., Baulieu, E. E. (eds.), pp. 227–243. New York: Raven Press 1977

Piacsek, B. E., Meites, J.: Effects of castration and gonadal hormones on hypothalamic content of luteinizing hormone releasing factor (LRF). Endocrinology *79*, 432–439 (1966)

Pietras, R. J., Szego, C. M.: Endometrial cell calcium and oestrogen action. Nature *253*, 357–359 (1976)

Pietras, R. J., Szego, C. M.: Specific binding sites for oestrogen at the outer surfaces of isolated endometrial cells. Nature *265*, 69–72 (1977)

Plant, T. M., Nakai, Y., Belchetz, P., Keogh, E., Knobil, E.: The sites of action of estradiol and phentolamine in the inhibition of the pulsatile, circhoral discharge of LH in the rhesus monkey (macaca mulatta). Endocrinology *102*, 1015–1018 (1978)

Plapinger, L., McEwen, B. S.: Ontogeny of estradiol-binding sites in rat brain. I. Appearance of presumptive adult receptors in cytosol and nuclei. Endocrinology *93*, 1119–1128 (1973)

Plapinger, L., McEwen, B. S., Clemens, L. E.: Onogeny of estradiol-binding sites in rat brain. II. Characteristics of a neonatal binding macromolecule. Endocrinology *93*, 1129–1139 (1973)

Pollow, K., L'ubbert, H., Boquoi, E., Kreuzer, G., Pollow, B.: Characterization and comparison of receptors for 17beta-estradiol and progesterone in human proliferative endometrium and endometrial carcinoma. Endocrinology *96*, 319–328 (1975)

Pollow, K., Schmidt-Gollwitzer, M., Nevinny-Stickel, J.: Progesterone receptors in normal human endometrium and endometrial carcinoma. In: Progesterone receptors in normal and neoplastic tissues. McGuire, W. L., Raynaud, J. P., Baulieu, E.-E. (eds.), pp. 313–338. New York: Raven Press 1977

Powell-Jones, W., Davies, P., Griffiths, K.: Specific binding of [^{3}H] oestradiol by cytoplasmic protein components of female rat liver. J. Endocrinol. *69*, 167–168 (1976)

Presl, J., Jirasek, J., Horsky, J., Henzl, M.: Observations on steroid-3β-ol dehydrogenase activity in the ovary during early postnatal development in the rat. J. Endocrinol. *31*, 293–294 (1965)

Presl, J., Horsky, J., Krabec, Z.: Development of the positive estrogen feedback in rat. Endocrinol. Exp. (Bratisl.) *2*, 161–165 (1968)

Presl, J., Rohling, S., Horsky, J., Herzmann, J.: Changes in uptake of [^{3}H]-estradiol by the female rat brain from birth to maturity. Endocrinology *86*, 899–902 (1970)

Price, D.: An analysis of the factors influencing growth and development of the mammalian reproductive tract. Physiol. Zool. *20*, 213–247 (1947)

Price, D., Pannabecker, J. R.: A study of sex differentiation in the fetal rat. In: The Chemical Basis of Development. McElroy, W., Glass, B. (eds.), pp. 774–778. Baltimore: Johns Hopkins Press 1958

Priore, R. L., Rosenthal, H. E.: A statistical model for the estimation of binding parameters in a complex system. Anal. Biochem. *70*, 231–240 (1976)

Puca, G. A., Bresciani, F.: Receptor molecules for oestrogens from rat uterus. Nature *218*, 967–969 (1968)

Puca, G. A., Bresciani, F.: Interactions of 6,7-[^{3}H]oestradiol-17β with mammary gland and other organs of the C3H mouse in vivo. Endocrinology *85*, 1–10 (1969)

Puca, G. A., Nola, E., Sica, V., Bresciani, F.: Estrogen-binding proteins of calf uterus. Partial purification and preliminary characterization of two cytoplasmic proteins. Biochemistry *10*, 3769–3780 (1971)

Puca, G. A., Nola, E., Sica, V., Bresciani, F.: Estrogen binding proteins of calf uterus. Interrelationship between various forms and identification of a receptor transforming factor. Biochemistry *11*, 4157–4165 (1972)

Puca, G. A., Sica, V., Nola, E.: Identification of a high affinity nuclear acceptor site for estrogen receptor of calf uterus. Proc. Natl. Acad. Sci. USA *71*, 979–983 (1974)

Puca, G. A., Nola, E., Hibner, U., Cicala, G., Sica, V.: Interaction of the estradiol receptor from calf uterus with its nuclear acceptor sites. J. Biol. Chem. *250*, 6452–6459 (1975)

Puca, G. A., Nola, E., Sica, V., Bresciani, F.: Estrogen binding proteins of calf uterus. Molecular and functional characterization of the receptor transforming factor: A Ca^{2+}-activated protease. J. Biol. Chem. *252*, 1358–1366 (1977)

Ramirez, V. D., Sawyer, C. H.: Differential dynamic responses of plasma LH and FSH to ovariectomy and to a single injection of estrogen in the rat. Endocrinology *94*, 987–993 (1974)

Rao, B. R., Meyer, J. S.: Estrogen and progestin receptors in normal and cancer tissue. In: Progesterone receptors in normal and neoplastic tissues. McGuire, W. L., Raynaud, J. P., Baulieu, E.-E. (eds.), pp. 155–169. New York: Raven Press 1977

Rao, B. R., Wiest, W. G.: Receptors for progesterone. Gynecol. Oncol. *2*, 239–248 (1974)

Rao, B. R., Wiest, W. G., Allen, W. M.: Progesterone "receptor" in rabbit uterus. I. Characterization and estradiol-17beta augmentation. Endocrinology *92*, 1229–1240 (1973)

Rao, B. R., Wiest, W. G., Allen, W. M.: Progesterone "receptor" in human endometrium. Endocrinology *95*, 1275–1281 (1974)

Rat, R. L., Vallet-Strouve, C., Erdos, T.: The properties of a uterine oestradiol-receptor after limited proteolysis. Biochimie *56*, 1387–1394 (1974)

Ratajczak, T., Hähnel, R.: Estradiol receptors: Influence of plasma proteins on detection and quantitation. J. Steroid Biochem. *7*, 741–744 (1976)

Raynaud, J. P.: Influence of rat estradiol binding plasma protein (EBP) on uterotrophic activity. Steroids *21*, 249–258 (1973)

Raynaud, J. P.: R5020, a tag for the progestin receptor. In: Progesterone receptors in normal and neoplastic tissues. McGuire, W. L., Raynaud, J. P., Baulieu, E.-E. (eds.), pp. 9–21. New York: Raven Press 1977

Raynaud, J. P., Mercier-Bodard, C., Baulieu, E.-E.: Rat estradiol plasma protein. Steroids *18*, 767–788 (1971)

Raynaud-Jammet, C., Cattelli, M. G., Baulieu, E.-E.: Inhibition of α-amanitin of the oestradiol-induced increase in α-amanitin insensitive RNA polymerase in immature rat uterus. FEBS Letters *22*, 93–96 (1972)

Raziano, J., Ferin, M., Vande Wiele, R. L.: Effects of antibodies to estradiol-17β and to progesterone on nidation and pregnancy in rats. Endocrinology *90*, 1133–1138 (1972)

Reddy, V., Naftolin, F., Ryan, K.: Conversion of androstenedione to estrone by neural tissues from fetal and neonatal rats. Endocrinology *94*, 117–124 (1974)

Reel, J. R., Shih, Y.: Oestrogen-inducible uterine progesterone receptors. Characteristics in the ovariectomized immature and adult hamster. Acta Endocrinol. (Kbh) *80*, 344–354 (1975)

Reel, J. R., Van Dewark, S. D., Shih, Y., Callantine, M. R.: Macromolecular binding and metabolism of progesterone in the decidual and pseudopregnant rat and rabbit uterus. Steroids *18*, 441–461 (1971)

Reiter, E. O., Goldenberg, R. L., Vaitukaitis, J. L., Ross, G. T.: A role for endogenous estrogen in normal ovarian development in the neonatal rat. Endocrinology *91*, 1537–1539 (1972)

Rennie, P.: Follicular activity in the maintenance of homotransplanted rabbit corpora lutea. Endocrinology *83*, 314–322 (1968a)

Rennie, P.: Luteal-hypophyseal interrelationship in the rabbit. Endocrinology *83*, 323–328 (1968b)

Reti, I., Erdos, T.: On the "native state" of the uterine estradiol "receptor". Biochimie *53*, 435–437 (1971)

Reynolds, S. R. M.: Determinants of uterine growth and activity. Physiol. Rev. *31*, 244–273 (1951)

Richards, J. E., Shyamala, G., Nandi, S.: Estrogen receptor in normal and neoplastic mouse mammary tissue. Cancer Res. *34*, 2764–2772 (1974)

Richards, J. S.: Estradiol binding to rat corpora lutea during pregnancy. Endocrinology *95*, 1046–1053 (1974)

Richards, J. S.: Content of nuclear estradiol receptor complex in rat corpora lutea during pregnancy: Relationship to estrogen concentration and cytosol receptor availability. Endocrinology *96*, 227–230 (1975)

Richards, J. S., Midgley, A. R., Jr.: Protein hormone action: A key to understanding ovarian follicular and luteal cell development. Biol. Reprod. *14*, 82–94 (1976)

Ringold, G. M., Yamamoto, K. R., Tomkins, G. M., Bishop, J. M., Varmus, H. E.: Dexamethasone-mediated induction of mouse mammary tumor virus RNA: A system for studying glucocorticoid action. Cell *6*, 299–305 (1975)

Ritzen, E. M., French, F. S., Weddington, S. C., Nayfeh, S. N., Hansson, V.: Steroid binding in polyacrylamide gels: quantitation at steady state conditions. J. Biol. Chem. *249*, 6597–6604 (1974)

Robertson, H. A.: Reproduction in the ewe and the goat. In: Reproduction in domestic animals. 3rd ed., Cole, H. H., Cupps, P. T. (eds.), pp. 477–498. New York: Academic Press 1977

Rochefort, H., Baulieu, E.-E.: New in vitro studies of estradiol binding in castrated rat uterus. Endocrinology *84*, 108–116 (1969)

Rochefort, H., Garcia, M.: Androgen on the estrogen receptor. I. Binding and in vivo nuclear translocation. Steroids *28*, 549–560 (1976)

Rochefort, H., Vignon, F.: Differences between the estradiol and testosterone receptors in rat uterus. Eur. J. Biochem. *48*, 503–512 (1974)

Rochefort, H., Vignon, F., Capony, F.: Effect of antiestrogens on uterine estradiol receptors. Gynecol. Invest. *3*, 43–62 (1972a)

Rochefort, H., Vignon, F., Capony, F.: Formation of estrogen nuclear receptor in uterus: Effect of androgens, estrone and Nafoxidine. Biochem. Biophys. Res. Commun. *47*, 662–670 (1972b)

Rodbard, D.: Mathematics of hormone-receptor interaction. Adv. Exp. Med. Biol. *36*, 289–326 (1973)

Rodbard, D., Feldman, H. A.: Theory of protein-ligand interaction. In: Methods in enzymology. Vol. XXXVI. O'Malley, B. W., Hardman, J. G. (eds.), pp. 3–6. New York: Academic Press 1975

Rosenthal, H. E.: A graphic method for the determination and presentation of binding parameters in a complex system. Anal. Biochem. *20*, 525–532 (1967)

Rosner, W., Christy, N. P., Kelley, W. G.: Partial purification and preliminary characterization of estrogen-binding globulins from human plasma. Biochemistry *8*, 3100–3108 (1969)

Ross, J. W., Shryne, J., Gorski, R. A., Marshall, J. R.: Effect of clomiphene citrate and its isomers on sexual behavior in ovariectomized rats. Endocrinology *92*, 1079–1083 (1973)

Rousseau, G. G., Baxter, J. D., Tomkins, G. M.: Glucocorticoid receptors: Relationship between steroid binding and biological effects. J. Mol. Biol. *67*, 99–115 (1972)

Rousseau, G. G., Baxter, J. D., Higgins, S. J., Tomkins, G. M.: Steroid-induced nuclear binding of glucocorticoid receptors in intact hepatoma cells. J. Mol. Biol. *79*, 539–554 (1973)

Rubin, R. P.: Calcium and the secretory process. New York: Plenum Press (1974)

Rudali, G., Apiou, F., Muel, B.: Mammary cancer produced in mice with estriol. Eur. J. Cancer *11*, 39–41 (1975)

Ruh, T. S., Baudendistel, L. J.: Different nuclear binding sites for antiestrogen and estrogen receptor complexes. Endocrinology *100*, 420–426 (1977)

Ruh, T. S., Katzenellenbogen, B. S., Katzenellenbogen, J. A., Gorski, J.: Estrone interaction with the rat uterus: In vitro response and nuclear uptake. Endocrinology *92*, 125–134 (1973)

Ruh, T. S., Wassilak, S. G., Ruh, M. F.: Androgen-induced nuclear accumulation of the estrogen receptor. Steroids *25*, 257–273 (1975)

Russell, J. T., Thorn, N. A.: Calcium and stimulus-secretion coupling in the neurohypophysis. In: Neurosecretion — the final neuroendocrine pathway. Knowles, F., Vollrath, L. (eds.), pp. 321. Berlin, Heidelberg, New York: Springer 1973

Russell, J. T., Thomas, G. H.: Oestradiol uptake and retention and high affinity binding sites in cultured rabbit uterus. Biochem. J. *144*, 99–106 (1974)

Saffran, J., Loeser, B. K., Haas, B. M., Stavely, H. E.: Binding of progesterone by rat uterus in vitro. Biochem. Biophys. Res. Commun. *53*, 202–209 (1973)

Saffran, J., Loesser, B. K., Bohnett, S. A., Faber, L. E.: Binding of progesterone receptor by nuclear preparations of rabbit and guinea pig uterus. J. Biol. Chem. *251*, 5607–5613 (1976)

Saiduddin, S.: [^{3}H]-estradiol uptake by rat ovary. Proc. Soc. Exp. Biol. Med. *138*, 651–653 (1971)

Saiduddin, S., Milo, G. E., Jr.: Effect of hypophysectomy and pretreatment on uptake and retention of estradiol by the ovary. Proc. Soc. Exp. Biol. Med. *146*, 513–517 (1974)

Saiduddin, S., Zassenhaus, H. P.: Estradiol-17beta receptors in the immature rat ovary. Steroids *29*, 197–213 (1977a)

Saiduddin, S., Zassenhaus, H. P.: In vitro and in vivo enhancement of progesterone binding to the uterine progesterone receptor by cortisol. Biochemistry *16*, 2829–2834 (1977b)

Sala-Trepat, J. M., Rete, E.: Dissociation studies of different forms of the estradiol-receptor complex from calf uterus: Higher stability of the complex bound to chromatin. Biochim. Biophys. Acta *338*, 92–103 (1974)

Sala-Trepat, J. M., Vallet-Strouve, C.: Binding on the estradiol receptor from calf uterus to the chromatin: Active forms. Biochim. Biophys. Acta *371*, 186–202 (1974)

Sanborn, B. M., Steinberger, E.: Androgen nuclear exchange activity in rat testis. Endocrinol. Res. Commun. *2*, 335–344 (1975)

Sanborn, B. M., Rao, B. R., Koreman, S. G.: Interaction of 17β-estradiol and its specific uterine receptor. Evidence for complex kinetic and equilibrium behavior. Biochemistry *10*, 4955–4960 (1971)

Sandberg, A. A., Slaunwhite, W. R.: Transcortin: A corticosteroid-binding protein of plasma. II. Levels in various conditions and the effect of estrogens. J. Clin. Invest. *38*, 1290–1296 (1959)

Sandberg, A. A., Rosenthal, H., Schneider, S. L., Slaunwhite, W. R., Jr.: Protein-steroid interactions and their role in the transport and metabolism of steroids. In: Steroid dynamics. Pincus, G., Nakao, T., Tait, J. F. (eds.), pp. 1–59. New York: Academic Press 1966

Sander, S.: The uptake of 17β-oestradiol in breast tissue of female rats. Acta Endocrinol. (Kbh) *58*, 49–56 (1968a)

Sander, S.: The capacity of breast tissue to accumulate oestradiol-17β in vitro. Acta Pathol. Microbiol. Scand. *73*, 20–36 (1968b)

Sanders, S., Attramadal, A.: An autoradiographic study of oestradiol incorporation in the breast tissue of female rats. Acta Endocrinol. (Kbh) *58*, 235–242 (1968)

Santi, D. V., Sibley, C. H., Perriard, E. R., Tomkins, G., Baxter, J. D.: A filter assay for steroid hormone receptors. Biochemistry *12*, 2412–2416 (1973)

Sanyal, M. D., Todd, R. B.: 5α-dihydroprogesterone influence on ovulation of prepuberal rats. Proc. Soc. Exp. Biol. Med. *141*, 622–624 (1972)

Sanyal, M. K., Villee, C. A.: Production and physiologic effects of progesterone metabolites in pregnant and pseudopregnant rats. Endocrinology *92*, 83–93 (1973)

Sar, M., Stumpf, W. E.: Neurons of the hypothalamus concentrate (^{3}H) progesterone or its metabolites. Science *182*, 1266–1268 (1973)

Sar, M., Stumpf, W. E.: Autoradiography of mammary glands and uteri of mice and rats after the injection of [^{3}H]-estradiol. J. Steroid Biochem. *7*, 391–394 (1976)

Sarff, M., Gorski, J.: Control of estrogen-binding protein concentration under basal conditions and after estrogen administration. Biochemistry *10*, 2557–2563 (1971)

Scatchard, G.: The attractions of proteins for small molecules and ions. Ann. N. Y. Acad. Sci. *51*, 660–672 (1949)

Schmidt, H., Noack, I., Walther, H., Voigt, K. D.: Einfluß von Zyclus und exogener Hormonzufuhr auf Stoffwechselvorgänge des Rattenuterus. Acta Endocrinol. *56*, 231–239 (1967)

Schrader, W. T., O'Malley, B. W.: Progesterone-binding components of chick oviduct. IV. Characterization of purified subunits. J. Biol. Chem. *247*, 51–59 (1972)

Schrader, W. T., Toft, D. O., O'Malley, B. W.: Progesterone binding protein of chick oviduct. VI. Interaction of purified progesterone receptor components with nuclear constituents. J. Biol. Chem. *247*, 2401–2407 (1972)

Schrader, W. T., Heuer, S. S., O'Malley, B. W.: Progesterone receptors of chick oviduct: Identification of 6S receptor dimers. Biol. Reprod. *12*, 134–142 (1975)

Schreiber, J. R., Ross, G. T.: Further characterization of a rat ovarian testosterone receptor with evidence for nuclear translocation. Endocrinology *99*, 590–596 (1976)

Schreiber, J. R., Reid, R., Ross, G. T.: A receptor-like testosterone binding protein in ovaries from estrogen stimulated hypophysectomized immature female rats. Endocrinology *98*, 1206–1213 (1976)

Schwartz, N. B., McCormack, C. E.: Reproduction: Gonadal function and its regulation. Ann. Rev. Physiol. *34*, 425–472 (1972)

Schwartz, R. J., Tsai, M. J., Tsai, S. Y., O'Malley, B. W.: Effect of estrogen on gene expression in the chick oviduct. V. Changes in the number of RNA polymerase binding and initiation sites in chromatin. J. Biol. Chem. *250*, 5175–5182 (1975)

Schwartz, B. E., Milewich, L., Johnston, J. M., Porter, J. C., MacDonald, P. C.: Initiation of human parturition. V. Progesterone binding substance in fetal membrane. Am. J. Obstet. Gynecol. *48*, 685–689 (1976)

Schwartz, B. E., Milewich, L., Gant, N. F., Porter, J. C., Johnston, J. M., MacDonald, P. C.: Progesterone binding and metabolism in human fetal membranes. Ann. N. Y. Acad. Sci. *286*, 304 (1977)

Scott, R. S., Rennie, P. I.: An estrogen receptor in the corpora lutea of the pseudopregnant rabbit. Endocrinology *89*, 297–301 (1971)

Segal, S. J., Nelson, W. D.: Antifertility action of chloramiphene. Anat. Rec. *139*, 273–279 (1961)

Segal, S. J., Scher, W., Koide, S. S.: Estrogens, nucleic acids and protein synthesis in uterine metabolism. In: Biology of the uterus. Wynn, R. M. (ed.), pp. 139–202. New York: Plenum Press 1977

Seiki, K., Hattori, M.: In vivo uptake of progesterone by the hypothalamus and pituitary of the female ovariectomized rat and its relationship to cytoplasmic progesterone-binding protein. Endocrinol. Jpn. *20*, 111–119 (1973)

Shaikh, A. A.: Estrone and estradiol levels in the ovarian venous blood from rats during the estrous cycle and pregnancy. Biol. Reprod. *5*, 297–307 (1971)

Shain, S. A., Barnea, A.: Some characteristics of the estradiol binding protein of the mature intact rat uterus. Endocrinology *89*, 1270–1279 (1971)

Shani, J., Givant, Y., Sulman, F. G., Eylath, U., Eckstein, B.: Competition of phenothiazines with oestradiol for oestradiol receptors in rat brain. Neuroendocrinology *8*, 307–316 (1971)

Shelesnyak, M. C.: Effect of estrogens on histamine release in the rat uterus. Proc. Soc. Exp. Biol. Med. *100*, 739–742 (1959)

Shelesnyak, M. C., Krarcer, P. F., Zeilmaker, G. H.: Studies on the mechanism of decidualization. I. The oestrogen surge of pseudo-pregnancy and progravidity and its role in the process of decidualization. Acta Endocrinol. (Kbh) *42*, 225–232 (1963)

Sheridan, P. J.: Is there an alternative to the cytoplasmic receptor model for the mechanism of action of steroids. Life Sci. *17*, 497–502 (1975)

Sheridan, P. J., Sar, M., Stumpf, W. E.: Autoradiographic localization of [^{3}H]-estradiol or its metabolites in the central nervous system of the developing rat. Endocrinology *94*, 1386–1390 (1974)

Sherman, M. R.: A combined steric and allosteric model of hormone-receptor interaction. In: Proc. 5th Int. Congr. of Endocrinology. James, W. H. I. (ed.), Vol. 402, pp. 572–577. 1977

Sherman, B. M., Korenman, S. G.: Inadequate corpus luteum function: A pathophysiological interpretation of human breast cancer epidemiology. Cancer *33*, 1306–1312 (1974)

Sherman, M. R., Corval, P. L., O'Malley, B. W.: Progesterone-binding components of chick oviduct. I. Preliminary characterization of cytoplasmic components. J. Biol. Chem. *245*, 6085–6096 (1970)

Sherman, M. R., Atienza, S., Shansky, J. R., Hoffman, L. M.: Progesterone receptors of chick oviduct: Steroid-binding "subunit" formed with divalent cations. J. Biol. Chem. *249*, 5351–5363 (1974)

Sherman, M. R., Tuazon, F. B., Diaz, S. C., Miller, L. K.: Multiple forms of oviduct progesterone receptors analyzed by ion exchange filtration and gel electrophoresis. Biochemistry *15*, 980–989 (1976)

Shin, S. H., Morris, A., Synder, J., Hymer, W. C., Milligan, J. V.: Subcellular localization of LH releasing hormone in the rat hypothalamus. Neuroendocrinology *16*, 191–201 (1974)

Shyamala, G.: Estradiol receptors in mouse mammary tumors: Absence of the transfer of bound estradiol from the cytoplasm to the nucleus. Biochem. Biophys. Res. Commun. *46*, 1623–1630 (1972)

Shyamala, G., Gorski, J.: Estrogen receptors in the rat uterus. Studies on the interaction of cytosol and nuclear binding sites. J. Biol. Chem. *244*, 1097–1103 (1967)

Shyamala, G., Nandi, S.: Interactions of [6,7-^{3}H]-17β estradiol with the mouse lactating mammary tissue in vivo and in vitro. Endocrinology *91*, 861–867 (1972)

Sibley, C. H., Tomkins, G. M.: Mechanism of steroid resistance. Cell *2*, 221–227 (1974)

Siegel, L. M., Monty, K. J.: Determination of molecular weights and frictional ratios of proteins in impure systems by use of gel filtration and density gradient centrifugation, application to crude preparations of sulfite and hydroxylamine reductases. Biochim. Biophys. Acta *112*, 346–362 (1966)

Siiteri, P. K., Schwartz, B. E., Moriyama, I., Ashby, R., Linkie, D., MacDonald, P. C.: Estrogen binding in the rat and human. In: Receptors for reproductive hormones. O'Malley, B. W., Means, A. R. (eds.), pp. 97–112. New York: Plenum Press 1973

Siiteri, P. K., Schwartz, B. E., MacDonald, P. C.: Estrogen receptors and the estrone hypothesis in relation to endometrial and breast cancer. Gynecol. Oncol. *2*, 228–238 (1974)

Singer, S. J.: The molecular organization of membranes. Ann. Rev. Biochem. *43*, 805–833 (1974)

Skipper, J. K., Hamilton, T. H.: Regulation by estrogen of the vitellogenin gene. Proc. Natl. Acad. Sci. USA *74*, 2384–2388 (1977)

Slaunwhite, W. R., Jr., Lockie, G. N., Back, N., Sandberg, A. A.: Inactivity in vivo of transcortin-bound cortisol. Science *135*, 1062–1063 (1962)

Sluyser, M., Van Nie, R.: Estrogen receptor content and hormone responsive growth of mouse mammary tumors. Cancer Res. *34*, 3253–3257 (1974)

Sluyser, M., Evers, S. G., DeGrocij, C. C. J.: Sex hormone receptors in mammary tumors in GR mice. Nature *263*, 386–389 (1976)

Smith, E. R., Barker, K. L.: Effects of estradiol and nicotinamide adenine dinucleotide phosphate on the rate of synthesis of uterine glucose 6-phosphate dehydrogenase. J. Biol. Chem. *249*, 6541–6547 (1974)

Smith, B. D., Bradbury, J. T.: Ovarian weight response to varying doses of estrogens in intact and hypophysectomized rats. Proc. Soc. Exp. Biol. Med. *107*, 946–949 (1961)

Smith, L. D., Ecker, R. E.: Role of the oocyte nucleus in physiological maturation in Rana pipiens. Dev. Biol. *19*, 281–309 (1969)

Smith, H. E., Smith, R. G., Toft, D. O., Neergaard, J. R., Burrows, E. P., O'Malley, B. W.: Binding of steroids to progesterone receptor proteins in chick oviduct and human uterus. J. Biol. Chem. *249*, 5924–5932 (1974)

Smith, J. A., Martin, L., King, R. J. B., Vertes, M.: Effects of oestradiol-17β and progesterone on total and nuclear-protein synthesis in epithelial and stromal tissues of the mouse uterus, and of progesterone on the ability of these tissues to bind oestradiol-17β. Biochem. J. *119*, 773–784 (1970)

Smith, R. G., Clarke, S. G., Zalta, E.: Two cytoplasmic estrogen receptors in reproductive tissue. J. Steroid Biochem. (in press) (1978)

Snow, L. D., Eriksson, H., Hardin, J. W., Chan, L., Jackson, R. L., Clark, J. H., Means, A. R.: Nuclear estrogen receptor in the avian liver: Correlation with biologic response. J. Steroid Biochem. *9*, 1017–1026 (1978)

Socher, S. H., O'Malley, B. W.: Estrogen-mediated cell proliferation during chick oviduct development and its modulation by progesterone. Dev. Biol. *30*, 411-417 (1973)

Soloff, M. S., Creange, J. E., Potts, G. O.: Unique estrogen-binding properties of rat pregnancy plasma. Endocrinology *88*, 427-432 (1971)

Sömjen, D., Kaye, A. M., Lindner, H. R.: Postnatal development of uterine response to estradiol-17β in the rat. Dev. Biol. *31*, 409-412 (1973)

Sömjen, G. J., Kaye, A. M., Lindner, H. R.: Demonstration of 8 S-cytoplasmic oestrogen receptor in rat müllerian duct. Biochim. Biophys. Acta *428*, 787-791 (1976)

Spaziani, E.: Relationship between early vascular responses and growth in the rat uterus: Stimulation of cell division by estradiol and vasodilating amines. Endocrinology *72*, 180-191 (1963)

Spaziani, E., Szego, C. M.: The influence of estradiol and cortisol on uterine histamine of the ovariectomized rat. Endocrinology *63*, 669-678 (1958)

Spaziani, E., Szego, C. M.: Early effects of estradiol and cortisol on water and electrolyte shifts in the uterus of the immature rat. Am. J. Physiol. *197*, 355-359 (1959)

Spelsberg, T. C.: Nuclear binding of progesterone in chick oviduct: Multiple binding sites in vivo and transcriptional response. Biochem. J. *156*, 391-398 (1976)

Spelsberg, T. C., Cox, R. F.: Effects of estrogen and progesterone on transcription, chromatin, and ovalbumin gene expression in the chick oviduct. Biochim. Biophys. Acta *435*, 376-390 (1976)

Spelsberg, T. C., Steggles, A. W., O'Malley, B. W.: Progesterone-binding components of chick oviduct. J. Biol. Chem. *246*, 4188-4197 (1971a)

Spelsberg, T. C., Steggles, A. W., O'Malley, B. W.: Changes in chromatin composition and hormone binding during chick oviduct development. Biochim. Biophys. Acta. *254*, 129-134 (1971b)

Spelsberg, T. C., Steggles, A. W., Chytil, F., O'Malley, B. W.: Progesterone binding components of chick oviduct. V. Exchange of progesterone binding capacity from target to nontarget tissue chromatin. J. Biol. Chem. *247*, 1368-1374 (1972)

Spelsberg, T. C., Webster, R., Pickler, G., Thrall, C., Wells, D.: Role of nuclear proteins as high affinity sites ("acceptors") for progesterone in the avian oviduct. J. Steroid Biochem. *7*, 1091-1101 (1976)

Spindler, B. J., MacLeod, K. M., Ring, J., Baxter, J. D.: Thyroid hormone receptors binding characteristics and lack of hormonal dependency on nuclear localization. J. Biol. Chem. *250*, 4113-4121 (1975)

Stabenfeldt, G. H., Shille, V. M.: Reproduction in the dog and cat. In: Reproduction in domestic animals. 3rd ed., Cole, H. H., Cupps, P. T. (eds.), pp. 499-528. New York: Academic Press 1977

Stancel, G. M., Leung, K. M. T., Gorski, J.: Estrogen receptors in the rat uterus. Multiple forms produced by concentration-dependent aggregation. Biochemistry *12*, 2130-2136 (1973a)

Stancel, G. M., Leung, K. M. T., Gorski, J.: Estrogen receptors in the rat uterus. Relationship between cytoplasmic and nuclear forms of the estrogen binding protein. Biochemistry *12*, 2137-2141 (1973b)

Steggles, A. W., King, R. J.: The use of protamine to study [6,7-^{3}H] oestradiol 17beta binding in rat uterus. Biochem. J. *118*, 695-701 (1970)

Steggles, A. W., Spelsberg, T. C., Glasser, S. R., O'Malley, B. W.: Soluble complexes between steroid hormones and target-tissue receptors bind specifically to target-tissue chromatin. Proc. Natl. Acad. Sci. USA *68*, 1479-1482 (1971)

Stephenson, R. P.: A modification of receptor theory. Br. J. Pharmacol. *11*, 379-393 (1956)

Stormshak, F., Leake, R., Wertz, N., Gorski, J.: Stimulatory and inhibitory effects of estrogen on uterine DNA synthesis. Endocrinology *99*, 1501-1511 (1976)

Strott, C. A.: Metabolism of progesterone in the chick oviduct: Relation to the progesterone receptor and biological activity. Endocrinology *95*, 826-837 (1974)

Stumpf, W. E.: Estradiol concentrating neurons: Topography in the hypothalamus by dry-mount autoradiography. Science *162*. 1001-1003 (1968)

Stumpf, W. E.: Nuclear concentration of [^{3}H]-estradiol in target tissues, dry mount autoradiography of vagina, oviduct, ovary, testes, mammary gland tumor, liver and adrenal. Endocrinology *85*, 31-37 (1969)

Stumpf, W. E.: Estrogen-neurons and estrogen-neurons systems in the periventricular brain. Am. J. Anat. *129*, 207-218 (1970)

Stumpf, W. E.: Estrogen, androgen and glucocorticoid concentrating neurons in the amygdala, studied by dry autoradiography. In: The neurobiology of the amygdala. Eleftheriou, B. E. (ed.), pp. 763-774. New York: Plenum Press 1972

Stumpf, W. E., Sar, M.: Steroid hormone target sites in the brain: The differential distribution of estrogen, progestin, androgen and glucocortico steroid. J. Steroid Biochem. *7*, 1163–1170 (1976)

Stumpf, W. E., Sar, M., Aumüller, G.: The heart: a target organ for estradiol. Science *196*, 319–321 (1977)

Styne, D. M., Goldsmith, P. C., Burstein, S. R., Kaplan, S. L., Grumbach, M. M.: Immunoreactive somatostatin and luteinizing hormone releasing hormone in median eminence synaptosomes of the rat: Detection by immunohistochemistry and quantification by radioimmunoassay. Endocrinology *101*, 1099–1103 (1977)

Sullivan, J. N., Strott, C. A.: Evidence for an androgen-independent mechanism regulating the levels of receptor in target tissue. J. Biol. Chem. *248*, 3202–3208 (1973)

Sutherland, R., Mester, J., Baulieu, E.-E.: Tamoxifen is a potent pure anti-oestrogen in chick oviduct. Nature *267*, 434–435 (1977)

Szego, C. M.: Role of histamine in mediation of hormone action. Fed. Proc. *24*, 1343–1352 (1965)

Szego, C. M.: The lysosomal membrane complex as a proximate target for steroid hormone action. In: The sex steroids. McKerns, K. W. (ed.), pp. 1–51. New York: Appleton-Century-Crofts 1971

Szego, C. M., Davis, J. S.: Adenosine 3'5'-monophosphate in rat uterus: Acute elevation by estrogen. Proc. Natl. Acad. Sci. USA *58*, 1711–1718 (1967)

Szego, C. M., Davis, J. S.: Inhibition of estrogen-induced elevation of cyclic 3',5'-adenosine monophosphate in rat uterus. I. By β-adrenergic receptor blocking drugs. Mol. Pharmacol. *5*, 470–480 (1969)

Szego, C. M., Lawson, D. A.: Influence of histamine on uterine metabolism: Stimulation of incorporation of radioactivity from amino acids into protein, lipid and purines. Endocrinology *74*, 372–381 (1964)

Szego, C. M., Roberts, S.: Steroid action and interaction in uterine metabolism. Recent Prog. Horm. Res. *8*, 419–469 (1953)

Taber, C. A., Karavolas, H. J.: Subcellular localization of LH releasing activity in the rat hypothalamus. Endocrinology *96*, 446–452 (1975)

Tachi, C., Tachi, S., Lindner, H. R.: Modification by progesterone of oestradiol-induced cell proliferation, RNA synthesis and oestradiol distribution in the rat uterus. J. Reprod. Fert. *31*, 59–76 (1972)

Takasugi, N., Bern, H. A.: Tissue changes in mice with persistent vaginal cornification induced by early postnatal treatment with estrogen. J. Natl. Cancer Inst. *33*, 855–865 (1964)

Talley, D. J., Li, J. J., Li, S. A., Villee, C. A.: Biochemical comparison of estrogen receptors of the hamster hypothalamus and uterus. Endocrinology *96*, 1135–1144 (1975)

Talwar, G. P., Segal, S. J., Evans, A., Davidson, O. W.: The binding of estradiol in the uterus: A mechanism for derepression of RNA synthesis. Proc. Natl. Acad. Sci. USA *52*, 1059–1066 (1964)

Talwar, G. P., Jailkhani, B. L., Narayahan, P. R., Narasimhan, C.: Oestrogen induced synthesis of a protein in avian liver. In: Karolinska symposia on research methods in reproductive endocrinology: Protein synthesis in reproductive tissue. Dicsfalusy, E. (ed.), pp. 341–356. Stockholm: Karolinska Institutet 1973

Tata, J. R.: The expression of the vitellogenin gene. Cell *9*, 1–14 (1976)

Tchernitchin, A.: Radioautographic study of the effect of estradiol 17β, estrone, estriol, progesterone, testosterone and corticosterone on the in vitro uptake of [2,4,6,7-^{3}H]-estradiol-17β by uterine eosinophils of the rat. Steroids *19*, 575–586 (1972)

Tchernitchin, A.: Effect of progesterone on the in vivo binding of estrogens by uterine cells. Experientia *32*, 1069–1071 (1976)

Tchernitchin, A., Tchernitchin, X., Galand, P.: Correlation of estrogen-induced uterine eosinophilia with other parameters of estrogen stimulation, produced with estradiol-17beta and estriol. Experientia *31*, 993–994 (1975)

Tchernitchin, A., Tchernitchin, X.: Characterization of the estrogen receptors in the uterine and blood eosinophil leukocytes. Experientia *32*, 1240–1242 (1976)

Tchernitchin, A., Tchernitchin, X., Galand, P.: New concepts on the action of oestrogens in the uterus and the role of the eosinophil receptor system. Differentiation *5*, 145–150 (1976)

Teng, C. S., Hamilton, T. H.: Role of chromatin in estrogen action in the uterus. II. Hormone-induced synthesis of non-histone acidic proteins which restore histone-inhibited DNA-dependent RNA synthesis. Proc. Natl. Acad. Sci. USA *63*, 465–472 (1969)

Teng, C. S., Hamilton, T. H.: The role of chromatin in estrogen action in the uterus. I. The control of template capacity and chemical composition and the binding of [^{3}H]-estradiol-17β. Proc. Natl. Acad. Sci. USA *60*, 1410–1417 (1968)

Teng, C. S., Teng, C. T.: Studies on sex-organ development: Ontogeny of cytoplasmic oestrogen receptor in chick Mullerian duct. Biochem. J. *150*, 191–194 (1975)

Teng, C. S., Teng, C. T.: Oestrogen receptor translocation in the developing chick Mullerian duct. Biochem. J. *154*, 1–9 (1976)

Terenius, L.: Oestrogen binding in the mouse uterus. Acta Endocrinol. *57*, 669–682 (1968)

Terenius, L.: Two modes of interaction between oestrogen and antioestrogen. Acta Endocrinol. *64*, 47–58 (1970)

Terenius, L.: Anti-oestrogens and breast cancer. Eur. J. Cancer *7*, 57–64 (1971)

Terenius, L., Ljungkvist, I.: Aspects on the mode of action of antiestrogens and antiprogestogens. Gynecol. Invest. *3*, 96–107 (1972)

Thompson, J. R., King, R. J.: A potential model for studying the control of oestradiol binding. Biochem. Soc. Trans. *4*, 772–774 (1976)

Toft, D. O.: The interaction of uterine estrogen receptors with DNA. J. Steroid Biochem. *3*, 515–522 (1972)

Toft, D., Gorski, J.: A receptor molecule for estrogens: Isolation from the rat uterus and preliminary characterization. Proc. Natl. Acad. Sci. USA *55*, 1574–1581 (1966)

Toft, D., O'Malley, B. W.: Target tissue receptors for progesterone: The influence of estrogen treatment. Endocrinology *90*, 1041–1045 (1972)

Toft, D., Shyamala, G., Gorski, J.: A receptor molecule for estrogens: Studies using a cell-free system. Proc. Natl. Acad. Sci. USA *57*, 1740–1743 (1967)

Tomkins, G. M., Martin, D. W., Jr., Stellwagen, R. H., Baxter, J. D., Mamont, P., Levinson, B. B.: Cold Spring Harbor Symp. Quant. Biol. *35*, 635–640 (1970)

Traish, A. M., Muller, R. E., Wotiz, H. H.: Binding of estrogen receptor to uterine nuclei. Salt-extractable versus salt-resistant receptor-estrogen complexes. J. Biol. Chem. *252*, 6823–6830 (1977)

Tsai, T. L., Katzenellenbogen, B. S.: Antagonism of development and growth of 7,12-dimethylbenz(a)anthracene induced rat mammary tumors by the antiestrogen W-23, 469 and effects on estrogen and progesterone receptors. Cancer Res. *37*, 1537–1543 (1977)

Tsai, S. Y., Tsai, M. J., Schwartz, R., Kalimi, M., Clark, J. H., O'Malley, B. W.: Estrogen effect on gene expression in chick oviduct: Nuclear receptor levels and initiation of transcription. Proc. Natl. Acad. Sci. USA *72*, 4228–4232 (1975)

Tseng, L., Gurpide, E.: Estradiol and 20 alpha-dihydroprogesterone dehydrogenase activities in human endometrium during the menstrual cycle. Endocrinology *94*, 419–423 (1974)

Tseng, L., Gurpide, E.: Effects of progestins on estradiol receptor levels in human endometrium. J. Clin. Endocrinol. Metab. *41*, 402–404 (1975)

Tseng, L., Gusberg, S. B., Gurpide, E.: Estradiol receptor and 17β-dehydrogenase in normal and abnormal human endometrium. In: Biochemical actions of progesterone and progestins. Gurpide, E. (ed.), pp. 190–198. New York: Ann. N. Y. Acad. Sci. 1977

Tsong, T. W.: Effect of phase transition on the kinetics of dye transport in phospholipid bilayer structures. Biochemistry *14*, 5409–5414 (1975a)

Tsong, T. W.: Transport of 8-Anilino-1-papthalenesulfunate as a probe of the effect of cholesterol on the phospholipid bilayer structure. Biochemistry *14*, 5415–5417 (1975b)

Tulchinsky, D.: Placental secretion of unconjugated estrone, estradiol and estriol into the maternal and fetal circulation. J. Clin. Endocrinol. *36*, 1079–1087 (1973)

Tyler, E. T., Winder, J., Gotlib, M., Olson, H. J., Nakabayashi, N.: Effect of MRL-41 in human male and female fertility studies. Clin. Res. *10*, 119–128 (1962)

Tytell, M., Peck, E. J., Jr., Clark, J. H.: The effects of estrogen on the subcellular distribution of luteinizing hormone releasing hormone (LHRH) in the ovariectomized rat. Endocrinology (in press) (1978a)

Ulfelder, H.: The stilbestrol-adenosis-carcinoma syndrome. Cancer *38*, 426–431 (1976)

Ullberg, S., Bengtsson, G.: Autoradiographic distribution studies with natural estrogens. Acta Endocrinol. (Kbh) *43*, 75–86 (1963)

Vaitukaitis, J. L., Bermudez, J. A., Cargille, C. M., Lipsett, M. B., Ross, G. T.: New evidence for an anti-estrogenic action of clomiphene citrate in women. J. Clin. Endocrinol. Metab. *32*, 503–508 (1971)

Van Tienhoven, A.: Reproductive physiology of vertebrates. Philadelphia: Saunders 1968

Vertes, M., King, R. J. B.: The mechanism of oestradiol binding in rat hypothalamus-effect of androgenization. J. Endocrinol. *51*, 271–282 (1971)

Vignon, F., Rochefort, H.: Regulation of estrogen receptors in ovarian-independent rat mammary tumors. I. Effects of castration and prolactin. Endocrinology *98*, 722–729 (1976)

Viladiu, P., Delgado, C., Pensky, J., Pearson, O. H.: Estrogen binding protein of rat liver. Endocrinol. Res. Commun. *2*, 273–280 (1975)

Vonderhaar, B. K., Kim, W. H., Mueller, G. C.: The subunit character of soluble estrogen receptors from the rat uteri and their modification in vitro. Biochim. Biophys. Acta *215*, 125–133 (1970)

Wade, G. N., Feder, H. H.: 1,2-[^{3}H]-progesterone uptake by guinea pig brain and uterus: Differential localization, timecourse of uptake and metabolism, and effects of age, sex, estrogen-priming and competing steroids. Brain Res. *45*, 525–543 (1972a)

Wade, G. N., Feder, H. H.: Uptake of [1,2-^{3}H] 20α-hydroxypregn-4-en-one [1,2-^{3}H] corticosterone, and [6,7-^{3}H] estradiol-17β by guinea pig brain and uterus: Comparison with uptake of [1,2-^{3}H] progesterone. Brain Res. *45*, 545–554 (1972b)

Wagner, R. K., Jungblut, P. W.: Differentiation between steroid hormone receptors CBG and SHBG in human target organ extracts by a single-step assay. Mol. Cell Endocrinol. *4*, 13–24 (1976)

Walker, M. D., Gozes, I., Kaye, A. M., Reiss, N., Littauer, U. Z.: The "estrogen-induced protein": Quantitation by autoradiography of polyacrylamide gels. J. Steroid Biochem. *7*, 1083–1089 (1976)

Walters, M. R., Clark, J. H.: Nuclear and cytoplasmic progesterone receptors in the rat uterus: Effects of R5020 and progesterone on receptor binding and subcellular compartmentalization. In: Progesterone receptors in normal and neoplastic tissues. McGuire, W. (ed.), pp. 271–285. New York: Raven Press 1977a

Walters, M. R., Clark, J. H.: Cytosol progesterone receptors of the rat uterus: Assay and receptor characteristics. J. Steroid Biochem. *8*, 1137–1144 (1977b)

Walters, M. R., Clark, J. H.: Cytosol and nuclear compartmentalization of progesterone receptors of the rat uterus. Endocrinology *103*, 601–609 (1978)

Warberg, J., Eskay, R. L., Barnea, A., Reynolds, R. C., Porter, J. C.: Release of luteinizing hormone releasing hormone and thyrotropin releasing hormone from a synaptosome-enriched fraction of hypothalamic homogenates. Endocrinology *100*, 814–825 (1977)

Warembourg, M.: Fixation de l'oestradiol-[^{3}H] au niveau des noyaux amygdaliens, Septaux et du systéme hypothalemo-hypophysaine chez la souris femelle. C. R. Acad. Sci. Paris *270*, 152–154 (1970a)

Warembourg, M.: Fixation de l'oestradiol-[^{3}H] dans le telencephale et le diencephale chez la souris femelle. C. R. Soc. Biol. *164*, 126–129 (1970b)

Warembourg, M.: Radioautographic study of the guinea pig uterus after injection and incubation with [^{3}H]-progesterone. Endocrinology *94*, 665–670 (1974)

Warembourg, M., Milgrom, E.: Radioautography of the uterus and after [^{3}H] progesterone injection into guinea pigs at various periods of the estrous cycle. Endocrinology *100*, 175–181 (1977)

Watson, C., Medina, D., Clark, J. H.: Estrogen receptor characterization in a transplantable mouse mammary tumor. Cancer Res. *37*, 3344–3348 (1977)

Watson, C. S., Clark, J. H.: Heterogeneity of estrogen binding sites in mouse mammary cancer. (Submitted)

Weil, P. A., Sidikaro, J., Stancel, G. M., Blatt, S. P.: Hormonal control of transcription in the rat uterus: Stimulation of deoxyribonucleic acid-dependent RNA polymerase III by estradiol. J. Biol. Chem. *252*, 1092–1098 (1977)

Weiss, G. B.: Cellular pharmacology of lanthanum. Ann. Rev. Pharmacol. *14*, 343–354 (1974)

Weniger, J. P.: Action feminisante de la gonade droite de l'embryon de Poulet fémelle en culture in vitro. Compt. Rend. Acad. Sci. Paris *246*, 1094–1096 (1968)

West, N. B., Verhage, H. G., Brenner, R. M.: Suppression of the estradiol receptor system by progesterone in the oviduct and uterus of the cat. Endocrinology *99*, 1010–1016 (1976)

West, N. B., Verhage, H. G., Brenner, R. M.: Changes in nuclear estradiol receptor and cell structure during estrous cycles and pregnancy in the oviduct and uterus of cats. Biol. Reprod. *17*, 138–143 (1977)

West, N. B., Norman, R. L., Sandow, B. A., Brenner, R. M.: Hormonal control of nuclear estradiol receptor content and the luminal epithelium in the uterus of the golden hamster. Endocrinology *103*, 1732–1741 (1978)

Westphal, U.: Steroid-protein interactions. Berlin, Heidelberg, New York: Springer 1971

Wetekam, W., Mullinix, K. P., Deeley, R. G., Kronenberg, H. M., Eldridge, J. D., Meyers, M., Goldberger, R. F.: Effect of estrogen on gene expression: Purification of vitellogenin messenger RNA. Proc. Natl. Acad. Sci. USA *72*, 3364–3368 (1975)

Whalen, R. E., Gorzalka, B. B.: Estrogen-progesterone interactions in uterus and brain of intact and adrenalectomized immature and adult rats. Endocrinology *94*, 214–223 (1974)

Whalen, R. E., Luttge, W. G.: Differential localization of progesterone uptake in the brain. Role of sex, estrogen pretreatment and adrenalectomy. Brain Res. *33*, 139–146 (1971a)

Whalen, R. E., Luttge, W. G.: Differential localization of progesterone uptake in brain: Role of sex, estrogen pretreatment and adrenalectomy. Brain Res. *33*, 147–155 (1971b)

Whalen, R. E., Maurer, R. A.: Estrogen "receptors" in brain: An unsolved problem. Proc. Natl. Acad. Sci. USA *63*, 681–685 (1969)

Wheaton, J. E., McCann, S. M.: Luteinizing hormone-releasing hormone in peripheral plasma and hypothalamus of normal and ovariectomized rats. Neuroendocrinology *20*, 296–310 (1976)

Whittaker, V. P.: The isolation and characterization of acetylcholine-containing particles from brain. Biochem. J. *72*, 694–706 (1959)

Whittaker, V. P., Michaelson, J. A., Kirkland, R. J. A.: The separation of synaptic vesicles from nerve ending particles (synaptosomes). Biochem. J. *90*, 293–303 (1964)

Wiest, W. G.: Progesterone and 20α-hydroxypregn-4-en-3-one in plasma, ovaries and uteri during pregnancy in the rat. Endocrinology *87*, 43–48 (1970)

Wiest, W. G., Rao, B. R.: Progesterone binding proteins in rabbit uterus and human endometrium. In: Advances in the biosciences No. 7. Raspe, G. (ed.), pp. 251–266. Vieweg: Pergamon Press 1971

Williams, D., Gorski, J.: A new assessment of subcellular distribution of bound estrogen in the uterus. Biochem. Biophys. Res. Commun. *45*, 258–264 (1971)

Williams, D., Gorski, J.: Equilibrium binding of estradiol by uterine cell suspensions and whole uteri in vitro. Biochemistry *13*, 5537–5542 (1974)

Wira, C. R., Munck, A.: Glucocorticoid-receptor complexes in rat thymus cells. J. Biol. Chem. *249*, 5328–5336 (1974)

Wittliff, J. L., Kenney, F. T.: Regulation of yolk protein synthesis in amphibian liver. I. Induction of lipovittellin synthesis by estrogen. Biochim. Biophys. Acta *269*, 485–492 (1972)

Wittliff, J. L., Gardner, D. G., Battema, W. L., Gilbert, P. J.: Specific estrogen receptors in the neoplastic and lactating mammary gland of the rat. Biochem. Biophys. Res. Commun. *48*, 119–125 (1972a)

Wittliff, J. L., Hilf, R., Brooks, W. F., Jr., Savlov, E. D., Hall, T. C., Orlando, R. A.: Specific estrogen-binding capacity of the cytoplasmic receptor in normal and neoplastic breast tissues of humans. Cancer Res. *32*, 1983–1992 (1972b)

Wittliff, J. L., Lee, K.-L., Kenney, F. T.: Regulation of yolk protein synthesis in amphibian liver. 2. Elevation of ribonucleic acid by estrogen. Biochem. Biophys. Acta *269*, 493–504 (1972c)

Wittliff, J. L., Beatty, B. W., Savlor, E. D., Patterson, W. B., Cooper, R. A., Jr.,: Estrogen receptors and hormone dependency in human breast cancer. In: Recent results in cancer research. St. Arneault, G., Band, P., Israel, L. (eds.), Vol. 57, pp. 59–77. Berlin, Heidelberg, New York: Springer 1976

Wittliff, J. L., Mehta, R. G., Kute, I. E.: Interaction of R5020 with binding sites in normal and neoplastic mammary tissues. In: Progesterone receptors in normal and neoplastic tissues. McGuire, W. L., Raynaud, J. P., Baulieu, E.-E. (eds.), pp. 39–57. New York: Raven Press 1977

Woosley, J. T., Muldoon, T. G.: Use of the direct linear plot to estimate binding constants for protein-ligand interactions. Biochem. Biophys. Res. Commun. *71*, 155–160 (1976)

Wotiz, H. H., Shane, J. A., Vigersky, R., Brecher, P. I.: The regulatory role of estradiol in the proliferative action of estradiol. In: Prognostic factors in breast cancer. Forrest, A. P. M., Kunkler, P. B. (eds.), pp. 368–376. Edinburgh: Livingstone Press 1968

Wotiz, H. H., Scublinsky, A., Walker, C.: Effect of estrogen antibodies on pregnancy in the rat. Gynecol. Invest. *3*, 124–131 (1972)

Wray, W., Stubblefield, E.: A new method for the rapid isolation of chromosomes, mitotic apparatus, or nuclei from mammalian fibroblasts at near neutral pH. Exp. Cell Res. *59*, 469–478 (1970)

Wyman, J.: Allosteric linkage. J. Am. Chem. Soc. *89*, 2202–2218 (1967)

Wynn, R. M.: Histology and ultrastructure of the human endometrium. In: Biology of the Uterus. Wynn, R. M. (ed.), pp. 341–376. New York: Plenum Press 1977

Yamamoto, K. R.: Characterization of the 4S and 5S forms of the estradiol receptor protein and their interaction with deoxyribonucleic acid. J. Biol. Chem. *249*, 7068–7075 (1974)

Yamamoto, K. R., Alberts, B.: On the specificity of the binding of the estradiol receptor protein to deoxyribonucleic acid. J. Biol. Chem. *249*, 7076–7086 (1974)

Yamamoto, K., Alberts, B.: The interaction of estradiol-receptor protein with the genome: An argument for the existence of undetected specific sites. Cell *4*, 301–310 (1975)

Yamamoto, K. R., Albert, B. M.: Steroid receptors: Elements for modulation of eukaryotic transcription. Ann. Rev. Biochem. *45*, 721–746 (1976)

Yen, S. S. C., Lein, A.: The apparent paradox of the negative and positive feedback control system on gonadotropin secretion. Am. J. Obstet. Gynecol. *126*, 942–954 (1976)

Yoshinaga, K., Hawkins, R. A., Stocker, J. F.: Estrogen secretion by the rat ovary in vivo during the estrous cycle and pregnancy. Endocrinology *85*, 103–112 (1969)

Young, P. C., Ehrlich, C. E., Cleary, R. E.: Progesterone binding in human endometrial carcinomas. Am. J. Obstet. Gynecol. *125*, 353–360 (1976)

Zarrow, M. X., Yochim, J. M., McCarthy, J. L.: Experimental Endocrinology: A sourcebook of basic techniques. New York: Academic Press 1964

Zava, D. T., Harrington, N. Y., McGuire, W. L.: Nuclear estradiol receptor in the adult rat uterus: A new exchange assay. Biochemistry *15*, 4292–4297 (1976)

Zava, D. T., Chamness, G. C., Horwitz, K. B., McGuire, W. L.: Human breast cancer: Biologically active estrogen receptor in absence of estrogen. Science *196*, 663–664 (1977)

Zeilmaker, G. H.: Experimental studies on the effects of ovariectomy and hypophysectomy on blastocyst implantation in the rat. Acta Endocrinol. (Kbh) *44*, 355–366 (1963)

Zigmond, R. E.: Binding, metabolism and action of steroid hormones in the central nervous system. In: Handbook of Psychopharmacology. Iversen, L.-L., Iverson, S. D., Snyder, S. H. (eds.), Vol. 5, pp. 239–327. New York: Plenum Press 1975

Zigmond, R. E., McEwen, B. S.: Selective retention of estradiol by brain cell nuclei in specific regions of the ovariectomized rat. J. Neurochem. *17*, 889–899 (1970)

Zor, U., Koch, Y., Lamprecht, S. A., Ausher, J., Lindner, H. R.: Mechanism of oestradiol action on the rat uterus: Independence of cyclic AMP, prostaglandin E_2 and β-adrenergic mediation. J. Endocrinol. *58*, 525–533 (1973)

Zumoff, B., Fishman, J., Cassouto, J., Gallagher, T. F., Hellman, L.: Influence of age and sex on normal estradiol metabolism. J. Clin. Endocrinol. Metab. *28*, 937–948 (1968)

Zumoff, B., Fishman, J., Bradlow, H. L., Hellman, L.: Hormone profiles in hormone dependent cancers. Cancer Res. *35*, 3365–3373 (1975)

Subject Index

Other Volumes of Interest from this Series

Volume 13: H. F. Deluca

Vitamin D – Metabolism and Function

1979. 14 figures. VIII, 80 pages
ISBN 3-540-09182-3

Volume 12

Glucocorticoid Hormone Action

Editors: J. D. Baxter, G. G. Rousseau
1979. 176 figures, 58 tables. Approx. 700 pages
ISBN 3-540-08973-X

Volume 11: S. Ohno

Major Sex-Determining Genes

1979. 34 figures, 6 tables. XIII, 140 pages
ISBN 3-540-08965-9

Volume 10: W. I. P. Mainwaring

The Mechanism of Action of Androgens

1977. 12 figures, 17 tables. XI, 178 pages
ISBN 3-540-07941-6

Volume 9: R. E. Mancini

Immunologic Aspects of Testicular Function

1976. 36 figures, 8 tables. IX, 114 pages
ISBN 3-540-07496-1

Volume 8: E. Gurpide

Tracer Methods in Hormone Research

1975. 35 figures. XI, 188 pages
ISBN 3-540-07039-7

Volume 7: E. W. Horton

Prostaglandins

1972. 97 figures. XI, 197 pages
ISBN 3-540-05571-1

Springer-Verlag
Berlin
Heidelberg
New York

Volume 6: K. Federlin

Immunopathology of Insulin

Clinical and Experimental Studies

1971. 53 figures. XIII, 185 pages
ISBN 3-540-05408-1

Volume 5: J. Müller

Regulation of Aldosterone Biosynthesis

1971. 19 figures. VII, 137 pages
ISBN 3-540-05213-5

Volume 4: U. Westphal

Steroid-Protein Interactions

1971. 144 figures. XIII, 567 pages
ISBN 3-540-05312-3

Volume 3: F. G. Sulman

Hypothalamic Control of Lactation

In collaboration with M. Ben-David, A. Danon, S. Dikstein, Y. Givant, K. Khazen, J. Mishkinsky-Shani, I. Nir, C. P. Weller

1970. 58 figures. XII, 235 pages
ISBN 3-540-04973-8

Volume 2: K. B. Eik-Nes, E. C. Horning

Gas Phase Chromatography of Steroids

1968. 85 figures. XV, 382 pages
ISBN 3-540-04277-6

Volume 1: S. Ohno

Sex Chromosomes and Sex-Linked Genes

1967. 33 figures. X, 192 pages
ISBN 3-540-03934-1

Springer-Verlag
Berlin
Heidelberg
New York

MIX
Papier aus verantwortungsvollen Quellen
Paper from responsible sources
FSC® C105338

If you have any concerns about our products,
you can contact us on
ProductSafety@springernature.com

In case Publisher is established outside the EU,
the EU authorized representative is:
Springer Nature Customer Service Center GmbH
Europaplatz 3, 69115 Heidelberg, Germany

Printed by Libri Plureos GmbH
in Hamburg, Germany